Water Resources in the Built Environment

Water Resources in the Built Environment

Management Issues and Solutions

Edited by

Colin A. Booth

Associate Professor of Sustainability
Associate Head of Research and Scholarship
Director of the Construction and Property Research Centre
University of the West of England, Bristol

Susanne M. Charlesworth

Reader in Urban Physical Geography
Department of Geography, Environment and Disaster Management
Coventry University

WILEY Blackwell

This edition first published 2014

Registered Office
John Wiley & Sons, Ltd, The Atrium, Southern Gate, Chichester, West Sussex, PO19 8SQ, United Kingdom.

Editorial Offices
9600 Garsington Road, Oxford, OX4 2DQ, United Kingdom.
The Atrium, Southern Gate, Chichester, West Sussex, PO19 8SQ, United Kingdom.

For details of our global editorial offices, for customer services and for information about how to apply for permission to reuse the copyright material in this book please see our website at www.wiley.com/wiley-blackwell.

Library of Congress Cataloging-in-Publication Data

Water resources for the built environment : management issues and solutions / edited by Colin Booth, Susanne Charlesworth.
pages cm
Includes bibliographical references and index.
ISBN 978-0-470-67091-0 (pbk.)
1. Water resources development. 2. Watershed management. 3. Water-supply. 4. Flood control. 5. City planning. I. Booth, Colin (Colin A.), editor of compilation. II. Charlesworth, Susanne, editor of compilation.
TC413.W375 2014
333.91–dc23
2013041154

A catalogue record for this book is available from the British Library.

Wiley also publishes its books in a variety of electronic formats. Some content that appears in print may not be available in electronic books.

Cover design by Garth Stewart
Cover image credit: Photos courtesy of iStock Photo, Alan Murray-Rust, Susanne Charlesworth and Kate Heal

Set in Sabon 9/11pt by SPi Publisher Services, Pondicherry, India
Printed and bound in Singapore by Markono Print Media Pte Ltd

1 2014

This book is dedicated to Esmée and Edryd

And in loving memory of Rónán John Coughlan Charlesworth
10/09/2011 – 25/09/2011

'So go and run free with the angels...'

Contents

Contents

Contributors

Joseph Akunna
Urban Water Technology Centre
Contemporary Sciences
University of Abertay
Dundee

Victoria Ashton
Natural Resources Wales
Tŷ Cambria
Cardiff

Joanne Bartie
Urban Water Technology Centre
School of Contemporary Sciences
University of Abertay
Dundee

David W. Beddoes
Drain Angels Ltd.,
Shropshire

Colin A. Booth
Faculty of Environment and Technology
University of the West of England
Bristol

Susanne M. Charlesworth
SUDS ARG, Coventry University
Coventry

Sean Churchill
Building Control
Wolverhampton City Council
Wolverhampton

Steve J. Coupe
SUDS ARG
Coventry University
Coventry

John W. Davies
Department of Civil Engineering
Coventry University
Coventry

J. Bryan Ellis
School of Health and Social Sciences
Middlesex University in London
London

Michael A. Fullen
Faculty of Science and Engineering
University of Wolverhampton
Wolverhampton

Trevor Goodhew
Faculty of Environment and Technology
University of the West of England
Bristol

Kate V. Heal
University of Edinburgh
School of GeoSciences
Edinburgh

Phil Harris
Faculty of Science and
Engineering
University of Wolverhampton
Wolverhampton

Emma Hatfield
Faculty of Science and Engineering
University of Wolverhampton
Wolverhampton

Augustine Ifelebuegu
GEDM
Coventry University
Coventry

Rotimi Joseph
Department of Architecture and
the Built Environment
University of the West of England
Bristol

Omolara O. Lade
Faculty of Science and
Engineering
University of Wolverhampton
Wolverhampton

Jessica E. Lamond
Faculty of Environment and
Technology
University of the West of England
Bristol

Craig Lashford
SUDS ARG
Coventry University
Coventry

John McEldowney
Warwick Law School
University of Warwick
Coventry

Sharron McEldowney
School of Life Sciences
University of Westminster
London

Neil McLean
MWH Global
Eastfield House
Edinburgh

Jay Millington
Faculty of Environment and
Technology
University of the West of England
Bristol

Peter Mills
Faculty of Science and Engineering
University of Wolverhampton
Wolverhampton

David Oloke
Faculty of Science and Engineering
University of Wolverhampton
Wolverhampton

Dayo Olugboye
Faculty of Science and Engineering
University of Wolverhampton
Wolverhampton

David G. Proverbs
Faculty of Environment and
Technology
University of the West of England
Bristol

Carly B. Rose
Department of Architecture and
the Built Environment
University of the West of England
Bristol

Miklas Scholz
School of Computing
Science and Engineering
University of Salford
Salford
Greater Manchester

Graham Squires
Faculty of Environment and Technology
University of the West of England
Bristol

Madhu Subedi
Food Security Monitoring Unit
UN World Food Programme
Kathmandu
Nepal

Kim Tannahill
Faculty of Science and Engineering
University of Wolverhampton
Wolverhampton

Rebecca Wade
Urban Water Technology Centre
School of Science Engineering and Technology
Abertay University
Dundee

Luke Walker
The Wilts & Berks Canal Trust
Wootton Bassett
Wiltshire

Frank Warwick
SUDS ARG
Coventry University
Coventry

Peter Wassell
Cunningham Lindsey
Wolverhampton

Simon Watkins
SUDS ARG
Coventry University
Coventry

About the Editors

Dr Colin Booth is an Associate Professor of Sustainability at the University of the West of England, Bristol, where he has been the Associate Head of Research and Scholarship since joining UWE in early 2012, and the Director of the Construction and Property Research Centre. He was previously a Reader in Construction Management, a Reader in Civil Engineering, an Overseas Tutor (Hong Kong), a Senior Lecturer in Water and Environmental Engineering, a European Research Project Manager and has also held several Post-doctoral Research Fellowship posts. He has a portfolio of external research income generation, a wealth of publications and a team of doctoral supervision successes. His previous books have focused on *Flood Hazards* and also on *Solutions to Climate Change Challenges for the Built Environment*.

Dr Susanne Charlesworth is a Reader in Urban Physical Geography at Coventry University in the Department of Geography, Environment and Disaster Management. She is Director of a Sustainable Drainage (SUDS) Applied Research Group at Coventry University, is the author of more than 50 peer reviewed journal articles on urban pollution and SUDS, many book chapters and has co-edited books on aquatic sedimentology and water resources. She collaborates with groups internationally and has given papers at international conferences worldwide.

Section 1 Introduction to the Book

1

Water Resources: Balancing too Little Versus too Much

Colin A. Booth and Susanne M. Charlesworth

1.1 Introduction

Why are we told to conserve water supplies and alter our attitudes towards water usage, yet our homes and businesses are increasing inundated with floodwaters? This is a question that is common to many communities in many counties around the world and is an ever-increasing issue being raised in the United Kingdom (Charlesworth and Booth, 2012).

1.2 Too Little Versus too Much

Being curtailed by a restricted water usage order, whilst standing knee-deep in floodwater inside your home is a confusing and perplexing scenario for society to comprehend – particularly when homeowners could be fined for using a hosepipe to clean out and sanitise their home after the destruction and devastation of a flood.

In the summer of 2010, with only ~300 mm of rain falling in several months and reservoirs at less than half their usual capacity, the water company for north-west England (United Utilities Plc) gained permission in early July 2010 for a drought order to restrict nonessential use of water for seven million homes in Cheshire, Lancashire, Greater Manchester, Merseyside and parts of Cumbria. However, within a matter of days of the restriction being imposed, residents in parts of Lancashire (Preston, Leyland, Ribbleton, Lostock Hall, Bamber Bridge and Coppull) and Merseyside (Bootle, Seaforth, West Derby and Bromorough) were inundated with floodwater after torrential rain (~50 mm in one hour) caused flash flooding. However, the drought order remained in place for many weeks later until mid August 2010 and, during which time, anybody caught breaching the ban would have be fined £1000 (~$1600 USD or ~ €1200 EUR). It is estimated that the water company saved about three billion litres of water during the drought order but the homes

Water Resources in the Built Environment: Management Issues and Solutions, First Edition.
Edited by Colin A. Booth and Susanne M. Charlesworth.

and businesses affected by the flooding were inconvenienced for many months later. This scenario highlights the problematic nature of attempting to report drought conditions to the general public so they will curb their water usage demands, when media reports are also screening the trauma and ruin of water excesses.

It would be wrong to have expected the water company to have envisaged or even anticipated the intensity of future rainfall events across its region and, furthermore, the rainfall did not bolster supplies because it fell in isolated places away from the main reservoirs. The company's decision to impose a drought order was an attempt to marry-up likely water demand with probable water availability, so as to maintain a regular and uninterrupted supply for its customers. However, the scenario clearly highlights the fact that water resources management decision-making is a complicated matter, which encompasses reliance upon nature to assist in the prediction of unknown rainfall events. Traditionally, it has been justifiable to assume that summer months will be warmer and drier than the other seasons. Unfortunately, for whatever reasons (and it is not our intention to persuade you to believe or disbelieve the climate change agenda; see Committee on Climate Change, 2012), there seems an ever-increasing shift in climate patterns towards extreme weather events with impacts that appear to be exacerbated by human activities in the built environment arena. As a result, this is causing widespread droughts and flooding to be commonplace for some countries. The following are examples.

Australia, the driest continent on Earth, is no stranger to drought conditions and through a host of measures they have dramatically reduced their water consumption over the last decade to address the issue. However, the Queensland floods in January 2011 served as a reminder that their highly variable climatic pattern of rainfall can have devastating effects (floodwater covered the equivalent area of France and Germany) on the coastal cities and towns, and their communities. In response, it has been proposed that new dams should be constructed to mitigate flooding and to provide a water resource for the growing population.

Many of the southern states of the United States are plagued by drought and flooding. New Orleans will always be remembered for the destruction caused by Hurricane Katrina (August 2005). However, by late 2011 and early 2012, much of the States of Louisiana and Mississippi were suffering extreme drought conditions. That was until heavy rainfalls brought ~180 mm to Louisiana State and ~250 mm to Mississippi State, causing flooding in many places. Elsewhere, in the State of Georgia, the City of Atlanta experienced its worst drought in living memory in 2007, yet within two years (September 2009) the city experienced an unprecedented 500-year flood event. Ever-demanding population growth and increasing urbanisation were highlighted as the precursors for these events.

Poorly maintained drainage systems were fundamental in causing flash flooding in Argentina's capital city, Buenos Aires, during February 2012, when torrential rains fell. However, the surrounding province, which suffered a lack of rainfall at this time, remained in drought for many months, until it rained ~200 mm in one night and caused extensive flooding (700 000 hectares) to the towns around Bolivar.

Many African nations are listed by the United Nations as being in a state of water stress (1700 m^3 per person) or scarcity (1000 m^3 per person). Ghana, like many other African countries, contends with the challenges of delivering a potable supply of water for its population, providing water for food production and growing its economy, confined within the constraints of its limited water resources. However, urban flooding in Ghana is becoming more frequent (February 2011, June 2011, October 2011) and with even greater impacts on communities and businesses.

India is a vast nation with extremes of water shortage and flooding. For instance, in 2002, more than 40 thousand villages in the State of Rajasthan were drought-stricken yet many millions of people in the States of Assam and Bihar were deep in floodwaters. Both scenarios caused suffering and the widespread destruction and failure of crops, together with associated poverty. Nearby, Pakistan was devastated by catastrophic floods (July 2010), which left some 20 million people homeless. Yet, in early 2012, many of the small communities neighbouring the River Indus were now suffering water shortages and were unable to irrigate their crops. Changes in glacial meltwater flows and upstream diversions were identified as compounding water resource issues and driving people into poverty. However, months later (August 2012) those same communities were once again forced from their houses when excessive rainfall caused the river to burst its banks and destroy many of their homes.

Elsewhere, also during August 2012, more than a million people were affected by flooding in the Philippine's capital, Manila, when two weeks' worth of rain fell in just 24 hours. As a consequence, about half the city was submerged with water up to 3–4 metres deep in places, which meant travel was impossible and some victims were stranded on the roofs of buildings. However, the memory of an earlier event in 2009 meant many people were well prepared and more organised when asked to evacuate.

Thailand was devastated in 2011 when it suffered its worst floods for several decades. Hundreds of people were killed, several millions of people were affected and the economic cost was estimated to be tens of billions of pounds (close behind Hurricane Katrina). However, several months later, 50 of its provinces were facing drought conditions. Nearby, severe drought also affected North Korea until heavy rains and flooding caused widespread damage, the deaths of >100 people, many thousands of people left homeless and a similar number of people in the City of Anju were left without potable water supplies (August 2012).

China has a water resources divide. The northern plain, with megacities such as Beijing and Tianjin, has endured severe water shortages to such an extent that reservoirs have diminished to only puddles (e.g. Shandong Province) and, as a consequence, to meet demand, groundwater aquifers are being abstracted faster than they can be replenished. In contrast, southern China is commonly afflicted by floodwaters. For instance, flooding in Sichuan, Guizhou, Hunan and Hubei Provinces (June 2011) caused enormous suffering and infrastructure damage, with many roads, bridges and buildings destroyed, and hundreds of thousands of people evacuated and many thousands of people left stranded. Recognising the imbalance of its water resources, the government is funding (~£37 billion) the North–South Water Project to build a series of massive pipes and canals to transfer water to where it is most needed.

Elsewhere, during June 2011, storms caused flooding in Hamburg, Germany, which inundated buildings and immobilised transport links. The rainfall, however, was welcomed because the country experienced its driest spring months on record. The previous year in Germany had brought extreme heat and drought (July 2010), yet it also brought the wettest August on record.

Spain has been a recent victim to both droughts and flash flooding. Following months of drought and scorching temperatures, the Andalucian Provinces of Almeria, Malaga and Murcia were inundated by a colossal amount of rainfall in only a few hours (September 2012). Such a large amount of rain in a short time meant streets were several metres deep with torrents of water that washed away cars and infrastructure, causing several deaths and mass evacuations.

The UK weather of 2012 can only be described as *topsy–turvy*. The early part of the year started with a second dry winter in succession, resulting in the implementation of drought orders across many parts of the country (affecting ~20 million people). Since then, the country has experienced some of the wettest periods since records began. Some places have reported up to 30 mm of rainfall in one hour and others have reported up to 100 mm in one day. As a consequence, flooding in June 2012 occurred in parts of Sussex (Bognor Regis, Bosham, Bracklesham, Earnley, Elmer, Felpham, Worthing, Middleton-on-Sea, Littlehampton and Hunston), West Wales (Dol-y-bont, Llandre, Machynlleth, Penrhyncoch and Talybont), the Midlands counties (Penkridge, Albrighton, Boningale, Frankley, Birmingham, Leicester, Kington, Kingsland and Eardisley), Greater Manchester (Wigan and Oldham), Lancashire (Croston, Darwen and Bacup), Cumbria (Kendal and Askam), Durham (Whitley Bay), Yorkshire (Mytholmroyd, Swillington, Todmorden and Hebden Bridge; during September 2012 in Boroughbridge, Catterick, Gilling and Tadcaster), Northumberland (Chester-le-Street, Durham, Morpeth, Newburn, Rothbury and Stockton-on-Tees) and Devon and Cornwall (Looe, Mevagissey, Bideford, Exmouth and Clovelly in October 2012).

The plethora of examples outlined above illustrate that droughts and flooding are concomitant global issues and, moreover, illustrate the necessity for water resources managers, water engineers and water policy-makers to ensure that they produce accurate and well-informed decisions to guarantee the sustained delivery of potable water supplies and the continued protection of society from floodwaters. Climate change may (or may not) transpire to be the root cause for droughts and flooding but perhaps there is also a need to reflect on a host of other reasons why these problems exist and concurrently learn to adapt the built environment and lifestyles for any predicted changes (Booth *et al*., 2012).

The foremost reasons for water scarcity include population growth, food production, water quality, water demand, plus a host of legislative, policy, social, economic, political and management decisions, while the primary reasons for flooding include natural reasons, such as excessive rainfall or storm surge, and anthropogenic reasons, such as restricted infiltration and excessive runoff from impervious landscapes, again brought about through a host of legislative, policy, social, economic, political and management decisions. Further and more fruitful insights into these issues and potential solutions are deliberated in the remaining chapters of this book.

1.3 Structure of the Book

This book comprises three parts and eight sections, which are collated into twenty-nine chapters. The first part of the book (Sections 2, 3 and 4) addresses management issues and solutions to minimise water shortages and provide water security for society, whilst the second part of the book (Sections 5 and 6) addresses management issues and solutions to control excessive rainfall and minimise flooding impacts. The latter part of the book (Section 7) contextualises the issues of the earlier sections within international case studies from the developing world.

Section 1 forms the *introduction to the book* and provides insights into issues and examples of the need to balance water resources from the extremes of having too little (drought) versus having too much (flooding). Section 2 introduces *water demand, policy and cost* and gives insights into water strategy, policy and legislation for meeting water demand, whilst also looking at the issues of regulating, privatising and economics of water. Section 3 concentrates on *water infrastructure and supply* and presents insights into issues of large-scale water storage, the impacts of powering the water industry, treatment of water to meet potable supply standards and delivering supplies in buildings. Section 4 assembles chapters dealing with *water conservation* and bestows insights into the concept of achieving water-neutral housing developments, building regulation attempts to reduce water usage, reaping water from rainwater and greywater harvesting, and an innovative approach to utilising inland waterways. Section 5 centres on *flooding responses and reinstatement* and furnishes insights into measuring and monitoring rainfall, engineered schemes for managing and protecting communities from floodwater, the economic cost of flooding, burdens on the insurance sector and a holistic approach to property flood protection. Section 6 ponders on *flood solutions in the urban landscape* and proffers insights into sustainable drainage systems, together with pavement drainage and green infrastructure benefits, the role of constructed wetlands and the treatment of wastewater. Section 7 contextualises *international case studies* with insights into water resources issues in Africa and Asia. Section 8 converges with a *summary of the book* and offers insights into the lessons that can be learnt for the future of water resources management.

1.4 Conclusions

The wealth of global examples and information communicated in this chapter have been randomly collated by the authors, from a host of media sources (television, radio, Internet and newspapers) throughout the last few years, and whilst the journalism reports have not been interrogated for absolute accuracy or scrutinised through a peer-review process, like the references used in the subsequent chapters, they are reported here to simply convey the scale of the water resources message of the need to balance too little with too much.

Balancing our water resources requirements and its management is clearly a complicated and multifaceted responsibility. Societies will complain when there is not enough water and the same communities will protest when they are flooded. The examples used portray a global problem of hardship and an obvious sense of frustration that must be so readily apparent to those affected. Whether you believe in climate change is a cause, or not, evidence suggests there is a shift towards more extreme weather events and the extent, frequency and repetition of droughts and floods illustrates a need to understand and adapt our lifestyles and behaviour, our homes and businesses, and our towns and cities to accommodate these events.

References

Booth, C.A., Hammond, F.N., Lamond, J.E. and Proverbs, D.G. (2012) *Solutions to Climate Change Challenges in the Built Environment*. Wiley–Blackwell, Oxford.

Charlesworth, S.M. and Booth, C.A. (2012) Water resources issues and solutions for the built environment: too little versus too much. In: Booth, C.A., Hammond, F.N., Lamond, J.E. and Proverbs, D.G. (eds) *Solutions to Climate Change Challenges in the Built Environment*. Wiley–Blackwell, Oxford, pp. 237–250.

Committee on Climate Change (2012) *Climate Change – Is the UK Preparing for Flooding and Water Scarcity*? Adaptation of Sub-Committee Progress Report. http://www.theccc.org.uk/reports/adaptation.

Section 2 Water Demand, Policy and Cost

2

Meeting Demand: Water Strategy, Policy and Legislation

Sharron McEldowney

2.1 Introduction

Water quality and availability is fundamental to human health and well-being and essential to economic productivity. Water plays a key role across a variety of commercial activities from food production and manufacture, to energy generation, industrial activity and for leisure activities. The sustainable use of water in urban landscapes involves the management of these diverse and often competing demands on a scarce resource. Policy-makers have to engage with managing water resources to ensure sustainability. It is a multidimensional problem which must somehow be resolved with the equitable distribution of water, while conserving water quantity and quality in the face of growing demand.

At present, ~18 billion litres of water are supplied daily by the UK water industry to residential and commercial users. The service sector is by far the largest single commercial user, accounting for 56% of nonresidential water use, with manufacturing using 28% and agriculture 12%. Some of this demand for water is met through direct abstraction by both agricultural and industrial businesses (Department of Environment, Food and Rural Affairs (DEFRA), 2011a, 2012a). Roughly 150 litres of water are used per day per person in the UK (www.waterwise.org.uk). Any water shortages will undoubtedly have diverse and substantial impacts. Policy development for sustainable water usage is faced with an expanding population, predicted to rise from just over 62 million to an estimated 70 million by 2027 in the UK (Office for National Statistics, 2011), and the impact of global warming. The UK 2012 Climate Change Risk Assessment (DEFRA, 2012a) sets out the vulnerability of the UK's water supply to climate change through reduced water availability and impacts on water quality.

There are a number of important contributions to the governance and management of water resources in England and Wales. DEFRA, together with the Welsh Assembly

Water Resources in the Built Environment: Management Issues and Solutions, First Edition.
Edited by Colin A. Booth and Susanne M. Charlesworth.

Government, is responsible for water policy. The Environment Agency (EA) for England and Wales contributes substantially to the sustainable development of the water resource. Nature Resources Wales, formed from the merger of the Environment Agency for Wales, the Countryside Council for Wales and the Forestry Commission for Wales in 2013, has similar responsibilities. OFWAT is the economic regulator for public water providers in England and Wales; it ensures water companies maintain a balance between supply and demand. The Drinking Water Inspectorate (DWI) oversees the quality of drinking water from public and private supplies.

This chapter will begin by reviewing the current legislative and regulatory framework for managing water quality and demand in the built environment. Strategies to foster greater water use efficiency in the future will be examined and current policy directions for water management in England and Wales are outlined.

2.2 Legislative and Regulatory Framework for Managing the Water Resources

The current regulatory framework for UK water resources reflects two major influences. Firstly, the European Union has substantially contributed to water protection, particularly through the Water Framework Directive (2000/60/EC) and its Daughter Directives. Secondly, the bulk of water supply and the treatment of wastewater, some 16 billion litres per day (DEFRA, 2012b), are provided for by the privatised public utility companies (see Chapter 3). The previous Labour Government's water strategy (2008) made clear the need to value water more highly and the necessity of conserving water resources for the future (DEFRA, 2008; Howarth, 2008). The new Coalition Government's recent White Paper, *Water for Life* (DEFRA, 2011b), has followed a similar approach, which underlines the value of water and emphasises the need to tackle pollution, overabstraction and improve water efficiency.

2.2.1 OFWAT and the Protection of the Water Supply

The Water Industry Act 1991 provides for water companies to supply water and sewage services and established a regulator for the industry. A modified Water Services Regulation Authority (the acronym OFWAT) was set up by the Water Act 2003, with a new emphasis on regulation of a service rather than the economic regulation of a company. This has brought greater cooperation with other agencies. OFWAT is under a primary duty 'to protect the interests of consumers, wherever appropriate by promoting effective competition'. Consumers are defined to include not only current but future users of water. The environmental consequences of water use also lie within the competences of OFWAT. There is an additional duty linked to the EU's Water Framework Directive (WFD) that requires the development of common principles 'to promote sustainable water use'. The Water Industry Act 1991 imposes a duty on water companies to promote the efficient use of water by its customers supplementing the powers of OFWAT.

The need to repair an ageing, Victorian water infrastructure is a major problem for OFWAT and the water industry. There is an ongoing major programme of infrastructure repairs carried out by the water companies and required by OFWAT. Undoubtedly, there has been improvement in leakage levels but, after an initial period of some success, current leakage reduction is disappointing. There were an estimated 2494 megalitres of losses from water companies' distribution networks and 787 megalitres from supply pipes on customers' properties per day in 2009–2010 (DEFRA, 2011a). These figures have remained fairly constant since 2000. Individual water companies have been set yearly targets to reduce leakage by OFWAT until 2015, and must publish data on annual leakage on their web sites. In the past, OFWAT has taken enforcement action over companies not reaching water leakage targets in breach of their statutory obligations and may do so again in the future. Water companies are expected, however, to balance the costs of reducing leakage against managing supply and demand in other ways (e.g. metering, promoting efficient use of the water resource and also against exploiting a new water resource). Guidelines were published by OFWAT as part of the 2009 Price Review on best practice in the calculation of a sustainable economic level of leakage (OFWAT, 2009). The calculation is intended to include consideration of social and environmental costs. There have been some difficulties in the interpretation of this calculation by water companies and OFWAT, the EA and DEFRA are currently reviewing the guidance. OFWAT has also set requirements for the improvements of sewage infrastructure at substantial cost to water companies and their customers. Again the age of the infrastructure is a fundamental problem and improvements are essential if EU Water Directives are to be achieved.

The water companies are expected to present an annual report to OFWAT, which in the past have been much criticised for being too onerous. The requirements for reporting have been simplified and from July 2012 the report has taken the form of a 'risk and compliance statement'. OFWAT believe this will provide more proportionate, targeted and risk-based regulation and incorporate a degree of horizon scanning for risk (OFWAT, 2012a). The statement includes confirmation that the company has met its statutory obligations, as well as licensing and regulatory obligations. The risk assessment element of the statement maps risks and sets out plans to manage or mitigate them. The companies must also report on their progress against a number of indicators for environmental impact, reliability and availability, customer experience and finance. The reliability and availability category include the serviceability of the water and sewage infrastructure and leakage. The environmental impact factors are primarily related to pollution incidence, safe sludge disposal and meeting discharge consents, all of which affect water quality. There is also a reporting requirement on greenhouse gas emissions. The indicators were developed in consultation with the EA, who will use them to help assess water companies' performance related to their environmental impact. The water companies can include other indicators that are not relevant to OFWAT and there seems to be the intention to include indicators of drinking water quality (OFWAT, 2012b).

OFWAT would undoubtedly claim there have been improvements in the efficiency of supply and service provided by the water companies. It is difficult to gauge, however, how much the improvements have been driven by the regulatory activity of OFWAT and how much is the result of the improving standards driven by the European Union. OFWAT has recently been pushing to increase the speed of introducing water metering in England and Wales, with a target of 90% of supplies metered by 2030. The regulator claimed this would not only have benefits for customers but would have substantial environmental benefits in reducing demand (OFWAT, 2011). However, this initiative does not seem to have been reflected in the recent Coalition Government's White Paper, *Water for Life* (DEFRA, 2011b) or the Water Bill 2013/14. The water companies through their distribution and supply of water are key players in the conservation of water resources. It is likely that there will be continued pressure on them to reduce environmental impacts relating to water quality, conservation and use.

2.2.2 The Drinking Water Inspectorate and Drinking Water Supply

The provision of safe, good-quality drinking water is regulated by the Drinking Water Inspectorate (DWI). The Chief Inspector of the DWI has statutory powers of inspection under the Drinking Water Act 2003 (Section 57). It is an offence under the Water Industry Act 1991 to sell water that is unfit for consumption (Section 70). The DWI has enforcement powers under the 1991 Act and has recently cautioned Thames Water, and Essex and Suffolk over drinking water supplies that were unfit for human consumption (DWI, 2011). The Drinking Water Directive (Directive 98/83/EC) sets the quality standards (chemical, microbiological and organoleptic) for drinking water and requires regular monitoring of supplies. There must also be public access to information on drinking water quality (see Chapter 9). Local Authorities keep records of private supplies in their area including reporting on compliance with the Drinking Water Directive. The Inspectorate prepares an annual report on the state of drinking water quality for both public supplies and private supplies (see, for example, Chief Inspector of Drinking Water, 2011).

The percentage of drinking water supplies failing tests in England and Wales was 0.04% in 2010, a slight improvement on 2009 where the rate of failure was 0.05%. There has been a rise in the incidence of failure since 2005, which has caused some concern (ENDS Report, 2009) and appears due to two factors, overfluoridisation and pesticides. There is a requirement for water companies to undertake risk assessments for each water treatment works and linked supply system, which is designed to prevent unhealthy water entering the water supply system. Water companies are expected to have an appropriate risk management system in place.

2.2.3 The Environment Agency and Protecting Water Resources

The EA is the major environmental regulator in England and plays a key role in protecting the quality and quantity of the water resource. It was established under the Environment Act 1995. There are a number of major pieces of legislation relevant to maintaining water resources. In particular the Water Resources Act 1991 provides comprehensive powers for the management and regulation of water resources. There are general duties on the EA to conserve, augment, redistribute and secure proper use of water resources in England and Wales. The 1991 Act provides a water pollution control system and powers under this Act also relate to making drought orders. The Environment Act 1995 gives the EA an extensive remit over water resources, including water management, water pollution control, abstraction, flood defences, conservation and fisheries management.

The EA is the competent authority in England for the Water Framework Directive (WFD) (Directive 2000/60/EC), which is implemented through the Water Environment (England and Wales) Regulations 2003. The Directive essentially applies sustainable development principles to the water industry and to water resources. It provides for comprehensive phased improvements in the status of the water resource. The WFD is intended to be the main regulatory system for covering surface water and groundwater in a common framework. The new Daughter Directives also contribute to the overall regulatory framework. One Daughter Directive is specifically designed to protect the groundwater resource (The Ground Water Directive 2006/118/EC), while a new Environmental Quality Standards Directive (Directive 2008/105/EC), also known as the Priority Substance Directive, sets environmental quality standards for chemical substances that are identified as presenting a risk to or via the aquatic environment. A 'priority list' of substances is set out in Annex X to the WFD and includes industrial chemicals, biocides, metals and metal compounds. These substances are selected on the basis of their persistence, bioaccumulation, toxicity or as agents of equivalent concern (e.g. endocrine disrupting chemicals). They must be monitored and exceedances of environmental quality standards (EQSs) reported. Under the WFD, discharges of these substances to the aquatic environment should be progressively reduced or stopped.

The WFD provides a ground-breaking holistic approach to the protection of water resources. Its technical complexity is matched by its ecological complexity and innovation (Josefsson and Baaner, 2011). The Directive employs river basin management plans (RBMPs) and ensures integrated management across river catchments from the point where the rivers rise to the coastal waters and incorporates all the catchment's water bodies (i.e. lakes, canals, reservoirs, wetlands, amongst others). England and Wales are divided into 11 river basin districts. There is a six-year cycle of planning; the next cycle is from 2015. The plans include objectives for each water body and, where relevant, reasons for not achieving the objectives. The programme of actions in order to achieve the objectives must also be included. The plans are available to the public on the EA website (http://www.environment-agency.gov.uk/research/planning/33240.aspx). Part of this planning specifically includes consideration of the likely effects of climate change on the water resources. The Floods Directive, essentially a sister Directive to WFD, also requires management across river basin districts and over a six-year planning cycle, which must be coordinated with RBMPs.

There is multilayer governance under the WFD and, because of the technical complexity of the legislation, a need to ensure some commonality in implementation across the European Union. There are working groups of experts and stakeholders from 'Member States' that produce documents on key areas of implementation as part of a 'Common Implementation Strategy'. These documents are guidelines rather than legally binding. Similarly, the UK Technical Advisory Group (UKTAG) on the WFD produces guidance for implementation of the scientific and technical aspects of the Directive within the United Kingdom to help ensure consistency of approach across the country (UKATG, 2012). Despite the attempt at common implementation, Member Sates actually have substantial implementation discretion and, indeed, can use broadly worded provisions to derogate from achieving good status (Howarth, 2009). Derogation can come from the application of an economic decision-making process, which forms part of the Common Implementation Strategy (European Commission, 2003a). This is intended to provide a process not only to judge the most cost-effective combination of measures to attain good-quality status as part of the planning process but also provides for a disproportionate cost analysis. The latter allows for time derogations and alternative objectives (Wright and Fritsch, 2011). There are concerns that some Member States are likely to claim derogations and adopt implementation targets less than ambitious, resulting in the Directive loosing effectiveness across the European Union as a whole (Keessen *et al.*, 2010).

Water bodies are assessed under WFD using 30 different criteria (dependent on the water body) under the general headings of biological (including ecological) quality, physical and chemical quality, environmental quality standards (priority substances) and physical quality (e.g. hydromorphological quality). The status of each water body is classified according to these criteria as high, good, moderate or bad, and by water type (e.g. river, estuary or coastal – one nautical mile from shore). Annex V to the Directive classifies waters into different ecological categories. Good ecological status comes when the biological elements of the water bodies deviate only slightly from undisturbed conditions and is the target status for water bodies by 2015. This involves Members States in establishing 'reference conditions' that represent the point at which the water body began to deviate from undisturbed conditions and against which the ecological status of the water body is assessed. This has been controversial, not least because identifying such historic conditions when most water bodies will have gone through substantial change over time is probably unachievable (Josefsson and Baaner, 2011). High-status water bodies must not be allowed to deteriorate from this status. In fact, minimising anthropogenic impacts on 'natural' waters is a legally binding obligation under the Directive (Howarth, 2006). In a sense, the status quo is not accepted for water bodies below good status, with the WFD intended to foster continual improvement until they achieve the best possible status. Designing techniques to establish and monitor ecological quality objectives have been demanding. There are a diversity of approaches with minimal coherence, as yet, in the selection of tools or sampling requirements to monitor water body status (Birk *et al.*, 2012). Monitoring is an important part of the WFD, but existing monitoring networks may not be sufficient or effective in supporting the integrated management process envisaged by the Directive (Collins *et al.*, 2012).

The target date to achieve 'good status' is 2015. Approximately 27% of surface water bodies in England and Wales are of 'good ecological status', leaving a considerable distance to travel. The reasons for failure of targets at the beginning of the planning cycle (in 2009) include: a physical modification typical of the built environment (31%); diffuse pollution (22%), with nonagricultural sources making up 9% of this total; point sources (16%), of which sewage treatment works are a major contributor; and abstraction (4%) (National Audit Office, 2010a, 2010b). The EA has had some success in controlling polluting discharges, especially through a substantial investment by water companies in sewage treatment infrastructure (Howarth, 2008), but controlling diffuse sources remains stubbornly problematic even in urban environments (see Thames River Basin Management Plan, 2009, at http://www.environment-agency.gov.uk/research/planning/125035.aspx). There are very real concerns about the problem of achieving 'good status' and questions about the 'one out–all out' approach (i.e. if a water body fails any one of the 30 assessment criteria it does not reach good status) (Josefsson and Baaner, 2011). There is considerable pressure to drop this approach (Phippard, 2012) and redefine ecological status (Josefsson and Baaner, 2011). A review of the WFD is currently underway and there is a consultation on policy options to safeguard EU waters (European Commission, 2012). A recent House of Lords European Union Committee Report (2012) has urged a reconsideration of the 'one out–all out' approach in the review, describing it as 'a blunt and rigid method which fails to capture effectively the ecological as well as the chemical quality of water'.

WFD specifically draws the public as well as listed statutory consultees into the decision-making process. The public includes 'users of the water' and this group can represent a diverse range of activities from leisure to commercial. Public participation is voluntary and agencies implementing the plan must seek public engagement from the beginning, as part of the planning process, through 'shared decision-making'. There is a 'self-determination' element in the engagement, which means local water management is handed to stakeholders and the public (Howarth, 2009). Guidance on public participation is provided under the Common Implementation Strategy (European Commission, 2003b). Outcomes from an EA pilot project on public participation in river basin planning emphasised the importance of the public understanding of the planning and implementation process and their limitations (Fox *et al.*, 2004). Crucially, fostering public engagement at the local level and encouraging the public to take 'ownership' of measures to protect and improve local water resources is likely to be far more effective than top-down action (Wright and Fritsch, 2011). Moreover, water conservation and resource protection are both likely to be supported by greater availability of environmental information (Doron *et al.*, 2011), as part of the WFD.

The EA went through a long consultation period in the development of the RBMPs as part of public engagement. It can be criticised for the complicated nature of the consultation and the number and complexity of the annexes (see, for example, the Thames River Basin Plan; see Environment Agency, 2008a), which may have had the effect of reducing full public engagement. Today, public and stakeholder participation is fostered at national and regional levels with a DEFRA National Stakeholder Group for England, an EA National Liaison Panel for England and also EA River Basin Liaison Panels. These may not, however, have moved beyond traditional methods of engagement dominated by centralised policy-making with minimal incorporation of stakeholder knowledge (Howarth, 2009; Collins *et al.*, 2012). DEFRA is currently heavily promoting a 'Catchment Based Approach' for WFD implementation that is intended to identify key issues in a catchment and, specifically, involve 'local groups in decision making' (DEFRA, 2012b). There are a total of 25 pilot catchments for this approach, 10 of which are hosted by the EA, while the remaining are hosted by a mix of Nongovernmental Organisations (NGOs) and water companies (Environment Agency, 2012c).

Groundwater is deemed so important in providing drinking water, sustaining agriculture and natural ecosystems under the WFD that there is a Daughter Directive on Groundwater (2006/118/EC). The aim of the WFD is to achieve good groundwater status and, as with surface water bodies, once achieved there should be no deterioration. Good status encompasses both chemical and quantitative objectives. Pollutant ingress should be prevented or at least limited and measures should be put in place to reverse downward trends in groundwater quality.

As mentioned above, abstraction of groundwater and surface water has major implications for the quantity and conservation of water resources. It has significant impact on achieving 'good status' in surface water (House of Lords, 2012) and, of course, groundwater. Water companies are responsible for ~50% of the water volume abstracted. The EA operates a licensing system for both groundwater and surface water abstraction, with ~21 000 abstraction licenses granted in England and Wales. Abstraction licensing was originally introduced under the Water Resources Act 1963 and at that time existing abstractors were given licenses of right, and volumes to be extracted were set on the basis of previous use or the capacity of equipment used for abstraction. There was no consideration of environmental impacts and few conditions to limit abstraction at times of low flow. The licences were not time-limited. Subsequent legislation consolidated the position and it was not until the drought of 1995–1996 that abstraction licensing was reviewed and a White Paper, *Taking Water Responsibly*, was published. The outcome was the Water Act 2003. This provides that:

- All new abstraction licences should be time limited.
- If abstraction results in serious environmental harm then the license can be revoked with no compensation from 2012.
- Licensing abstraction limits can be changed.
- Small abstractors no longer require a licence (i.e. some deregulation).
- Significant abstractors outside the original regime are brought into licensing requirements (e.g. dewatering of excavations).

This Act also made drought plans and water resource management plans prepared by water companies a statutory requirement.

The EA has established 'Catchment Abstraction Management Strategies' setting water availability for abstraction on a catchment basis and intended to inform the permitting system. The EA is attempting to control historic abstraction licences if they result in environmental damage through a 'Restoring Sustainable Abstraction Programme' (Environment Agency, 2010a). This is primarily done through negotiation and voluntary agreements encouraged by cost incentives. The EA has only limited enforcement powers, however, and controlling abstraction has proved difficult. At present, 600 licences are under review. OFWAT consulted on incorporating an 'Abstraction Incentive Mechanism' in its regulatory toolkit for water companies, in order to provide an economic incentive for efficient abstraction management (OFWAT, 2012c). The recent White Paper, *Water for Life*, has added further impetus to controlling abstraction to conserve the water resource (DEFRA, 2011b).

2.3 Water Management and Conservation for the Future

Demands on water resources in urban areas are many and varied. The quality and status of water bodies has implications for leisure users, water companies and other commercial enterprises in urban environments. The security of drinking water resources, the efficiency of their use and their availability are fundamental to health and well-being. Developing policy, legislation and strategies to protect a precious resource into the future is a considerable challenge and involves balancing competing demands. It is, however, increasingly recognised as key to the countries economic, social and environmental prosperity (House of Lords, 2012; Institution of Civil Engineers, 2012).

There are a number of new policy and regulatory developments that will be important for the future of our water resources. The new 'National Policy Planning Framework' has a requirement for local plans to incorporate water management issues. This, however, may not sufficiently encourage Local Authorities to act as key players at the local level in implementation of the WFD (House of Lords, 2012), but nevertheless it is an encouraging development. The recent White Paper, *Water for Life* (DEFRA, 2011b), and the new Water Bill will also influence the future management of water in England. The European Commission has also published an important policy document, *A Blueprint to Safeguard Europe's Water Resources* (European Commission, 2012), on responses to the present and predicted challenges to water resources across the European Union. This is a time when risks to our water resources are becoming ever more apparent and workable responses to meet the challenges ever more critical. Achieving a balance between supply and demand will increasingly be put to the test by climate change, a rising population and land use intensification. Pollution pressures on our water bodies are likely to increase in urban conurbations as well as rural environments, especially from diffuse pollution (Environment Agency, 2008b).

It is clear that government and regulators will have to be proactive and collaborative in ensuring sustainable water supplies and resources for water users. The current overlap between the responsibilities of regulators is clear and the need for active partnership between regulators as a strategy has not gone unrecognised (see Environment Agency, 2010b). A coordinated response is a necessity for a workable strategic approach to water resource management (Council for Science and Technology, 2009). The WFD puts to the fore stakeholder and public participation and again this is an essential strategy for the conservation of water resources and for implementation of the WFD. The role of academics as actors in the protection of resources should not go unmentioned. The 'Thames River Basin Management Plan' recognises the need to engage more fully with academic knowledge in implementing the WFD (Environment Agency, 2009) and other commentators have pointed out the significance of academic contributions (Collins *et al.*, 2012). This is not necessarily easy and knowledge exchange can be affected by poor communication and limited resources, which act as substantial barriers to a constructive dialogue between policy-makers, regulators and academics (Slob *et al.*, 2007). It is similarly important to stimulate innovation in the water sector and explore novel technological solutions to specific problems (House of Lords, 2012). There are many examples in the built environment where technology may be able to contribute workable strategies towards ensuring water resource efficiency and limiting urban water pollution. These include energy and cost-effective technologies (see Chapter 7) to reuse wastewater, rainwater and greywater harvesting (see Chapter 13); water saving devices and products (see Chapter 12); designing technologies to reduce water use in industrial processes; designing technologies such as SUDS (see Chapter 23) to protect the quality of water bodies (European Environment Agency, 2012).

The WFD has served as an excellent illustration of the importance of an integrated and holistic approach to water management and the multidimensional characteristics of successful management. It could be argued that actually protecting the quality of water resources is primarily about land use management and planning, while protecting the quantity of water is all about managing the demand for water and promoting an understanding of the value of water. At present the full cost of producing potable water or wastewater treatment has not been reflected in pricing regimes, and this undoubtedly has led to a disconnect between water users and the true value of the resource (see Chapter 4). The concept of ecosystem services has drawn considerable interest in recent years as a useful method of placing an economic (i.e. monetary) value on the services provided by natural environments such as water bodies (Brown *et al.*, 2007). In fact, the WFD does not incorporate the concept. Ecosystem services are not only those that directly relate to a commercially valuable product but also less tangible benefits that come from an improvement in human well-being. The more indirect the benefits the more difficult they are to value. It seems unlikely that relying on market valuation of water ecosystem services will, on its own, provide sustainable water resources (Watson and Albon, 2011). However, it will undoubtedly provide another valuable tool for the protection of water resources and is certainly worthy of consideration in WFD review (European Commission, 2012; House of Lords, 2012).

Article 9 of the WFD is intended to contribute to better water pricing in a number of ways. There are the principles of 'recovery of the costs of water services' and 'the polluter pays principle', as well as incentive pricing. The 'polluter pays principle', of course, requires a polluter to bear the financial burden of pollution reduction. The principle of 'recovery of costs of water services' brings a focus on the monetary, societal and environmental costs of water supply and that these should be reflected in water pricing. Water pricing should also contain an element of 'incentive pricing' to provide water users with incentives to use water efficiently. It is important that any pricing through tariffs and metering takes into account times of scarcity and seasonal variations in setting charges (European Environment Agency, 2012). There are a number of alternative economic regulatory mechanisms that can be used to reduce water use by domestic and commercial consumers as well as the water industry. It is possible to apply environmental taxes to alter prices or introduce environmental subsidies, perhaps to encourage development and adoption of new water efficient technologies by industry or change consumer behaviour through green purchasing schemes. Tradeable permit schemes could be applied effectively to abstraction or pollution, in a similar way to carbon trading. Economic instruments may not work in isolation but would form part of a coterie of regulatory mechanisms aimed at protecting and conserving the water resource (European Environment Agency, 2012).

The White Paper, *Water for Life* (DEFRA, 2011b), focused on a number of key areas particularly relevant to this discussion. Firstly, tackling water pollution through the catchment based approach intended to protect water quality. The efficacy of this policy is being tested at present through a number of pilot catchments (see above) and should not need any further legislation beyond the implementation of WFD. Secondly, tackling overabstraction through the reform of the abstraction licensing regime through new legislation, which is a much needed initiative. The intention is also to reduce barriers to trading abstraction licences. Thirdly, changing the way we use and value water. This is to take a number of forms but draws back from legislation. Water efficiency is to be encouraged and incentivised, with the provision of water efficiency advice but universal water metering is not to be a requirement. The House of Lords European Union Committee (2012) thought this an important gap in the armoury of measures to manage demand. The White Paper again turned away from regulation in the context of sustainable drainage systems, which are to be encouraged but not required. *Water for Life* is perhaps a missed opportunity for the sustainable management of the water resource, rather concentrating on competition in the water sector. The resulting legislation, in the form of the Water Bill, will not be passed through the Houses of Parliament until 2014.

2.4 Conclusions

Future sustainable water management in the built environment will require an imaginative mixture of regulation, financial incentives, technology and societal contributions. Fundamental to success is collaboration between different actors, including government and regulators, private enterprises, the public and NGOs, with a real integration of their activities and a dialogue that recognises the merit of each contribution and is willing to make trade-offs. Measuring the worth of water as a financially valuable commodity through ecosystem services or setting tariffs on potable water would undoubtedly result in changes to consumer and producer attitudes. It may control demand and protect water quantity. However, water quality may be better protected not just by 'valuing' water as a commodity for whatever use but also through the public acceptance of stewardship and feeling ownership of their local resource.

References

Birk, S., Bonne, W., Borja, A., Brucet, S., Courrat, A., Poikane, S., Solimini, A., van de Bund, W., Zampoukas, N. and Hering, D. (2012) Three hundred ways to assess Europe's surface water: an almost complete overview of biological methods to implement the Water Framework Directive. *Ecological Indicators*, **18**, 31–41.

Brown, T., Bergstrom, J. and Loomis, J. (2007) Defining, valuing and providing ecosystems goods and services. *Natural Resources Journal*, **47**, 329–376.

Chief Inspector of Drinking Water (2011) *Drinking Water 2010. Private Water Supplies in England.* Drinking Water Inspectorate.

Collins, A., Ohandja, D.-G., Hoare, D. and Voulvoulis, N. (2012) Implementing the Water Framework Directive: a transition from established monitoring networks in England and Wales. *Environmental Science and Policy*, **17**, 49–61.

Council for Science and Technology (2009) *Improving Innovation in the Water Industry: 21st century challenges and opportunities*. Available at: http://webarchive.nationalarchives.gov.uk/+/http://www2.cst.gov.uk/cst/reports/files/water-report.pdf.

Department of Environment, Food and Rural Affairs (DEFRA) (2008) *Future Water*. London.

Department of Environment, Food and Rural Affairs (DEFRA) (2011a) *Environmental Statistics – Key Facts*, December 2011. London.

Department of Environment, Food and Rural Affairs (DEFRA) (2011b) *Water for Life*. Cm 8230. London.

Department of Environment, Food and Rural Affairs (DEFRA) (2012a) *The UK Climate Change Risk Assessment: Government Report – Water*, January 2012. London.

Department of Environment, Food and Rural Affairs (DEFRA) (2012b) *Catchment Based Approach*. Available at: http://www.DEFRA.gov.uk/environment/quality/water/legislation/catchment-approach/.

Doron, U., The, T.-H., Haklay, M. and Bell, S. (2011) Public engagement with water conservation in London. *Water and Environment Journal*, **25**, 555–562.

Drinking Water Inspectorate (DWI) (2011) *Objectionable Taste and Odour in Water Supplies in North–East London between January and March 2010*. Ref 2010–2661/2663, 25 October 2011. Available at: http://dwi.DEFRA.gov.uk/stakeholders/information-letters/2011/08-2011-annexa.pdf.

ENDS Report (2009) Drinking water incidents rise. *The ENDS Report*, **414**, 20.

ENDS Report (2012) Tussle over plans to merge Welsh Regulator. *The ENDS Report*, **448**, 30–31.

Environment Agency (2008a) *Water for Life and Livelihoods. A consultation on the Draft River Basin Management Plan, Thames River Basin District*. Available at: http://www.environment-agency.gov.uk/research/planning/33250.aspx.

Environment Agency (2008b) *Water Resources in England and Wales – current state and future pressures*. Available at: http://publications.environment-agency.gov.uk/PDF/GEHO1208BPAS-E-E.pdf.

Environment Agency (2009) *Water for Life and Livelihoods. River Basin Management Plan, Thames River Basin District*. Available at: http://publications.environment-agency.gov.uk/PDF/GETH0910BSWA-E-E.pdf.

Environment Agency (2010a) *Managing Water Abstraction*. Available at: http://publications.environment-agency.gov.uk/PDF/GEHO0310BSBH-E-E.pdf.

Environment Agency (2010b) *Creating a Better Place 2010–2015. Water Supporting Strategy*. Available at: http://publications.environment-agency.gov.uk/PDF/GEHO1109BQXG-E-E.pdf.

Environment Agency (2012a) *Catchment Based Approach for a Healthier Environment*. Available at: http://www.environment-agency.gov.uk/research/planning/131506.aspx.

Environment Agency (2012b) *How the Environment Agency Manages Water Resources in England and Wales*. Available at: http://www.environment-agency.gov.uk/business/topics/water/133013.aspx.

Environment Agency (2012c) *Water Situation Report. Monthly Bulletin for England and Wales*, March 2012.

European Commission (2003a) *Common Implementation Strategy for the Water Framework Directive (2000/60/EC)*. Guidance Document No. 1. Economics and the Environment: the Implementation Challenge of the Water Framework Directive. Office of Official Publications of the European Communities, Luxembourg.

European Commission (2003b) *Common Implementation Strategy for the Water Framework Directive (2000/60/EC)*. Guidance Document No. 8. Public Participation in Relation to the Water Framework Directive. Office of Official Publications of the European Communities, Luxembourg.

European Commission (2012) *A Blueprint to Safeguard Europe's Waters and a Consultation on Policy Options for the Blueprint to Safeguard Europe's Water Resources*. COM (2010) 673 final. Available at: http://ec.europa.en/environment/water/blueprint.

European Environment Agency (2012) *Towards Efficient Use of Water Resources in Europe*. EEA Report/No. 1/2012.

Fox, P., Bond, D., Orr, P., Madeira, N., Riley, C., Rees, Y. and Davis, M. (2004) *Ribble Pilot River Basin Public Participation and River Basin Planning – Early Experiences*. Environment Agency. Available at: http://www.merseybasin.org.uk/archive/assets/48/original/MB15_Ribble_Pilot_River_Basin_Public_Participation_and_River_Basin_Planning_Early_experiences.pdf.

House of Lords (2012) *An Indispensable Resource: EU Freshwater Policy*. European Union Committee, 33rd Report of Session 2010–12, HL Paper 296.
Howarth, W. (2006) The progression towards ecological quality standards. *Journal of Environmental Law*, **18**, 3–35.
Howarth, W. (2008) Visions, strategies and realization. *Environmental Law Review*, **10**, 310–318.
Howarth, W. (2009) Aspirations and realities under the Water Framework Directive: Proceduralisation, Participation and Practicalities. *Journal of Environmental Law*, **21**, 391–417.
Institution of Civil Engineers (2012) *The State of the Nation. Water 2012*. Available at: http://www.ice.org.uk/Information-resources/Document-Library/State-of-the-Nation-Water.
Josefsson, H. and Baaner, L. (2011) The Water Framework Directive – A Directive for the Twenty–First Century? *Journal of Environmental Law*, **23**, 463–486.
Keessen, A.M., van Kempen, J.J.H., van Rijswick, M., Robbe, J. and Backes, C.W. (2010) European River Basin Districts: are they swimming in the same implementation pool? *Journal of Environmental Law*, **22**, 197–221.
National Audit Office (2010a) *Environmental Protection Briefing for the House of Commons Environmental Audit Committee*. National Audit Office.
National Audit Office (2010b) *Tackling Diffuse Water Pollution in England*. HC 188 Session 2010–2011.
National Planning Policy Framework (2012) Available at: http://www.communities.gov.uk/publications/planningandbuilding/nppf.
Office for National Statistics (2011) *National Population Projections 2010 – Based Statistical Bulletin*.
OFWAT (2009) *Price Review 2009*. Available at: www.OFWAT.gov.ukj/pricereview/pr09faqs.
OFWAT (2011) *Water Today, Water Tomorrow*. Exploring the costs and benefits of faster, more systematic water metering in England and Wales. Available at: http://www.OFWAT.gov.uk/future/customers/metering/pap_tec201110metering.pdf.
OFWAT (2012a) *Delivering Proportionate and Targeted Regulation – OFWAT's Risk-Based Approach*. Available at: http://www.OFWAT.gov.uk/regulating/compliance/pap_pos1203regcomp.pdf.
OFWAT (2012b) *Key Performance Indicators – Guidance*. Available at: http://www.OFWAT.gov.uk/regulating/compliance/gud_pro1203kpi.pdf.
OFWAT (2012c) *The Case for Change – Reforming Water Abstraction Management in England*. Available at: http://www.OFWAT.gov.uk/future/markets/waterrights/pap_pos20111205abstraction.pdf.
Phippard, S. (2012) A DEFRA perspective on the Blueprint. Paper given in *A Blueprint to Safeguard Europe's Water – Highlighting the Critical Issues for the UK*. SOAS 11 January 2012. Available at: http://www.ciwem.org/media/474436/Sonia%20Phippard.pdf.
Slob, A.F.L., Rijnveld, M., Chapman, A.S. and Strosser, P. (2007) Challenges of linking scientific knowledge to River Basin Management Policy: AquaTerra as a case study. *Environmental Pollution*, **148**, 867–874.
Thames River Basin Management Plan. Available at: http://www.environment-agency.gov.uk/research/planning/125035.aspx.
UKTAG (2012) *The United Kingdom Technical Advisory Group for the Water Framework Directive*. Accessible at: http://www.wfduk.org.
Watson, R. and Albon, S. (2011) *UK National Ecosystem Assessment. Understanding nature's value to society. Synthesis of key findings*. Information Press, Oxford.
Wright, S.A.L and Fritsch, O. (2011) Operationalising active involvement in the EU Water Framework Directive: why, when and how? *Ecological Economics*, **70**, 2268–2274.

3

Water Privatisation and Regulation: The UK Experience

John McEldowney

3.1 Introduction

Water privatisation is a global phenomenon and refers to nonstate actors involved in the delivery of water (World Bank, 2012). The UK model, as it has become known, involves the entire water system from abstraction to sewage treatment being sold to private firms (Schofield and Shaol, 1997). Monitoring and regulation oversight of privatised water companies becomes the responsibility of regulatory agencies supported by legislative and licensing frameworks. Privatisation may take a different form than the UK approach. Different types of private sector partnerships may be formed with state-owned water entities (Pinsent Masons, 2011). There is also a form of privatisation, common in France, where the actual operations and planning of water services are undertaken by private entities (Wollmann and Marcou, 2011).

On the basis of this broader definition and taking into account the growth of both population and water privatisation between 2007 and 2011, Pinsent Masons (2011) estimates that 909 million people in 62 countries, or 13% of the world population, were served by the private sector in one form or another. A World Bank report (2007a), which took account of the growth of both population and water privatisation, estimated that by 2015 almost 21% of the population will be served by the private sector engaged in some form of water delivery. The report did not include estimates of the number of people served by private companies on the wastewater side (World Bank, 2007b).

The organisation of this chapter is firstly to provide consideration of the structure of privatisation, followed by a discussion of the water industry and the domestic market. Then the water industry and the market for business customers are detailed, before outlining the challenges facing the future regulation of the water industry.

Water Resources in the Built Environment: Management Issues and Solutions, First Edition.
Edited by Colin A. Booth and Susanne M. Charlesworth.

3.2 The First Country to Fully Privatise its Water and Sewerage Business

After a hesitant start, utility privatisation took root in the early 1980s. The Littlechild Report (Littlechild, 1986) identified the absence of competition in the water industry as an issue, suggesting that regulation and privatisation might be an effective replacement for the 10 Regional Water Authorities and 20 private water companies brought in under the Water Act 1973. The 10 Regional Water Authorities replaced a mixture of local councils, regional water and river boards, taking responsibility for all water-related functions in river basins including operational and regulatory responsibilities. Littlechild found that the Regional Water Authorities were ineffective and largely underfunded, particularly for sewage treatment. Regulation was weak as the Regional Water Authorities acted as 'poacher and gamekeeper' with a general decline in standards and an absence in competition (Byatt, 2001). Hence water became one of many public utilities to be privatised in the United Kingdom during the 1980s (Byatt, 2001). The organisational model, including the regulatory structure, followed a similar pattern set by other privatisations, such as gas, telecommunications and electricity (Vogel, 1986; Foster, 1992). Within the privatisation legislation are a complex set of legal rules and regulatory techniques, such as licences and economic instruments (e.g. pricing mechanisms). Bundled in with these are various statutory duties and regulatory powers to promote competition and protect consumers, including lengthy technical utility contracts and licences. Private companies were formed under a plethora of licences and binding contracts and agreements. Strategies to protect the environment and competition policy as well as transparency in costs and access to the utility were also introduced (Graham, 2000).

The Water Act 1989 established the privatisation of the water companies (Littlechild, 1986). A mixture of regulatory bodies has oversight of the privatised water companies. OFWAT was established as an independent regulatory body and is the economic regulator with oversight of the water companies, including their functions and finance as well as consumer interests and promoting industry competition. Overall, OFWAT has to ensure that there is economy and efficiency, as well as sustainable development of the water industry. The Department for Environment Food and Rural Affairs (DEFRA) sets the water and sewage policy framework for England. In Wales, water and sewerage policy is set by the Welsh Assembly Government. The Environment Agency (www.environment-agency.gov.uk) has responsibility for environmental policy and the reduction of flood risk as well as sustainable development. Drinking water quality is specifically the responsibility of the Drinking Water Inspectorate.

Financial mechanisms, such as auditing, and reporting responsibilities over government departments concerned with utilities involve the National Audit Office (NAO), which also contributes to value-for-money studies. Regulation involves different subject disciplines including law, economics and political science and, in particular instances, science. It has created its own distinct subject regulation specialists, mainly lawyers and economists. Through drafting and regulatory experience, UK institutions and lawyers have rapidly gained an international expertise for their work, which forms a global market for economic and legal consultancy. The expertise of regulating the utilities provides a major income stream for leading London based city firms of solicitors and financial institutions (Guislain, 1997).

Water privatisation and subsequent regulation in the United Kingdom provide an example of an increasing degree of political policy-making in this area, adding considerably to the complexity and volatility of the regulatory system (UK Groundwater Forum, 1995; McCrudden, 1999; McEldowney, 1999). In that context, the importance of international markets, the impact of globalisation and change beyond purely domestic or European perspectives cannot be ignored. The current financial crisis, since 2008, has added financial uncertainty, especially in relation to large investment projects.

Local concerns in England and Wales also have to be taken into account. Drought conditions in early 2012 throughout the South of England have exposed the vulnerability of the water industry. The absence of competition within the water and sewerage industry is difficult to address, especially the problem of the high costs of water abstraction, the treatment of sewage and the provision of various retail services.

Since privatisation, >£90 billion has been invested to improve drinking water quality and upgrade existing pipes and sewage systems (DEFRA, 2008). The water industry has been the subject of numerous reviews. One of the most significant is the review of water industry competition, concluded in 2009 by Professor Martin Cave, a leading economist at the University of Warwick. Its recommendations are considered in the Coalition Government's White Paper, *Water for Life*, which was published in December 2011 (DEFRA, 2011). The White Paper is the result of the Coalition Government's policy and strategy for the future of water resources. It identifies some of the future issues and concerns over water resources, taking account of climate change (HM Government, 2008). The White Paper is strongly in favour of competition but falls short of direct government deregulation. In November 2011, a review of OFWAT concluded that changes are required in its role in the promotion of competition. A Water Bill 2013/14 is currently being considered by Parliment.

3.3 Water Privatisation and Structure

Water privatisation under the Water Act 1989 created a model of vertical integration (Ogus, 1994). Previously, public ownership of water was seen as a natural resource, publicly owned for the public benefit. Water privatisation was one of the most controversial of any of the Government's privatisation strategies because of the view that a natural resource ought not to be turned into a profit-making structure. The United Kingdom was the first country to privatise water and privatisation broke new ground, as at the time of privatisation there was no other country that operated a fully privatised water and sewerage business (Milman, 1999). Unsurprisingly, the privatisation of water proved a greater problem than the previous privatisations of gas or telecommunications (Ogus, 1994). Water privatisation had to address one of the most complex legal arrangements for water services due to ownership, financial and competition issues.

Before privatisation there were 10 public water authorities and 29 private water companies that made up the water industry. Many companies owed their origins to the Victorian period of municipal government. Small private water companies had been successful, often centred on spa towns and traditional Roman excavations. Underinvestment was a legacy of the past. Post Second World War, the expansion of new towns and cities had developed apace, without much thought being given to the infrastructure. Old waterworks and sewerage treatment plants struggled to meet enhanced EU requirements for drinking and bathing water. Limited competition regionally left 'natural monopolies' untouched by broader national issues.

Privatisation was eventually carried out under the Water Act 1989, which created a new public body, the National Rivers Authorities (NRA), later subsumed into the Environment Agency, under the Environment Act 1995, with the rights and liabilities of the existing water authorities divided between the NRA and successor companies. The 29 statutory water companies are retained as water undertakers for their areas. The successor companies inherited the responsibilities of sewerage and water subject to the terms of the instruments of appointment. The various commercial companies, as water and sewerage undertakers, received powers under the Water Industry Act 1991. The duty to maintain and develop an efficient and economical water supply and sewerage system falls under Sections 37 and 94 of the 1991 Act. The supply of water for domestic purposes 'must be wholesome and of adequate quality'.

The current law is complicated and housed within six major pieces of legislation: the Water Resources Act 1991, the Water Industry Act 1991, the Statutory Water Companies Act 1991, the Land Drainage Act 1991, the Water Consolidation (Consequential Provisions) Act 1991 and the Water Act 2003. As noted above, the Environment Agency established under the Environment Act 1995 is the major environmental regulator and takes responsibility for water pollution enforcement. There are equivalent responsibilities for other parts of the UK held by the Department of the Environment, Transport and Regions, Northern Ireland Office (DETR NI) (www.ehsni.gov.uk) and the Scottish Environment Protection Agency (SEPA) (www.sepa.org.uk).

In the area of water resources one significant change is the Water Framework Directive 2000 (Directive 2000/60/EC).This is an ambitious part of an ongoing strategy to develop and improve water quality in Member States. The Water Act 2003 amends the previous regulatory system over water set up at the time of water privatisation. This has resulted in OFWAT becoming (in 2006) a regulation authority as part of a panel of regulators. OFWAT's role as the water industry regulator is a key element in ensuring water standards. There is equivalent legislation in Scotland, the Water Environment and Water Services (Scotland) Act 2003, and for Northern Ireland, the Water Environment (WFD) Regulations (Northern Ireland) 2003.

The Water Act 1989 broadly followed some of the legal characteristics of the post-privatisation arrangements of the other main utilities (Prosser, 1997, 2004). The main regulatory instrument is the licence that contains a regulatory mechanism, which 'caps' the price companies may charge their customer. The annual increase is restricted to the 'RPI plus an additional factor K allocated to the companies on an individual basis for each of the next 10 years'. This is designed according to the first Director-General of Water Services, Ian Byatt, 'to off-set the significant investment programmes, which have been necessary to achieve the higher standards which we all seek' (Byatt, 2001).

In the case of water, reference to the Monopolies and Mergers Commission (MMC), now the Competition Commission, may be made by the Secretary of State for Trade and Industry following advice from the Office of Fair Trading (OFT). The water companies may appeal to the MMC if they wish to contest the action of the Director-General of Water Services in respect of determining the 'K' factor in the price cap (a pricing mechanism for setting consumer prices), amendments to their licences and accounting guidelines. The actual management of the industry is, subject to the legal framework identified above, left to the individual water companies to develop. OFWAT's remit includes a 10-year periodic review of the company, investment programme, management plan, efficiency standards and the regulatory regime in general.

The Water Act 2003 follows the broad framework contained in many utility privatisations, gas, electricity and telecoms, and also found in a number of features common to the arrangements put in place for the Office of Energy (OFGEM). Originally, it was thought that a single regulator would be retained and thus OFWAT would have survived without any amendment. In fact, a new regulatory authority replaces the post of Director-General of Water Services with new duties to provide sustainable development and to safeguard consumer interests. The legislation covers the following:

- Changes in the regulation for supervising the financial probity of water companies.
- A new duty on OFWAT to take into account sustainable development and environmental and social matters.
- A limited protection for water companies of confidentiality to ensure that competition between the companies is encouraged.
- Resolution of the problems of liability for water companies involved in abstracting water that causes damage.

There are additional powers that supplement existing powers granted to the Secretary of State to issue social and environmental guidance to the regulator. There are increased penalties for water companies who fail to perform to certain standards. There is also a new consumer council for water with increased powers to obtain information from water companies. Another important feature of the new arrangements is increased competition for new entrants into the water industry with a water consumption threshold of 50 m/year. Water licences are closely monitored through new revised powers shared between the Secretary of State, the Welsh Assembly, the Water Regulator and the Drinking Water Inspectorate.

The Water Act 2003 also contains detailed regulations covering water abstraction licences and conditions for the monitoring and supervision of abstraction in terms of water quality and standards and the requirements set by the Environment Agency. Increased penalties for abstraction offences are included and the removal of certain civil immunities from civil claims is intended to strengthen regulation of the industry. The Drinking Water Inspectorate gains new powers for the direct enforcement of legal proceedings in the area of providing water that is unfit for human consumption. Taken together, the Water Act 2003 sets tougher enforcement powers and provides a strengthened system for the regulation of the water supply. At stake are issues about the increase in competition intended in the Act and the mechanisms to monitor new entrants into the water industry.

3.4 The Water Industry and the Domestic Market

The previous Labour Government (1997–2010) undertook a distinctive approach to the rights of domestic consumers (House of Lords Select Committee, 2003–2004). The statutory duty to supply domestic premises was conditional upon only those premises connected to an existing water main, under Section 52 of the Water Industry Act 1991. Any connection to the water mains must be paid for by the person requiring the connection. One of the remarkable features of the current water arrangements concerns metering of the water supply. Nearly 80% of the water supply in England and Wales is unmetered (OFWAT, 2011). The charging system is one that must be approved by the regulatory authority, OFWAT, subject to licence conditions that may change once every five years. Water companies also have to produce a Code of Practice for the payment of charges that addresses the problem of those having difficulties making payments. There is no power to disconnect a water supply for the nonpayment of a bill and the use of pre-payment meters, common in the energy sector, is not permitted.

The previous government also took steps to address the ability of those to pay water charges. OFWAT's latest analysis is that 23% of households pay >3% of their income after housing costs on water and sewage charges and 11% pay >5% of their income; this amounts to ~2.4 million customers (OFWAT, 2011). The main initiative to address ability to pay came from WaterSure (Water Metering Standard, House of Commons, 2008). This recognises vulnerable groups and there is financial assistance from the water company to pay their bills, provided they meet a qualification category dependent on their means. This system of relief is linked to social payments and the number of children under 16, as well as any pre-existing medical conditions in the household. The scheme is narrowly defined and coverage is limited. Take-up amongst the vulnerable groups is also limited, with only 45 336 households benefitting from the scheme (in 2010–2011). The limited scope of the scheme has given cause for concern. Currently, the water companies are unable to interrogate social payments schemes to find out whether more eligible customers might benefit from the scheme. It is estimated that 60% of customers with affordability problems are unable to access the scheme.

The new Coalition Government (since 2010) plans to encourage a social tariff for the most vulnerable groups and plans to amend the current 'Vulnerable Groups Regulations' for that purpose. There is also a long-held view that metering of the water supply is most likely to be of benefit to single adults and pensioners, who probably pay larger bills than appropriate for their water use at present. The view of the Coalition Government is to pursue a metering policy with the water companies; however, they have drawn back from compulsory metering (House of Commons Standard Note, 2008).

There is also a need to provide, in certain circumstances, assistance to a water and sewerage company to secure a reduction in household bills. The Water Industry (Financial Assistance) Act 2012 provides the Government with powers to reduce water bills in certain regions and provide financial assistance in suitable cases. There are also powers to address one of the critical problems facing the water industry in the South of England, which has an ongoing shortage of water. Similarly, the government have powers to make an adjustment to the already high bills in South West England. The aim is to cut bills by £50 per year for all household customers until the end of the next spending review. This is an attempt to mitigate some of the problems facing South West Water (www.southwestwater.co.uk).

Finally, there is a major issue about the future regulation of the water supply when facing profound drought conditions. A recent risk assessment, UK 2012 Climate Change Risk Assessment, identified the demand and supply of water as vulnerable to change (HM Government, 2012). The environmental impact of drought requires a response and there are a number of potential mechanisms for change, including incentives to reduce water consumption and to encourage trading between water companies to compensate for variations in supply.

3.5 The Water Industry and the Market for Business Customers

The previous government commissioned Professor Martin Cave to undertake a review into water competition and innovation. His review, *Competition and Innovation in Water Markets* (2009) dealt with some of the concerns about how a free market in water services might be compatible with the environment and make the water industry more competitive (Cave, 2008, 2009). It was proposed that competition throughout the sector should help improve innovation and that the abstraction and licensing system should be reformed to bring to bear full costs on the industry. It was thought that this might help to takes steps to avoid water scarcity. Increased competition has not really been embraced under the Water Act 2003 and has been very limited, reflecting the way privatisation was undertaken.

Water is a natural resource with regionally privatised companies and, consequently, it has proved difficult to bring effective competition in the water sector. This is a reflection of the economic challenges facing the water industry in terms of investment. Company mergers are also difficult under existing rules. To date, in the UK, the limited opportunity for competition in the water industry has been subject to criticism. The main findings of the review are:

- Reducing the current threshold limit of 50 megalitres of water per year to allow users to choose their water supplier.
- Unbundling the water companies retail part of water services from the wholesale part and allow OFWAT to set the wholesale price of water and wastewater treatment at a level to allow efficient retailers to cover costs.
- Where there is a balance between water resources and abstraction licenses, all abstraction licences should be tradeable.
- In cases of shortfall in the water supply then the Environment Agency might fund buy-outs of abstraction licences through negotiated agreements.
- Currently any water company merger for firms with a turnover exceeding £10 million must be referred to the Competition Commission. One recommendation is to allow mergers between firms of up to a turnover of £70 million subject with no referral to improvements in OFWAT's comparative calculation of the effects of the mergers.
- The introduction of upstream competition for wholesale water supply and sewerage systems. This would require transparency of costs and a requirement for suppliers to obtain the best value for water, wastewater and infrastructure services (Cave, 2009).

The approach taken is intended to move the institutional organisation of the water industry in the direction of other utilities, such as gas, electricity and telecommunications. This is clearly in a direction that favours greater transparency through increased competition. The Cave Review (2009) is innovative and forward thinking but marks a remarkable change in the culture of the water industry. Unfortunately, the financial crisis has overtaken policy-making in terms of future funding possibilities. Nevertheless, the Cave Review points the way for the future direction of the water industry around increased competition, leading to efficiency savings. However, it will take some time for the proposals to become implantable.

Since the Cave Review, there has been a change of government with a refocus in direction. The Coalition Government's White Paper, *Water for Life*, in December 2011 provided suggestions for changes in the way a more vibrant and competitive market in competition might be introduced. One suggestion is to build on the changes made in the Water Act 2003, which allow a limited retail market that enables large customers to switch suppliers and for new entrants to enter the water industry as competitive companies. The application of the 2003 Act has been very limited and only one commercial customer has managed to shift supplier in the past six years (Cave, 2009). The law allows a nonhousehold consumer using 50 million litres of water to shift supplier in England. In Scotland, an alternative regulatory system has allowed >40% of the market to renegotiate the terms of their supply. The Coalition Government is set to allow new market entrants and encourage competition. However, the Cave Review (2009) had suggested there should be a separation of the retail businesses of the water companies and this has been rejected. The Government's view is that competition will be sufficiently encouraged without the break-up of the retail industry from the main water companies.

The Government's proposal to reduce the threshold allowing a customer to switch (from 50 million litres to 5 million litres) will allow a larger number of customers to change supplier. There are also plans to work with the 'Water Industry Commission' in Scotland, to establish a new market for retail water and sewerage services in partnership. This is intended to generate more diversity in supply and provide customers with better options. There is also a major initiative to make it easier to change supplier and allow greater flexibility. Particular attention is being given in a new Water Bill (2012–2013) to allow a transparent wholesale price access to be made available to facilitate new entrants. The retail market will remain regulated but subject to prevention measures intended to ensure that incumbent companies are not able to exploit their position against new entrant companies. OFWAT is expected to introduce new statutory market codes and regulated wholesale charges. There is a pre-existing special mergers arrangement in place for water that requires a merger to be referred to the Competition Commission if the turnover of the acquiring or acquired company is greater than a threshold of £10 million. The Coalition Government is considering increasing the threshold to £70 million. OFWAT is likely to be given an increasing role in fostering competition and engaging in fair and transparent access for new entrants. This may also require changes in the licensing arrangements to ensure the availability of information and transparency in the development of competition. Standards of service will be carefully regulated and monitored with an increase in overall levels of transparency. Emphasis is also placed on new developments and investments are to be encouraged.

The role of OFWAT was considered in a review carried out by David Gray in July 2011 (www.defra.gov.uk/environment/quality/water/industry). It was accepted that regulation had worked reasonably well and achieved stability in investment and forward planning. There were additional findings, however, that considered there were too many burdens on the water companies and priority should be given to reducing costs and introducing cultural changes within the companies. It was also felt there is a need for greater cooperation between the various agencies including OFWAT, the Environment Agency and the Drinking Water Inspectorate, which should engage more effectively with companies and other stakeholders. There needs to be greater clarity on the longer-term objectives for the sector to provide a coherent framework, which OFWAT can regulate. Strategies to accommodate a new regulatory framework under a Regulatory Compliance project to introduce risk-based regulation into the industry are calculated to increase effective regulation. The 'Future Price Limit Proposals' provide a new focus on outcomes and also on incentivising companies. Particular attention will be given to regulating leakages into the future (OFWAT, 1991, 1993, 1994).

3.6 Conclusions

Water privatisation under the Water Act 1989 created a model of vertical integration. Public ownership of water was seen in terms of a natural resource, publicly owned for the public's benefit. Privatisation of water was seen as the most controversial of any of the Government's privatisation strategies. It broke new ground as, at that time, there was no other country that operated a fully privatised water and sewerage business. Today, water privatisation and regulation is going through a complex and politically difficult experience with competing and, at times, contradictory policies. The Coalition Government published a Water Bill 2013 that includes provisions on competition. The Government has yet to have the legislation passed into law (House of Commons Standard Note, 2013).

The drive for competition in the sector has to be tempered by demands for value for money, affordable water charges and updating of the sewerage and water transport system. Large investments to bring the water industry up to standard have to be seen in the development of longer-term strategies. At times, this may be at odds with specific problems of climate change and drought conditions. Water regulation has to manage with financial pressures on consumers who have higher bills than other areas. Existing arrangements to provide help for vulnerable customers are hardly adequate to meet current needs. Environmental standards are increasingly being strengthened, including the quality of drinking water, and this places additional pressures on water companies to respond. Water companies face increasing fines for failing to meet stringent standards of performance.

The water industry is likely to remain a highly regulated industry with a minimal prospect of this being changed in the near future. The potential for increased competition and consumer choice will largely depend on the vitality of OFWAT and its ability to engage with water companies. The politicisation of many of the issues surrounding the regulation of the water industry is also a major influence (Regulatory Impact Unit, 1999). This creates regulatory tensions in terms of dialogue and debate about the most appropriate form of regulation (Veljanovski, 1991; Black, 2001). There is no single or ideal type of regulation that will deliver the expectations of all stakeholders. Day-to-day decisions have to be made and justified and legislative changes often need considerable time before they are made effective (Maloney and Richardson, 1995).

Domestic and commercial sectors are not easily reconcilable and this is likely to be the cause of considerable tensions within the industry (Veljanovski, 1993). The multidisciplinary nature of regulation and the role of legal analysis sets many challenges in the dynamics of water regulation after privatisation (Cabinet Office, no date).

References

Black, J. (2001) Decentring Regulation: Understanding the Role of Regulation and Self Regulation in a 'Post-Regulatory' World. *Current Legal Problems*, **54**, 103.

Byatt, I. (2001) The Water Regulation Regime in England and Wales. In: Henry, C., Matheu, M. and Jeunemaître, A. (eds), *Regulation of Network Utilities: The European Experience*. Oxford University Press, Oxford, pp. 79–93.

Cabinet Office (no date) London. Available at: www.cabinet-office.gov.uk/regulation.

Cave, M. (2008) Interim review of competition and innovation in water markets. *ENDS Report*, **407**, December, 50–51.

Cave, M. (2009) *Competition and Innovation in Water Markets*.

Department of Environment, Food and Rural Affairs (DEFRA) (2008) *Future Water*. Cm 7319. London.

Department of Environment, Food and Rural Affairs (DEFRA) (2011) *Water for Life*. Cm 8230. London.

Foster, C. (1992) *Privatization, Public Ownership and the Regulation of Natural Monopoly*. Blackwell, Oxford.

Graham, G. (2000) *Regulating Public Utilities*. Hart Publishing, Oxford.

Guislain, P. (1997) *The Privatization Challenge*. World Bank, Washington.

HM Government (2008) *Future Strategy, The Government's Water Strategy for England*, February 2008, Cm 7319.

HM Government (2012) *UK Climate Change Risk Assessment*. DEFRA, London.

House of Commons Standard Note (2008) *Water Metering*, SN/SC/1509.

House of Commons Standard Note (2013) SN06638, The Water Bill.

House of Lords Select Committee on the Constitution (2004) *The Regulatory State Ensuring Its Accountability*. 6th Report Session 2003–2004, HL Paper 68–I. London.

Littlechild, S.C. (1986) *Economic Regulation of Privatised Water Authorities: A Report Submitted to the Department of the Environment*. HMSO, London.
Maloney, W. and Richardson, J. (1995) *Managing Policy Change in Britain: The politics of water*. Edinburgh University Press, Edinburgh.
McCrudden, C. (1999) *Regulation and Deregulation*. Oxford University Press, Oxford.
McEldowney, J.F. (1999) *National and International Perspectives on Law and Privatisation*. British Institute of International and Comparative Law, London.
Milman, D. (1999) *Regulating Enterprise*. Hart Publishing, Oxford.
OFWAT (1991) *Paying For Water: A Time for Decisions*. OFWAT.
OFWAT (1993) *Paying for Quality: The Political Perspective*. OFWAT.
OFWAT (1994) *Future Charges for Water and Sewerage Services: The Outcome of the Periodic Review*. OFWAT.
OFWAT (2011) *Water Services Regulation Authority (OFWAT) Annual Report and Accounts 2010–11*. HC 978. OFWAT.
Ogus, A. (1994) *Regulation*. Clarendon Law Series, Oxford.
Pinsent Masons (2011) *Water Yearbook 2010–2011*. London.
Prosser, T. (1997) *Law and Regulators*. Clarendon Press, Oxford.
Prosser, T. (2004) Regulation, markets and legitimacy. In: Jowell, J. and Oliver, D. (eds), *The Changing Constitution* 5th edition. Oxford University Press, Oxford, pp. 351–374.
Regulatory Impact Unit (1999) *Modernising Government*. Cm 4310.
Schofield, R. and Shaol, J. (1997) Regulating the water industry, swimming against the tide or going through the motions. *The Ecologist*, **27** (1), January/February.
UK Groundwater Forum (1995) *Groundwater in the UK: a strategic study*. UK Groundwater Resources.
UK Parliament (no date) www.uk.parliament.
Veljanovski, C. (1991) The Regulation Game. In: Veljanovski, C. (ed.), *Regulators and the Market*. IEA, London.
Veljanovski, C. (1993) *The Future of Industry Regulation in the UK*. London European Policy Forum.
Vogel, D. (1986) *National Styles of Regulation*. Cornell University Press, Ithaca and London.
Water Metering Standard (2008) Note SN/SC/1509, 6 November 2008. House of Commons Library.
Wollmann, H. and Marcou, G. (2011) *The Provision of Public Services in Europe*. Edward Elgar, Cheltenham.
World Bank (2007a) *Strategy and Policies*, Washington, DC.
World Bank (2007b) *IDA at Work: Water Management*. Washington, DC.
World Bank (2012) *Achieving Financial Sustainability and Recovering Costs in Bank Financed Water Supply and Sanitation and Irrigation Projects*. Washington, DC.

4

Urban Water Economics

Graham Squires

4.1 Introduction

This chapter looks at 'environmental resource' issues in connection with water. Externalities as the third-party costs and benefits are analysed using water pollution as an applied example. This external cost is then examined in relation to the introduction of water pollution control. Natural resource economics is applied with regards to water production and consumption in urban areas, which have increasing wants for water resources that are scarce. Sustainable development objectives are further introduced in this chapter as both a theme and as applied water examples in practice. Valuing the environment using the example of water as a resource will also be demonstrated. International cooperation and agreement on water issues require economic thinking on both local and global scales, especially as water has less regard for the national administrative boundaries they cross.

4.2 Externalities

Externalities can be more broadly defined as the spillover or third-party effects arising from the production and/or consumption of goods and services in the marketplace. In order to improve capture of full costs in using water resources that are produced or consumed within urban areas, models can be used to conceptualise and value the external costs in using water resources. The degree of cost will depend on the sustainability in the production and consumption of water as a good or service. A more sustainable water provision would be one that has less external costs at a higher volume of output.

Water Resources in the Built Environment: Management Issues and Solutions, First Edition.
Edited by Colin A. Booth and Susanne M. Charlesworth.

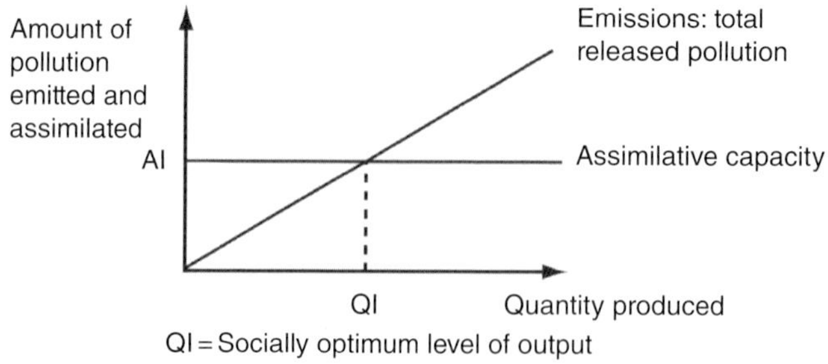

Figure 4.1 Socially optimal level of output.

4.2.1 Socially Optimal Level of Output for Water

The socially optimal level of output model (Figure 4.1) considers the assimilative capacity of external water costs, such as water pollution costs, to determine the point at which water resource costs stop being assimilated by the environment and society by beginning to detrimentally impact upon it. In using Figure 4.1 as a model, and the example of chemicals, at a global level the optimum amount of chemical by-products generated by the chemical plant that can be assimilated and absorbed by the water system are at quantity Q1, with a level of pollution emitted at a level of A1. This point is regarded as the socially optimal level of output – any further output would be directly affecting society through environmental degradation. If there is any further production beyond Q1, this would take the amount of water pollution beyond the finite assimilative capacity (expressed as a straight line) of the water system, and as a result generate a surplus external cost that is the difference between the total released pollution and the assimilative capacity.

4.3 Pollution Control (of Water Resources at a Market or Zero Price)

The firm could internalise some of these external costs of pollution if some intervention is applied by governing authorities. This means that incentives provided within the marketplace can alter and direct change in production and consumption patterns. For environmental goods and services, price incentives centre around restricting and regulating production and consumption, as at their most basic the price of inputs for water resources (without extraction and processing costs) are zero. Water resources are, in essence, zero-priced or free goods and services, such as the price of fresh drinking water (prior to collection and distribution) being free at source (depending on the ownership of the land from which it is being sourced). Without market incentives and regulation, an *unfettered* or uncontrolled price mechanism will use too much of zero-priced goods. In the fresh water example, a zero source cost of water will mean an overproduction of the product if it is to be turned into a commodity and sold, as the cost to produce is a noncontrolling factor, or, more simply, the zero price provides no incentive to produce less of the good. If a surplus and oversupply of free goods is produced there will be a cost to society above the private internal cost to the firm producing the fresh water for consumption. This excess cost and divergence away from the private cost is referred to as the 'social cost' and/or 'environmental cost'.

Market inefficiency of zero-priced goods can be reduced if individuals and organisations control their behaviour though some nonprice incentives. Behaviour can be considered a nonprice determinant of demand function with regards to changes in tastes and preferences, altering the level of demand at the same price point. An example of nonpriced determinants of demand that alter the behaviour (tastes and preferences or fashion) of zero-priced goods is through education. Unfettered markets could have improved water quality if consumers change demand through education, awareness and choice of less-polluting products. Incentives to environmental goods and services can therefore shape markets or correct market failure. These incentives can be viewed as either direct or indirect in the way in which they operate. Complete allocation control to the free market can be created via direct full privatisation. In the United Kingdom and in most other countries, the supply of water utility has been in control of the government. In a process of privatisation this relaxation of governmental control has meant that competing water companies can lease parts of the infrastructure and sell water as a commodity at a price determined by the quantity of units consumed by the end-user. The movement in viewing water provision from a public good to one that is private mirrors the transformation of value of the environmental resource as zero-priced to one that has a price attached to it. Other direct regulatory changes could see a shift in the opposite direction from a deregulated free market to that of a command-and-control approach to water distribution. For instance, the government approach could be to decide what can and cannot be produced and polluted with respect to water. For instance, water standards can be directly commanded and enforced via legislation, such as setting and enforcing water purity standards for firms that are operating near water outlets.

As well as direct regulation as an enforced incentive to value water resources, indirect incentives are another mechanism that can alter their functioning. Indirect incentives could be regulations such as taxes and charges that financially incentivise the value and allocation of water as a good and service. A higher service charge for water for residential use may discourage the quantity of water used by the households subject to the service charge. As a result of such indirect manipulation of the market, this will adjust the price in existing markets for water goods and services. A radical intervention into the market for water, which will change their economic functioning, is by creating 'new markets'. Governments can create new markets for environmental services by direct regulation and creation of new tradeable entities that act as a new 'currency' in the marketplace. A classic example is the creation of environmental carbon credits that are traded and regulated. Water credits could similarly run if water scarcity becomes an even more pressing issue.

4.3.1 Pollution Control: Charges, Permits and Deposits for Environmental Use

Indirect economic mechanisms can therefore shape and influence the market price and resultant value of water resources that are used and produced in urban areas. Table 4.1 consolidates five key instruments that can be used to influence the price of water resources on the market. Firstly, emission charges can be used to ensure that the emitter has to pay for any polluting activity, so with the charge they are encouraged to pollute less and thus be charged less. The charge is often related to the quantity (and quality) of pollutant damage. Therefore, in the case of water pollution by a firm in proximity to a water source, the cubic feet of water polluted could be charged, plus the intensity of pollution affecting its regenerative capacity to return to pure water will equally be charged more intensely.

Table 4.1 Five instruments using charges to price water use.

Instrument	Descriptor
1. Emission charges	Charge on discharge of pollutants into water – related to the quantity and quality of pollutant damage
2. User charges	For revenue raising in relation to water treatment, collection, disposal and administrative costs
3. Product charges	Harmful environmental products have charges attached when used in water production, consumption or disposal
4. Marketable permits	Permits given that match a water quota or allowance for water pollution levels and can later be traded
5. Deposit–refund systems	Deposit paid on a potentially water polluting product; authorities return deposit (or part of the deposit) depending on the level of water pollution on the returns

Secondly, user charges are another instrument to influence the price of environmental goods. As they are different from emission charges, a user charge is one where the more of a resource is used the greater the charge allocated. For instance, the more units of chemicals used in a chemical plant the greater the charge attached to the use of the input materials. Product charges are the third instrument, and are based upon the particular product that is being produced in order to gauge how much pollution is caused and hence charged by a regulatory force. The harm to a water system in producing cars may be more polluting than the creation of bicycles (if compared at equivalent values, e.g. one car and ten bicycles), and therefore the charge will be more heavily felt in the former product.

Marketable permits are the fourth instrument and draw on the carbon emissions trading example discussed earlier. Here a certain quota of water for a firm or nation could be set and any production under the set quota will generate a surplus water credit that can be traded to firms or nations that overproduce goods and services that pollute water. The fifth and final instrument that can affect the price of water resources, particularly in affecting price in controlling their pollution discharge, is the use of deposit–refund systems. A deposit–refund system is where a firm pays a deposit up-front to an authority if they are potentially likely to emit pollutants in their process. At a certain point the authority will return the deposit if no pollution has occurred or retain part of the deposit if there are partial pollution emissions. An example could be a drinking water bottling firm that pays a deposit to extract a certain amount of water, and if at a point of audit they have overextracted, they will lose part of the initial deposit.

4.3.2 Pollution Control Example: Water Pollution Tax

A specific and simple indirect regulatory approach to influencing market behaviour on water pollutants through price is via a tax. Here the mechanisms that occur through a water pollution tax are now demonstrated. A tax on production or consumption of water may be a strong incentive that is fed through into the price mechanism. More complex mechanisms can be attached to taxing water; for instance, a water tax could be increased at a graduated rate if the pollution content of water increases. The effect is also not entirely clear because the impact of a tax will affect the consumer as well as the producer, and it is the consumer who may substitute to a less-expensive (less-taxed) water source such as transported bottled water for drinking.

As well as these complexities associated with a water tax (as an example of indirect taxes as an instrument of environmental resource use), there are very clear disadvantages associated with a pollution tax instrument being used to affect the market for water resources. One disadvantage of a water tax is that it is socioeconomically regressive in terms of economic development. This is because the poor will pay a larger percentage of their income in a water tax and, inversely, wealthier social strata will pay less of a proportion of their income on a water tax. Secondly, a water tax may be detrimental as the tax adds to the tax base of the taxing country and hence, paradoxically, may encourage the use of excessive water use as public revenue can be extracted from it. Thirdly, difficulties of water taxes are that the outcome achieved in the taxing country may not be achieved in another country, which may be emitting an extremely high level of pollution. This will mean that, on a global scale, if there is not unilateral agreement between countries there will be a negligible impact on dealing with water pollution.

4.4 Natural Resource Economics and Water

Resources are the backbone of every economy. In using resources and transforming them, capital stocks are built up that add to the wealth of present and future generations. However, the dimensions of our current resource use are such that the chances of future generations (and the developing countries) having access to their fair share of scarce resources are endangered. Moreover, the consequences of our resource use in terms of impacts on the environment may induce serious damages that go beyond the carrying capacity of the environment (European Commission, 2011). Natural resource economics (NRE) is a branch of economics that is of interest in this field, and is one that can be tailored to suit interests involving the use and management of water.

Natural resource economics deals with the supply, demand and allocation of the Earth's natural resources. One main objective of natural resource economics is to be able to understand the role of natural resources in the economy better in order to develop more sustainable methods of managing those resources, such as water, to ensure their availability to future generations. Natural resource economists study interactions between economic and natural systems, with the goal of developing a sustainable and efficient economy. Natural resource economics aims to address the connections and interdependence between human economies and natural ecosystems. For instance, economic connection and modelling is made with fisheries, forestry and minerals (i.e. fish, trees and ore), air, the global climate and, of importance here, water.

The value of NRE has also moved into the management and value of recreational use as well as commercial use of resources including water. Of particular interest to NRE is the understanding via an economic lens as to differences between perpetual resources and exhaustible resources – what can be referred to as renewable and exhaustible when discussing the difference between ecological economic (EE) and environmental resource economic (ERE) paradigms. For urban areas and natural resource economics involving water, several features can be applied. For instance, natural water resources available within and used by urban areas will have an impact on its economy, and thus develop more sustainable methods of managing water resources. Cities with close proximity to natural resources may have advantages, such as low transportation costs of water availability, although the strength of the urban economy may enable it to command a greater consumption of natural resources from locations that have no direct ecological and environmental connection to the urban space under investigation, such as wealthy cities enabling transportation of water over great distances.

4.5 Resource Valuation and Measurement

Approaches for valuation and measurement in urban water economics apply certain techniques in order to measure and value resources and it is these techniques that will be introduced and explained. Key measurement and valuation techniques can be both positive and normative, asset-based, contingent or stated preference, market priced, opportunity cost and replacement cost orientated and involve charging instruments.

4.5.1 Valuation in Positive and Normative Economics

Economic costs and benefits will have different approaches depending on whether a positive or normative economic approach is taken. Positive economics is an approach that explains economic activity in a more factual and testable manner. Positive economics is more aligned to the natural sciences by appropriating disciplines such as mathematics in order to attribute relationships and correlations precisely between various components and factors. Using such a 'hard science' approach, positive economics attempts to describe *what is, what was, or what will be* in respect to time. This enables current, past and future economic activity to be retrospectively analysed for improved foresight in decision-making. Within positive economics the emphasis on facts means that disagreements in arguments, particularly over scarce resources and infinite wants can be resolved by an appeal to the facts. For instance, if it is argued that an urbanising geographical space consumes more water, the addition of statistics to provide evidence that urban areas consume more water compared to rural areas will be of prominence. Positive economics is also an economic approach where statements of 'truth' can be tested. As per the urbanising water example, facts revealing that there is greater water use in urban areas or rural areas can be tested, by recording and measuring how much water is consumed by all of the households within both urban and rural neighbourhoods.

Distinct to positive economics is 'normative economics'. Rather than provide testable facts as per positive economics, normative economics is more interested in dealing with what *ought to be* in society. It is thought in this approach that even if tested evidence provides some universal 'truth', it does not necessarily mean that the 'fact' is of merit to progress in society. For instance, the high correlation of higher water use per unit of space in urban areas does not necessarily mean that all urban areas should be deurbanised if water scarcity is to be dealt with. The 'should be' is more of interest to normative economics, where any disagreements, such as arresting the amount of water used or polluted, can be resolved by negotiation. Any negotiation will involve an element of value judgements being traded between parties in order to reach a clearer 'truth' through dialogue. This normative economic approach can therefore provide guidance on the desirability of a programme before or after implementation, such as whether water pollution in urban areas is the key contributor to poor health, and therefore leads to justification that a water efficiency programme should be in operation.

Positive and normative economics are not always diametrically opposed approaches, as aspects of both approaches can help to determine an understanding of what economic forces are operating on a particular event. Both applications are useful for the economic analysis of water. If the relationship between the economy and water consumption is considered, both positive and normative economic approaches will provide some input. Positive economics would be more interested in describing relationships of water consumption and the economy. For normative economics, this example would generate value judgements as to whether water consumption was the principle issue to be tackled in the first instance when considering the economy.

4.5.2 Water as an Asset

So there are different approaches to economics, as positive and normative. In considering these approaches, differing measuring and valuation techniques are used in scenarios involving water. The first technique to look at is an environmental asset approach that can be used to explain and quantify phenomena by urban areas. An asset approach to resources is one where something tangible has value now as it has the potential to generate future economic benefits. The asset approach also means that some past economic activity has provided its current asset value. It is also an asset resource that is controlled by an entity with respect to the result of these past transactions or events from which future economic benefits have been obtained. If 'life' itself is seen as a future benefit that can be generated from water, then water is, as a result, an asset that provides future life-support services. As such, in thinking of water as a measured and valued asset, any future fall in the future life-sustaining services will be reducing the asset value of water as the environmental entity. Inversely, it is therefore important, if the goal is to protect the environment, that individuals and groups try to prevent depreciation of the asset (in this case water) so that it can continue to provide aesthetic and life-sustaining services.

To value water as an environmental asset, measurement of its future benefits needs to be conceptualised and recorded. In doing so, thought and questioning need to be focused on what the environment provides as a benefit (and cost) for the economy into the future. Firstly, future benefits by natural resources such as water are as a raw material that can be transformed into goods and services in the production process (e.g. a soft drink). The future benefits of water as an asset are therefore the economic benefits of the goods and services produced from the natural resources, such as the future economic benefits of lakes to provide fresh water as a raw material that will last *X* amount of years.

Water asset values can appreciate (go up in value) or depreciate (go down in value) within a closed internal economic system. Otherwise, the value is externalised as water goods and services become waste. The water asset depreciates as the future benefit starts to generate fewer future benefits if there is a greater output of waste. An excessive provision of waste depreciates the water asset at a point where the absorptive capacity of waste is exceeded – to the point of pollution. To illustrate, a coal-fired power station that overheats the water supply in its cooling process will risk the deoxygenation of the water supply that will then not allow any further reproduction of natural resources in the food chain as benefits for other goods and services.

4.5.3 Contingent Valuation (Stated Preference): Willingness to Pay (WTP) and Willingness to Accept (WTA)

Water as an asset introduces the idea that economic concepts such as assets and revenues can be attributed to environmental resources and urban spaces. In addition, it has been intonated that asset approaches can begin to measure and value such concepts. Another valuation method in the economics of water is through the willingness to pay (WTP) technique. WTP is a standard measurement of benefit in economics, and is, as the term suggests, what an individual or group is willing to pay for a particular good or service that is not directly sold on the market. The WTP is the price, sacrifice or exchange that a person will enter into in order to avoid a detrimental consequence, such as incurring pollution or a reduced quality of life. An example of a WTP value would, for instance, be the £5 charge for visitors to enter a wetland national park and, as a proxy, provide some value as to the wetland national park's economic worth, in addition to the donations and public funds directed towards its conservation and preservation.

As it is a willingness to pay, the WTP concept has strong connection with demand rather than supply. Supply connections in this type of valuation method would be a willingness to accept (WTA) technique. In connection to demand a WTP therefore means that it is the purchaser that has a WTP that is equal to or exceeds the price. This means that the consumer has a WTP at a particular maximum (rather than minimum) value. To attach some illustration, if a dam was constructed to provide water for an urban area, an economic analysis using WTP could be attached. The WTP for such a dam would be to generate what the average price would be that consumers would be willing to pay for such a scheme in order to improve the quality of life. If, as an average, the urban areas' population (of, say, 50 000 people) were willing to pay a minimum of £1 per day and a maximum of £3 per day to use the service, the WTP principle would suggest a charge of £2 and an estimated annual valuation of the scheme at £36.5 million ($2 \times 50\ 000 \times 365$). Note that WTP is constrained by an individual's level of wealth (a function of demand). If applying WTP to quantifying one's own life, the WTP to not suffer the detriment of death (e.g. die of thirst) would be the maximum amount of wealth the individual owns (assuming that all beings wish to survive, notwithstanding suicide). Conversely, then, the WTA life would be an accepted value at any price (again, notwithstanding suicide) if all people wish to live irrespective of the 'price tag'.

4.5.4 Replacement Cost Approach

Another economic technique used to value environmental (built and natural) resources such as water is the replacement cost approach (RCA). The RCA operates in a similar way to the opportunity cost principle, with opportunity cost being the value of the next best opportunity forgone. For instance, the value of a river basin is the opportunity cost of it being a reservoir – or whatever next best alternative has the highest cost. Similarly, the RCA is a measure of the benefit formed through avoiding damage as a result of improved environmental conditions. For instance, the asset value of property close to the waterfront will benefit in value if flood resilience and resistance measures are taken to counter sea levels rising in response to global warming. As a specific measurement of replacement cost, an approximated market value can be attached in terms of the cost to prevent, restore or replace the damage in question. In the flooding and property example, the cost to restore flooded property will be the value, and therefore the aggregate cost of all potential flood replacement costs, of the environmental phenomena in question.

The replacement cost approach can be based on what it would cost at current prices to build an equivalent new structure (e.g. a reservoir) to replace the old one (which was built in a previous time on an entirely different cost structure). As an example, the occurrence of acid rain due to changes in vegetation dynamics from overcultivation can be valued using RCA. Acid rain deteriorates a nation's infrastructure, such as highways, bridges and historical monuments, and therefore has a cost to infrastructure if damage occurs; the reduction in acid rain can, as a consequence, also have a benefit value. If the replacement cost of repairing built structures from acid rain is $1 million, the benefit value of reducing acid rain to a level where no damage is incurred is therefore also $1 million. If a bill is passed by government to reduce sulphur and nitrate emissions by 50%, this would mean that from the replacement cost valuation of built structures, a benefit of 50% of the replacement cost, in this example $0.5 million, is attributed to sulphur and nitrite emissions reduction. In short, the replacement cost is the current market-value benefit of reduced costs in restoration or replacement damage of entities that can be valued, such as real estate and infrastructure. The valued benefits of reduced damage from environmental emissions as pollution are the savings realised from reduced expenditure on repairing, restoring and replacing the nation's valued goods and services, such as infrastructure and property.

This abstract costing concept has some difficulties, despite its ability to measure and value entities that may be described as intangible. Firstly, the environmental damage by 'abnormal' flooding or acid rain on to structures may not be able to be completely repaired or replicated. For instance, a historical monument has artistic value that may not be able to be (a) measured or (b) restored to its original form. If this replacement cannot be completed it cannot be given a full cost, and thus the benefit of reducing the environmental damage cannot be quantified. Even if a replica is put in the place of the original structure at a certain cost, the replica may have less worth than the original. As a result of this incomplete replacement, the replacement cost approach is a tool to consider for economic analysis of urban areas, particularly in the built environment that constitutes urban areas, but should be used with caution and expertise as to whether the full replacement cost has been attributed.

4.5.5 The Market Pricing Approach (Revealed Preference)

Cost, benefit, price and value can now be formed from an abstract of economic concepts rather than a simple acceptance of a number being applied to water. The standard approach to deconstructing what is meant by price is the use of the market of a good or service that incorporates a combination of supply, demand and quantity. Water as a resource is no exception and the market mechanism can be applied to provide an equilibrium price for water. In returning to basic neoclassical (and thus classical) market diagrams, valuation of water pollution can also be shown in market terms of price and quantity. In Figure 4.2, the increase in crop yield from improved water quality in irrigation means that the supply of

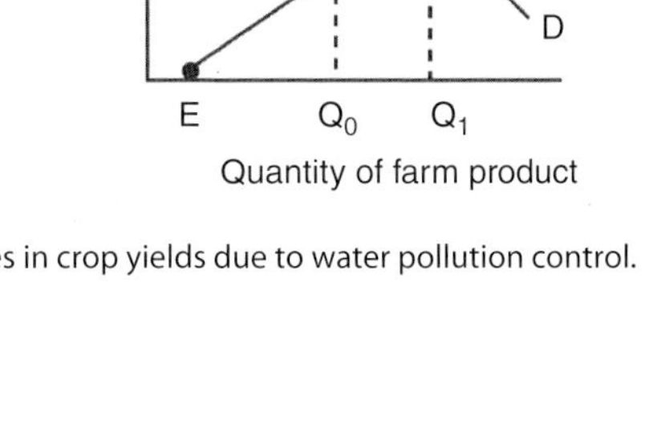

Figure 4.2 Market changes in crop yields due to water pollution control.

crops can be increased, as demonstrated by a shift in the supply curve. The previous revenue would have been represented by the rectangle formed by P0Q0, but now with increased production from water quality improvement the new revenue is Q1P1. The difference between these two revenues, before and after, is therefore the revenue gained from the increased yield and thus the value that can be attached to water pollution control.

4.6 International Issues and Development in Water

International perspectives are paramount when studying the allocation of water resources, within and between shared urban spaces. Water policy in response to environmental problems are seen to have developed at an international platform for two key reasons: (1) many of the world's threatened natural resources are shared resources or 'common property' (such as the oceans) and (2) because water pollution travels, as actions in one part of the world will affect the quality of life in another part of the world (Turner *et al.*, 1994). For the economics of water, difficulty may arise if a pollution control is implemented in one country when it is several other countries that are contributing to the pollution. In terms of governance, there may not be a transnational authority with the power to impose a water tax unilaterally. It is widely thought that poverty is the greatest 'cause' of environmental and water degradation. For instance, the poorer someone is, the less likely they are to worry about tomorrow, with the immediate concern of food today. This suggests that poorer communities will show little concern for water quality provision and will not undertake conservation practices to prevent water pollution. It is true that poor people are often trapped in just such situations. However, it is not so much that they do not care about tomorrow, but that they have limited ability to do anything about conservation if it diverts resources from the process of meeting today's needs. Nor is it universally true that poor communities fail to conserve resources such as water. Many have fairly elaborate structures of rules and regulations that are aimed at conserving water resources. As such, initial solutions may rest on rights and land resources that must be clearly defined to enable monitoring and regulation.

The urban poor, who are unable to compete for scarce water resources or protect themselves from harmful environmental conditions, are most affected by the negative impacts of urbanisation. The growth of large cities, particularly in developing countries, has been accompanied by an increase in urban poverty, which tends to be concentrated in certain social groups and in particular locations (United Nations Environment Programme, 2002). As for the environment, in cities with human and capital resource imbalances (i.e. developing country urban areas), further impacts are experienced. Effects can be felt further afield, such as pollution of waterways, lakes and coastal waters by untreated effluent. Water pollution from cities has an impact on residents' health as well as on vegetation and soils at a considerable distance. Urbanisation in coastal areas often leads to the destruction of sensitive ecosystems and can also alter the hydrology of coasts and their natural features, such as mangrove swamps, reefs and beaches, which serve as barriers to erosion and form important habitats for species. Water is a key issue in urban areas. The intensity of demand in cities can quickly exceed local supply. The price of water is typically lower than the actual cost of obtaining, treating and distributing it, partly because of government subsidies. Poor sanitation creates environmental and health hazards, particularly by direct exposure to faeces and drinking water (ibid.). The study of urban water economics (UWE) therefore has the opportunity to include concepts such as price, cost and value of water in many contexts, whilst ensuring that the normative economic concerns of meeting water 'wants' with water scarcity are secured for future generations.

4.7 Conclusions

The commodification of water has enabled the natural resource to become an economic good and service in itself, as well as providing a significant input into the production and consumption process. Water at source is 'in theory' at a zero (or free) price and could be considered a public good where it 'in theory' is nonexcludable to everyone on the basis of price and is nonrivalrous to everyone – in that the use of water by one person does not affect the use of another. The use of water in economic terms can be both internalised into the market mechanism and, due to any third-party 'spillover' effects in water resource use, many external costs and benefits are generated. Negative externalities are those such as water pollution, where the socially optimum level of absorption by the environment is exceeded. Water pollution costs can be measured, valued and conceptualised in economic terms – and pollution controls can consider economic benefits if intervention provides an improvement of the factors that are measured and valued (e.g. the benefit value of clean water in the productive process as opposed to polluted drinking water). Pragmatic economic tools and instruments can play some part in incentivising economic behaviour and direct water use to pollute less or internalise the external costs (and generate benefits). Instruments include: water permits, water emission charges, water user charges, product charges and water deposits.

The role of natural resource economics (NRE) clearly has a more complex task of dealing with unequal and often nonsystemic economies, with natural resources that operate within a fragile but dynamic ecosystem. The power of urban wealth, for instance, can reallocate and transport even the heaviest of water supplies over many miles. Water resources still have value and are measurable even though markets cannot easily put a price on them. Techniques via a revealed market preference for water are those in contingent valuation, where a stated preference through a willingness to pay (WTP) and a willingness to accept (WTA) is formed at a central point where a price can be stated. The asset value of water, in contrast to revenue, can also be applied, as holding stocks of water in a lake or reservoir enables future water benefits over many years rather than just in the immediate short term. Cost concepts of water can also be viewed in economic terms, for instance, with regards to opportunity cost, the next best alternative of land use for a reservoir may be for agriculture, the opportunity cost here is the economic value hypothetically extracted from agricultural activity. The replacement cost of replacing a particular water resource can also determine a measured economic approach – the cost to replace the supply of timber and its supply chain is one such 'replacement cost' that can be determined in demonstrating the total cost of water pollution (e.g. through acid rain). Environmental concerns involving water are both local and global in scale, with human induced changes and its effects crossing national boundaries (e.g. climate change). International as well as intranational solutions will be needed if weaker national and urban economies are less well equipped to engage with sustainable water use and to be resilient to water-based problems such as floods and droughts.

References

European Commission (EC) (2011) *Sustainable Use of Natural Resources*. European Commission: Environment.

Turner, R., Pearce, D. and Bateman, I. (1994) *Environmental Economics: An Elementary Introduction*. Harvester Wheatsheaf, New York.

United Nations Environment Programme (UNEP) (2002) *State of the Environment and Policy Retrospective: 1972–2002*.

Section 3 Water Infrastructure and Supply

5

Impacts and Issues of Dams and Reservoirs

Kim Tannahill, Peter Mills and Colin A. Booth

5.1 Introduction

Increased water demand is an inevitable consequence of an expanding population. This has provided the impetus and necessity to impound vast volumes of water behind large, physical barriers. It is estimated there are >45 000 large dams (i.e. reservoirs storing >4000 km^3 of water) in over 140 countries (mostly in China, the United States, India, Spain and Japan) (World Commission on Dams, 2000). As a consequence, it is predicted that human-made reservoirs are land-locking ~10 800 km^3 of water worldwide, which retains water from the oceans and lowers the sea-level to the equivalent of a 3-cm displacement (www.nationalgeographic.com).

Dams and the water they store are closely coupled to the economic fabric of a nation (Graf, 1999) and, besides ensuring a regular supply of water for potable use, their waters are also chiefly used for agricultural and industrial purposes, such as crop irrigation, waste product removal, cooling thermal power plants and generating hydroelectric power (see Chapter 7), plus they have provided improved navigation, offered flood protection and enhanced recreational activities (Blake, 1989; Schilt, 2006; Zhao *et al.*, 2012). Thus, dams offer a vital security to communities and, concomitantly, generate sizeable economic returns. However, global estimates suggest several tens of millions of people have been displaced due to the construction of dams and ~60% of the Worlds Rivers have been affected by dams and diversions.

It is perhaps not surprising, nowadays, that decisions to build new dams are being contested increasingly as human knowledge evolves, technologies advance and decision-making becomes more open and transparent (WCD, 2000). For instance, in South East England, where water supply shortages indicate there is less water available per capita than in many Mediterranean countries (Environment Agency, 2007), campaigners recently claimed a victory (March 2011) when the Secretary for State (on behalf of the Government)

Water Resources in the Built Environment: Management Issues and Solutions, First Edition.
Edited by Colin A. Booth and Susanne M. Charlesworth.

rejected Thames Water's plans to build a much-needed £1 billion reservoir near Abingdon, in Oxfordshire. Despite the localised community triumph, the regional water shortage issue remains an unresolved predicament.

Lessons learnt from past experiences indicate that dams can contribute to the delivery of a sustainable water resource but they need to be carefully planned and managed, and unlike the past, there needs to be due consideration of environmental and social issues through meaningful participation that leads to negotiated outcomes. This chapter briefly outlines the types of dams frequently designed and utilised by civil engineers and then moves on to outline the environmental and socioeconomic impacts and issues associated with several case study dams, before closing with a discussion on the insights of the United Nations Environment Program (UNEP) Dams Project.

5.2 Building Dams

Once the need for a dam has been established, the process of deciding which type of dam, where it is to be located and what materials are to be used begins. Typically, it is an iterative process, the materials to be used depending on the location and the preferred location depending upon suitable materials and local geography. Large deep dams develop huge water pressures, which tend to preclude the use of local soils and the choice is limited to steel reinforced concrete. The Three Gorges Dams and the Lesotho Highlands Scheme are prime examples of such large concrete structures.

The local geography and geology are also important technical factors. A narrow deep gorge allows for a relatively narrow dam wall to be constructed, but this is a compromise on storage capacity; a wider wall at another location is likely to provide a much larger storage capacity, but with increased construction costs.

It is usual practice for a number of proposed locations to be established, typically four or five. For each proposed location the technical feasibility is evaluated from some preliminary investigations. Such investigations are intended to give a clearer indication of the local geology and identify any major issues such as fault lines. Increasingly the technical feasibility is combined with environmental and socioeconomic assessments (discussed further in Sections 5.3 to 5.9), which together produce the selected site. Once the site is selected, detailed investigations are required to enable detailed design works to proceed, covering numerous aspects including foundation sealing requirements (grouting) and the geometrical arrangement of the dam.

There are numerous different types of dam construction, but a simple classification is either embankment dams or masonry/concrete dams. These two broad classifications are expanded upon in Sections 5.2.1 and 5.2.2.

5.2.1 Embankment Dams

An accepted definition of embankment dams is those 'made from nonorganic particulate material excavated from the Earth's surface local to the dam site and used more or less as excavated' (Blake, 1989). Commendable examples include Fort Peck (United States), Mangla (Pakistan) and Nurek (Tajikistan).

The traditional arrangement, sometimes referred to as 'zoned earth' and 'rockfill', consists of a clay core (which provides the water barrier) supported on both sides by shoulders of much higher permeability soils and rocks. In an effort to prevent failures such as those described in Section 5.10, filter material is provided between the clay core and the shoulders. Water will flow through the clay core but at a very slow rate. This seepage must be managed to avoid the build-up of pressure and the wash-out of material.

Many other arrangements are available: homogeneous earthfill and earthfill with various arrangements of toe drains, some of which combine natural materials with man-made layers (concrete face rockfill) (Fell *et al.*, 2005).

5.2.2 Concrete Dams

Concrete dams still utilise excavated material but it is bound together and strengthened by hydraulic cement. There are various stability mechanisms associated with concerete dam construction and these mechanisms dictate subdivision, which are defined below (Blake, 1989):

(i) Gravity dams: 'these are the simplest because they rely on gravitational force to oppose the overturning movement caused by the pressure of the reservoir water on their upstream faces' (e.g. Grand Coulee (US), Bratsk (Russia) and Solina (Poland)).
(ii) Hollow gravity dams: 'these require less concrete and therefore cost less to construct. Foundation requirements are more critical.'
(iii) Buttress dams: 'these also require less concrete than gravity dams. The buttresses support the upstream face of a buttress dam. The upstream edges of the buttresses are commonly widened so that they join, forming contiguous buttress dam. As an alternative, the upstream face may consist of small arches between buttresses, forming a multi-arch dam' (e.g. Daniel Johnson (Canada), Idukka (India) and Topolnitza (Bulgaria)).
(iv) Arch dams: 'these may be constructed as a whole in one large arch, spanning the valley sides and relying on them to carry the very large thrusts caused by reservoir water pressure. This type is the most sophisticated of the concrete dams and may be subdivided into single curvature and double curvature, according to whether the vertical section is straight, or is curved to further reduce bending moments in the concrete' (e.g. Glen Canyon (US), Kurobegawa (Japan) and Bhumiphol (Thailand)).

5.3 Historical and Global Context

It is believed that the first dams were constructed over 5000 years ago and existing records suggest that the oldest dam in the world was constructed in Jawa in Jordan around 4000 BCE (Blake, 1989; Tullos *et al.*, 2009). The earliest known dam remains are located in Egypt in Sadd El-Kafara and these have been dated back to 2600 BCE (Tullos *et al.*, 2009). Both these dams were embankment type dams using locally sourced materials. The Romans built the first concrete and mortar dams much later, around 100 AD, followed by arch dam construction in Mesopotamia around 1280 AD. In the 1600s, the Spanish took their expertise in dam building to the New World, initiating a global spread in river regulation (Tullos *et al.*, 2009). It should be noted, however, that despite the Roman's introduction of concrete and mortar dams as early as 100 AD, prior to the 1800s, the majority of the world's dams were of the embankment type. It was not until concrete technology and methods of structural analysis were well enough developed during the 19th century that this became the more popular choice.

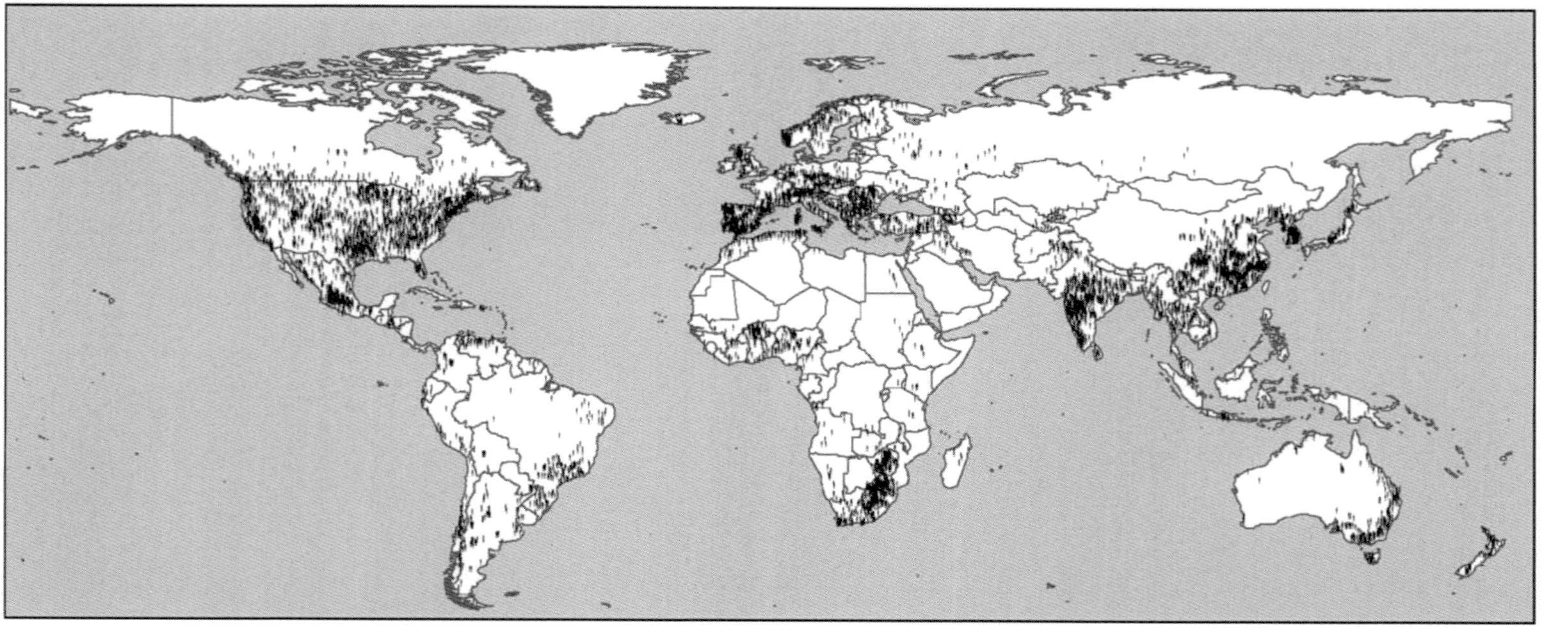

Figure 5.1 Distribution of dams and reservoirs that have been digitally georeferenced as part of an ongoing collaborative development of a comprehensive global database on dams and reservoirs (Tullos *et al.*, 2009).

Early dams were constructed to provide storage for irrigation water, but the birth of industrialisation shifted the focus for large-scale water use. In the United Kingdom, the industrial revolution saw an increase in demand for water for industrial processes, transport and the support of a growing population. During the 18th century, dams were built primarily to store water for canals, during the 19th century, water supply was a priority and by the early 20th century, they were used to supply power to the aluminium smelting industry in Scotland and Wales (Blake, 1989).

The latter half of the 20th century saw a peak in dam construction on a global scale due to an increase in population and rapid economic development increasing the demand for electrical power. According to the International Commission on Large Dams (www.icold-cigb.org), current figures indicate that there are >45 000 large dams worldwide (with large dams being defined as those greater than 15 m in height or having a storage capacity greater than 3 million m^3). Figure 5.1 is an illustration of the global distribution of dams and has been derived from a digitally georeferenced database of all dams and reservoirs worldwide that is being developed as part of an ongoing research effort.

Although there was a rapid rise in the amount of dams being constructed, the number of embankment dams being built fell during the second half of the 19th and early 20th centuries. This period saw an increase in production of concrete dams, with the early half of the 20th century seeing the proportion of embankment-style dams falling to around 30%. Conversely, the latter half of the century saw an improved understanding of embankment dams due to advances in soil mechanics and an increased capacity of earth-moving machinery that led to a rise in their popularity and a shift in proportion to around 80% (Blake, 1989).

According to Tullos *et al.* (2009), the fragmentation of large river basins (those with a historical mean annual discharge of greater than 350 m^3) by dams is a global problem. Approximately half are fragmented on the main channel and two-thirds fragmented on both the main stem and major tributaries (Nilsson *et al.*, 2005).

While dams allow us to create reservoirs of water for a variety of uses, such as irrigation, commercial consumption, household use, etc., the end result can often be an overexploitation of water resources. According to Tullos *et al.* (2009), there are at least six out of the world's 292 large river basins that have reservoirs where storage exceeds the annual discharge (Manicougan, Colorado, Volta, Tigris–Euphrates, Mae Khlong and Rio Negro) (Nilsson *et al.*, 2005). Of the remaining river basins, another 14 major rivers have more than 50% of their annual flow diverted for reservoir storage and some major rivers, including the Colorado and the Nile, no longer reach the sea year round. As a result, the impacts of damming rivers are significant and it is important to understand how these impacts affect the environment and affect us socially on both a temporal and spatial basis.

5.4 Environmental Impact

Modification in fluvial channel hydrologic dynamics, such as flow regimes and the ability to transport sediment, can alter the response of many variables and, ultimately, upset a river's natural cycle. The result will be a modification of deposition and transportation processes, which can adversely impact the river's ecology and biodiversity (Stevaux *et al.*, 2009) through altering habitats. The barriers created by dams alter the natural flow of the water and can interfere with ecological processes and change the spatial configuration, surface water temperature and quality of water in aquatic habitats (Xiaoyan *et al.*, 2010). In fact, according to Zhao *et al.* (2012), the construction of dams for flood control, navigation and hydroelectric power, etc., are often cited as having the most significant impact on rivers on a global scale by reducing connectivity, fragmenting watersheds, eroding downstream channels and altering key hydrological processes.

The most obvious impacts of dam construction can be seen through landscape changes (i.e. land use and land cover change) in the outer river area, and these have a fundamental reciprocal relationship with ecological processes (Ouyang *et al.*, 2010). However, the hydrologic regime in the inner river area is more immediately affected through blocks in the continuity of hydrology, disruptions in sediment flow, alteration in surface and subsurface water levels, changes in magnitude, duration, frequency, timing, predictability and variability of flow events and modification in seasonality of flows. The result of these changes is a loss of biological diversity and a decrease in the ecological functions of the aquatic ecosystem (Nilsson and Berggren, 2000; Ouyang *et al.*, 2011). Studies carried out on the aquatic habitat communities in the middle reaches of the Lancang River in China after the construction of the Manwan Dam support this and Xiaoyan *et al.* (2010) state that 'Manwan Dams across Lancang River changed habitat of landscape and interfered with aquatic habitat, including flow regime, surface water temperature and water quality that could influence species number and composition. The cover area of water body increasing was converted from the reduction of forest, shrub, cultivated land and other habitats. Discharge between pre-dam period and post-dam period had a great variation in dry season and wet season.' They also found that surface water temperature variation and flow rates accelerated the growth of green algae and ultimately changed the number and composition of fish species.

5.5 Socioeconomic Impact

The formation of the World Commission on Dams in 1998 was in response to growing global concern about the social costs of large-scale dam projects and how to solicit meaningful participation from those most affected. This prompted a number of social impact assessments (SIAs) on various large dam projects to evaluate the short- and long-term effects of dam construction (Tilt *et al.*, 2009).

The impacts can be substantial and, according to Tullos *et al.* (2009), can include 'migration and resettlement, changes in household size and structure, changes in employment and income-generating opportunities, alteration of access and use of land and water resources; changes in social networks and community integrity, and often a disruption of the psychosocial well-being of displaced individuals'. According to the World Commission on Dams (2000), approximately 80 million people worldwide have been displaced due to dam construction, resulting in loss of land and increased unemployment. Relocation efforts have also been shown to lead to higher population densities and greater struggles over available land (Brown *et al.*, 2009).

Many studies have been conducted on large-scale dam projects to assess the social impacts. Among those studied are the Lesotho Highlands Water Project (LHWP) and the Manwan Dam in China.

5.6 Socioeconomic Impacts of the Lesotho Highlands Water Project

The project was designed as a water delivery scheme between the governments of South Africa and Lesotho and was designed to include five dams linked to cross-national tunnels constructed in four phases over a period of 30 years (1987–2017). The largest reservoir, a

185-m double curvature arch dam with a capacity to hold 1950 million m^3 was constructed at Katse in 1993, followed by the Mohale Dam, with a capacity to hold 947 million m^3. To date, three of the dams have been completed. The aims of the project are to sell, transfer and deliver water from Lesotho to South Africa and to create a hydroelectric power station that would allow Lesotho to generate its own electricity (Tilt *et al.*, 2009).

The communities disrupted were primarily rural with subsistence-oriented agriculture being the main way of life. The effects of the construction project on these communities were varied and included:

- change in their relationship to the environment and their resource base;
- loss of potable water sources and natural springs;
- decrease in access to wild vegetables and herbs necessary for food and medicine;
- loss of access to forests and wooded areas that were submerged in the reservoir; and
- loss of some of their most productive arable land for agriculture.

There were also changes to the employment structure of the surrounding areas and, while it was intended that the project would create jobs locally, few local people were actually employed by the LHWP and jobs that did go to the local population were usually given to men. As a result, most of the jobs taken by the female population were informal, unregulated and poorly paid.

There were also negative impacts on the infrastructure, transport and housing that affected almost 20 500 residents in the surrounding areas. The worst of the adverse effects could arguably be found in the Khokhoba village, which found itself centrally located by a main traffic intersection. The main road through the middle of the village was expanded and paved and the pasture lands surrounding the village were appropriated so that the Lesotho Highland Development Agency (LHDA) so that they could build a hotel, shopping centre, offices and a residential village for employees.

It should be noted that it is a documented obligation of the project that current standards of living for those impacted by the project cannot be adversely affected. Development agencies are required to have mitigation and/or compensation strategies in place, but as the dam is a development project and development programmes are justified as poverty reduction methods, it can be argued that dam infrastructure projects are, in themselves, mitigative measures for those impacted (Tilt *et al.*, 2009).

5.7 Socioeconomic Impacts of the Manwan Dam, Upper Mekong River, China

The exploitation of the upper Mekong River for hydroelectric purposes began around 60 years ago and has had measurable negative impacts on the surrounding communities. The Manwan Dam was completed in 1986 with a height of 132 m and a reservoir area of 23.6 m^2. Its construction resulted in the destruction of 411 ha of farmland and 562 ha of woodlands and displaced 7260 farmers and their families. Economic productivity was significantly impacted due to:

- decline in agricultural productivity;
- shortage of irrigation water;
- increasing electricity costs; and
- depletion of forest resources (Tilt *et al.*, 2009).

Compensation for lost housing is a key part of China's 'Developmental Resettlement Policy', which mandates that the social and economic impacts on communities must be minimised. However, compensation offered was too little for many people to build comparable housing as property prices were fluctuating due to the fact that China was undergoing the transition from a centrally planned economy to a market economy (Bartolome *et al.*, 2000).

5.8 Environmental and Socioeconomic Impacts of the Three Gorges Dam, China

The Three Gorges Dam is the world's largest hydroelectric scheme (Jackson and Sleigh, 2000) and was built with three main purposes:

- to reduce flood damage;
- improve river transport; and
- generate electricity (Jackson and Sleigh, 2000; Zhang and Lou, 2011).

Construction of the concrete gravity dam began in 1993 and was completed in 2009. It is 185 m high and has a reservoir capacity of 39.3 billion m^3. Its construction was controversial due to the massive scale of the project and the potential environmental and social implications.

Many of the predicted environmental problems documented in the Environmental Impact Assessment (EIA) during the 1980s have been realised and ongoing monitoring has shown that there are environmental changes downstream. The area is biologically sensitive due to its complex topography and climatology and distinctive geographic location. It is home to 6400 plant species and 3400 insect species, some of which are endemic to the region (Zhang and Lou, 2011). It was predicted that the river would change course and seasonal flows would be altered (Jackson and Sleigh, 2000). As discussed in Section 5.4, this would have an adverse effect on the ecology and biodiversity. As the reservoir is located in an ecologically fragile region, the environmental impacts are sizeable. It has necessitated ongoing environmental monitoring since 1997 and the implementation of conservation strategies to ensure the survival of endangered and rare species. According to Jackson and Sleigh (2000), ex situ programmes have successfully preserved known endangered plant species and projects on the artificial breeding for fish continue to increase populations in the Yangtze River. Success has been limited, however, during attempts to preserve aquatic mammals through establishing nature reserves.

Levels of waste in the reservoir also remain problematic and reduction of pollution is a necessary step to ensure good water quality. However, there is a lack of funding available to deal with treatment that may help alleviate the pollution loading and help stabilise the river banks to reduce the internal sources of pollution loading due to sedimentation.

There are also many adverse social implications associated with the construction of the Three Gorges Dam project. The lives, habitat and economy of at least 20 million people upstream of the dam were affected and at least 300 million people downstream. This resulted in the relocation of 1.3 million people, most of whom were farmers who lost their livelihoods through the loss of 34 000 ha of agricultural land. It has been suggested that of those people relocated, some may never recover economically or socially and those areas with unexpected increased populations may suffer from increased conflict (Jackson and Sleigh, 2000).

Unfortunately, the resources are not available to deal with these problems effectively as they occur. In order to reduce pollutants from agricultural runoff, this requires a shift from a land-based to a knowledge-based economy and waste materials from internal sources of pollutants and shipping will only increase as navigation of the river has been improved.

5.9 Dam Risks: Incidents and Failures

Aside from the documented environmental and socioeconomic problems, there is also the impact to local populations and ecology from dam failures. Embankment dams made with earthfill and/or rockfill are the most common type globally, and they are most likely to fail through unrelenting water pressure seeking out cracks and fissures through the embankment or under the embankment. According to Warren (2011), 'By creating a reservoir, the potential of water to be released in a sudden uncontrolled manner is also created, posing a small but finite risk to those living in the path of a dam break flood wave.'

In 1864, the Dale Dyke Dam burst killing 244 people during the first filling of the reservoir. This is one of the most devastating historical dam failures in the United Kingdom. The outcome of the inquiry into the disaster was a statement from the jury that suggested the need for action that would result in 'governmental inspection of all works of this character; and that such inspections should be frequent, sufficient and regular'. After two further dam failures and more lives lost, the United Kingdom saw the introduction of reservoir safety legislation. The Reservoirs (Safety Provisions) Act 1930 introduced regular reservoir inspections by specialist engineers. This act was further developed in the Reservoirs Act 1975 (Warren, 2011) (recently amended under the provisions of the Flood and Water Management Act 2010 (Warren, 2011)).

The most recent potential catastrophe in the United Kingdom was the Ulley Reservoir, near Rotherham, South Yorkshire. The reservoir was created in 1873 by the construction of a 16 m high embankment dam. The spillway channel, which was designed and constructed to avoid overtopping in the event of a flood, failed (June 2007) when there was unexpected severe rainfall. The failure of the wall led to rapid erosion of the dam earthfill material and threatened the overall stability of the dam embankment. According to Hinks *et al.* (2008), the investigators were unable to determine with any certainty the physical mechanisms by which the spillway wall collapsed. Research was carried out to establish the pressures that could be exerted on masonry blocks as it was thought that the spillway collapsed due to pressure differentials on this structure; recent research has confirmed that this is a feasible mode of failure (Winter *et al.*, 2010). Thankfully, the failure did not cause the dam to breach and the potential catastrophe was averted.

The generic causes of structural failures are listed in Table 5.1. However, insufficient investigations into site geological conditions are the cause of many catastrophic dam failures. Although there is generally a very low probability of failure, they do happen globally on an annual basis and are investigated, with the findings becoming public knowledge. Noteworthy examples of geological failures include the Malpasset Dam (France) and Vajont Dam (Italy); a human-related failure includes the Austin Dam (United States).

Table 5.1 Generic causes and examples of structural failures (adapted from Warren, 2011).

Causes	Examples
Project planning	• Lack of clear scope • Conflicting client expectations
Site investigation	• Inadequate scope or extent of ground investigations • Misinterpretation of information
Design errors	• Conceptual design errors • Lack of redundancy • Failure to identify all loads and load combinations • Calculation errors • Detailing deficiencies • Specification deficiencies • Failure to consider surveillance, monitoring and maintenance
Construction errors	• Inappropriate temporary works • Improper sequencing • Improper methods or timing of construction • Excessive construction loads
Material deficiencies	• Material inconsistency • Premature deterioration • Fabrication defects
Operational errors	• Structural alterations • Operation beyond the scope of the design • Change in structure use • Inadequate surveillance, monitoring or maintenance

The Malpasset Dam, in Provence–Alpes–Cote d'Azur, southern France, was an arch dam (60 m high, 220 m long) created in 1954 to supply water and irrigation for the region. Unfortunately, lack of funding meant a thorough geological survey was not conducted and/or the local lithology was not fully appreciated because the foliation in the schist and gneiss bedrock (relatively impermeable metamorphic rocks) aided the development of a large uplift pressure on the dam, which led to a wedge-type failure that caused the entire wall to collapse (2 December 1959). This resulted in the release of a wave of water (~40 m high) that destroyed infrastructure and buildings in its path, flooded vast areas of Frejus town and killed >400 people.

The Vajont Dam (260 m high) is a double arched dam and despite experts indicating the surrounding land was unstable the dam was built in 1959. The dam has never failed and remains standing in Erto e Casso, Province of Pordenone in northern Italy (albeit with an empty reservoir). However, a massive landslide (300 million m^3 of material) fell into the reservoir, causing a huge displacement of water and a major overtopping of the dam (9 October 1963). The resultant towering wave (~100 m high at the crest) swept away several downstream villages and was responsible for >2000 deaths.

The Austin Dam in Pennsylvania, United States, was a large concrete dam (15 m high, 160 m long) created in 1909. It was originally designed to be ~9 m thick when built but, for whatever reason(s), it was only constructed at two-thirds of its intended breadth. Therefore, it is perhaps not surprising that, soon after its completion, it was noted that pressures caused by holding back the water (760 megalitres) were forcng the dam to bow and the concrete construction to crack. Efforts were made to alleviate the issue but the dam failed (30 September 1911), with the loss of ~80 lives.

5.10 Insights into the UNEP Dams Project

The last two decades has seen an increase in awareness of the impacts and performance of dams and this has led to conflict and heightened the debate over the costs and benefits of large-scale dam projects. In April 1997, a representative of diverse interests came together in Gland, Switzerland, to discuss some of these controversial issues surrounding dam construction and its effects on the local community and environment, and from this workshop was founded the United Nations Environment Program (UNEP) and the World Commission on Dams (WCD).

The UNEP Dams Project was undertaken by the WCD and their task was 'to conduct a rigorous independent review of the development effectiveness of large dams, to assess alternatives and to propose practical guidelines for future decision-making'. They recognised that the design, construction and operation of dams themselves are not the only issues that need to be considered when assessing the cost/benefit implications of large- and small-scale projects, and that a variety of environmental, social and political implications on which our development and desire to improve our well-being are based must also be taken into consideration.

The Project aimed to evaluate a number of dams on a global scale and stated that of those studied, a considerable amount of projects fell short of their physical and economic targets. However, there were still a considerable amount of benefits and services accounted for, such as 12–16% of global food production and generation of ~19% of the world's electricity supply.

5.10.1 Environmental Performance

The Project findings largely supported the impacts that have been documented previously in this chapter. The WCD states that 'Large dams generally have extensive impacts on rivers, watersheds and aquatic ecosystems.' Loss of wildlife habitat was observed through inundation of reservoir areas and upstream catchment areas generally suffered from degradation. They further observed that increased greenhouse gas emissions due to rotting vegetation and carbon inflows from dam basins were evident. However, the study indicated that it could be possible to achieve some mitigation through early cooperation between ecologists, designers and those likely to be affected. Further advances can be made through improving policy and legislation, particularly those that promote site selection that will avoid building dams on main river stems. A number of countries, particularly the United States, have been attempting to restore ecosystem functions and fish populations through the decommissioning of both small- and large-scale dam projects. This proved successful after the decommissioning of the Grangeville Dam on Clearwater Creek in Idaho in 1963. The dam was built in 1903 and caused a blockage of the passage for migratory fish due to excessive sedimentation. Within 6 months of removal, the river has washed out the accumulated sediment with no recorded downstream effects and fish were able to resume their migratory pattern.

5.10.2 Social Performance

The WCD study found that in terms of social performance, 'past decision making and planning efforts have often neither adequately assessed nor accounted for the adverse social impacts of large dams'. As previous authors detailed in this chapter have found, there are serious and lasting effects on the lives, livelihood and health of the local communities affected. The project found that the ability or commitment to cope with displaced peoples was often lacking. Of the 40–80 million people that are estimated to be displaced on a global scale through large-scale dam projects, many are not recognised as such and thus not resettled or adequately compensated. Of those that were resettled, policy concentrated on physical relocation and, thus, there was a failure to consider the economic and social development that would have allowed them to regain their livelihood. However, while the report found that there were improvements in policy and legal requirements, the weak link when considering dam construction projects was during the planning and decision-making phase and it is through these stages that there were opportunities for reducing negative impacts.

5.10.3 The Way Forward

The Commission has attempted to develop a framework within which developers and those involved in construction can operate effectively to minimise negative impacts. Five key stages have been identified throughout the development process that provide a framework within which stakeholder groups and decision-makers can work to ensure that they comply with agreed procedures and commitments. The stakeholder involvement is essential to the process.

The process begins with the selection of the preferred development plan. This will establish what the needs are that need to be met and whether or not a dam is the preferred option.

Stage 1. Needs assessment. Establishing what the local and national energy and water needs are. The needs assessment is validated by a decentralised consultation process.

Stage 2. Selecting alternatives. This involves identifying the preferred development plan from among the full range of options. It is important during this stage that the assessment is carried out using criteria that gives equal consideration to social and environmental concerns as to technical, economic and financial aspects. If the feasibility study offers a dam as the preferred option, then there are three further key decision points that are required for project preparation, implementation and operation.

Stage 3. Project preparation. This involves verifying that agreements are in place before tender of the construction contract. This is a crucial planning stage and the tender is conditional upon reaching an agreement for mitigation measures.

Stage 4. Project implementation. This stage is designed to ensure compliance before commissioning and covers procurement and construction. Licenses will only be issued once compliance with benefit sharing and mitigation measures has been confirmed.

Stage 5. Project Operation. Adapting to change within the boundaries of mitigation strategies as agreed throughout previous stages.

The benefits of using this key stage approach are described in the Dams and Development documentation (www.dams.org) as: 'lowering risks to livelihoods and cost escalation, reducing the number of disputes, and encouraging local ownership. In the short term, additional financial resources for needs and options assessment will be required to achieve compliance with the Commission's policy principles, and efforts will be required to strengthen institutional capacity. In the longer term, the potential exists for major cost savings and increased benefits.'

5.11 Conclusions

An expanding population and a rise in industrialisation has facilitated the need for storage of large amounts of water to ensure a regular supply for potable, agricultural and industrial purposes (Shilt, 2006; Zhao *et al.*, 2012). However, dams offer vital security to communities and generate economic returns; they are also associated with various negative environmental and socioeconomic factors (World Commission on Dams, 2000).

While dams are a necessary part of our developing world, the findings of the WCD's Dams Project indicate that many mitigation strategies and policies are not properly implemented throughout the lifecycle of the construction project and generally prove ineffective. They suggest that the key to tackling the negative impacts of both large- and small-scale dam projects lies with improved communications between designers, decision-makers and stakeholder groups, particularly at the planning and decision-making stage. This allows for an open and transparent participatory dialogue to ensure that we address water shortage issues and deliver a sustainable water supply that will minimise the impacts on the environment and local communities without political intrusion.

References

Bartolome, L.J., de Wit, C., Mander, H. and Nagraj, V.K. (2000) *Displacement, Resettlement, Rehabilitation, Reparation, and Development: China Report*. World Commission on Dams Thematic Review 1.3. World Commission on Dams, Cape Town.

Blake, L.S. (1989) *Civil Engineer's Reference Book*, 4th Edition. Butterworth-Heinmann Ltd, Oxford.

Brown, P.H., Tullos, D., Tilt, B., Magee, D. and Wolf, A.T. (2009). Modeling the costs and benefits of dam construction from a multidisciplinary perspective. *Journal of Environmental Management*, **90**, 5305–5311.

Environment Agency (EA) (2007) *Making Better Use of Our Water Resources: Identifying Areas of Water Stress*. Environment Agency Publications, Bristol.

Fell, R. (2005) *Geotechnical Engineering of Dams*. A.A. Balkema Publishers, Leiden, The Netherlands.

Graf, W.L. (1999) Dam nation: a geographic census of American dams and their large-scale hydrologic impacts. *Water Resources Research*, **35**, 1305–1311.

Hinks, J.L., Mason, P.J. and Claydon, J.R. (2008) Ulley Reservoir and high velocity spillway flows. *Proceedings of the 15th British Dam Society Conference Ensuring Reservoir Safety into the Future*, pp. 227–237. Thomas Telford, London.

Jackson, S. and Sleigh, A. (2000) Resettlement for China's Three Gorges Dam: socio-economic impact and institutional tensions. *Communists and Post-communist Studies*, **33**, 223–241.

Nilsson, C. and Berggren, K. (2000) Alterations of riparian ecosystems caused by river regulation. *BioScience*, **50**, 783–792.

Nilsson, C., Reidy, C.A., Dynesius, M. and Revenga, C. (2005) Fragmentation and flow regulation of the world's large river systems. *Science*, **308**, 798–800.

Ouyang, W., Hao, F.H., Zhao, C. and Lin, C. (2010) Vegetation response to 30 years hydropower cascade exploitation in upper stream of Yellow River. *Communications in Nonlinear Science and Numerical Simulation*, **15**, 1928–1941.

Ouyang, W., Hao, F.H., Song, K.Y. and Zhang, X. (2011) Cascade dam induced hydrological disturbance and environmental impace in the upper stream of the Yellow River. *Water Resources Management*, **25**, 913–927.

Schilt, C.R. (2006) Developing fish passage ad protection at hydropower dams. *Applied Animal Behaviour Science*, **104**, 295–325.

Stevaux, J.C., Martins, D.P. and Meurer, M. (2009) Changes in a large regulated tropical river: the Parana River downstream from the Porto Primavera Dam, Brazil. *Geomorphology*, **113**, 230–238.

Tilt, B., Braun, Y. and He, D. (2009) Social impacts of large dam projects: a comparison of international case studies and implications for best practice. *Journal of Environmental Management*, **90**, S249–S257.

Tullos, D., Tilt, B. and Liermann, C.R. (2009) Introduction to the special issue: understanding and linking the biophysical, socioeconomic and geopolitical effects of dams. *Journal of Environmental Management*, **90**, S203–S207.

Warren, A.L. (2011) Investigations of dam incidents and failures. *Forensic Engineering*, **164** (FE1), 33–41.

Winter, C., Mason, P., Baker, R. and Ferguson, A. (2010) *Guidance for the Design and Maintenance of Stepped Masonry Spillways, Report SC080015*. Environment Agency, London.

World Commission on Dams (WCD) (2000) *Dams and Development: A New Framework for Decision-Making*. Earthscan Publishers, London.

Xiaoyan, L., Shikui, D., Qinghe, Z. and Shiliang, L. (2010) Impacts of Manwan Dam construction on aquatic habitat and community in Middle Reach of Lancang River. *Procedia Environmental Sciences*, **2**, 706–712.

Zhang, Q. and Lou, Z. (2011) The environmental changes and mitigation actions in the three Gorges Reservoir region, China. *Environmental Science and Policy*, **14**, 1132–1138.

Zhao, Q., Lui, S., Dong, S., Cong, Wang, Yang, Z. and Yang, J. (2012) Landscape change and hydrologic alteration associated with dam construction. *International Journal of Applied Earth Observation and Geoinformation*, **16**, 17–26.

6

Powering the Water Industry

Jay Millington

6.1 Introduction

The global water industry is being encouraged to develop new strategies to guarantee supplies and maximise efficiency, because of less-reliable rainfall patterns associated with climate change and growing consumer demands (e.g. Grafton and Hussey, 2011). In many countries water companies have the added issue of an ageing infrastructure, with the need to replace worn pipework and associated ancillaries. However, these challenges also present the opportunity to explore the potential for harvesting energy from flowing water wherever it may be possible, using novel approaches and new technologies.

This chapter begins by setting the context of hydropower within the water industry; examples of conventional schemes are given followed by a discussion of available opportunities. Issues in the United Kingdom with government subsidies that limit installed capacity, as well as the legislation that restricts water companies from generating with pre-pumped flows are explored. To ensure best practice, a number of other factors pertinent to the development of new hydropower schemes in the industry are also highlighted.

6.2 Conventional Approach

The transfer of potable water from where it is sourced to where it is required is explained in detail by Savić and Banyard (2011). In the United Kingdom it must arrive at the customer's property with a minimum flow rate of 9 l/min and minimum pressure head of 7 m, as defined by the Water Services Regulation Authority (OFWAT); its quality must also meet the standards set by the Drinking Water Inspectorate (DWI). Before it arrives, however, the water may have passed through many kilometres of pipework and numerous components within a typical distribution system. From the abstraction point, a river or reservoir for

Water Resources in the Built Environment: Management Issues and Solutions, First Edition.
Edited by Colin A. Booth and Susanne M. Charlesworth.

Example 6.1 Poundbury water supply (Dorset, UK)

Wessex Water supplies the recent development of 2000 houses in Poundbury on the edge of Dorchester, at an elevation of 110 m above datum, with drinking water. The town's existing water tower had insufficient head, so a new service reservoir was built 4 km to the west at Lambert's Hill with a top water level of 156 m, from where water flows by gravity to the urban development. However, to supply water to the service reservoir from the regional strategic main at Burton, 2 km to the north-east of Poundbury with an elevation of 59 m, a pumping station (with treatment works) is required to lift the water by 97 m at a flow rate of 7 Ml/day (peak week demand). Assuming a typical efficiency of 75%, the pumps require a total power of approximately 100 kW, plus further capacity for duty, standby and assist operation. One of these pumps is shown in Figure 6.1(a). From the service reservoir at Lambert's Hill, flow gravitates through the distribution main into the network at Poundbury. This main is designed to supply 120 l/s (peak hour on maximum day demand) so will have a suitable diameter to ensure it arrives at the customer above the minimum head level set by OFWAT. As flow for the most part will not be at the maximum level the head-loss will, therefore, be less and pressure-reducing valves (PRVs) are required to protect the system from excessive water pressures; Figure 6.1(b) shows an example of a PRV at Dorchester Depot.

Figure 6.1 (a) Pump 1 of 3 at Burton Pumping Station; (b) a pressure-reducing valve at Dorchester Depot (taken in 2011).

surface water or a borehole or spring for groundwater, raw-water mains take the flow to treatment. On leaving the works, water may be sent to area link mains, currently being installed to guarantee supply throughout whole regions, which now move the water much further than in the past; even in times of plentiful supply, the quality of water must be maintained in these large-diameter pipes. Next, trunk mains deliver the water to a service reservoir, with further treatment along the way. Finally, distribution mains transfer the water into the local distribution network.

In order to travel through the distribution system, it can be seen from the Bernoulli equation that water requires potential or pressure energy for conversion into kinetic energy. This is achieved either as a result of topography for gravity flow or by being mechanically lifted with a pump. Gravity flow is usually adopted for distribution networks and wherever else it may be possible (e.g. the Elan Aqueduct, see a later example), but it would be unlikely for a distribution system not to rely on pumping at some stage. A typical example of a water distribution system is given in Example 6.1.

6.3 Hydropower

By connecting a hydroturbine's rotating shaft to a generator, the mechanical energy of flowing water can be converted into electrical energy. In the water industry this could be used on site or exported to generate revenue, offsetting the high costs associated with pumping.

6.3.1 The Variables

As is widely published (e.g. Douglas *et al.*, 2005; Massey, 2006; Hamill, 2011), the output power of a turbine is directly proportional to the available volume flow rate and the effective head:

$$P = \eta \rho g Q H$$

where

P = output power (W)
η = overall efficiency
ρ = density of the fluid (water) (kg/m^3)
g = acceleration due to gravity (m/s^2)
Q = volume flow rate (m^3/s)
H = effective head (m)

Head is categorized as low, medium or high. The term low head generally applies to schemes <10 m and high head >100 m; usable heads range from 1 or 2 m to in excess of 1000 m. Any water course from a small stream up to the largest of rivers has been used, but the most flow a single turbine can cope with is around 1000 m^3/s. Definitions vary, but generally hydropower schemes are classified according to their size as:

$$\text{Pico} < 5\,\text{kW} < \text{Micro} < 100\,\text{kW} < \text{Small} < 5\,\text{MW} < \text{Medium} < 100\,\text{MW} < \text{Large}$$

The largest hydropower scheme is the Three Gorges project in China, completed in July 2012 with an installed capacity of 22 500 MW; the largest scheme in terms of electricity produced per annum is Itaipú on the border of Brazil and Paraguay, whose capacity is lower but it suffers from less seasonal variation.

There are two types of turbine: reaction and impulse. The rotating part through which the fluid passes in a turbine is called a runner; in reaction turbines the runner is completely enclosed and full of the working fluid, whereas impulse turbines are not fully submerged and would be better described as constant pressure (Massey, 2006). Reaction types include Propeller, Kaplan and Francis turbines and impulse types include Pelton, Turgo, Crossflow and Archimedes turbines, plus the water wheel. Examples of different turbines are shown in Figure 6.2.

Figure 6.2 (a) Kaplan turbine runner at Merikoski (Oulu, Finland, 2012); (b) Francis turbine runner at the University of the West of England (Bristol, UK, 2011); (c) Pelton turbine runner from Mary Tavy No. 2 (Devon, UK, 2011); (d) Crossflow turbine being installed at Catcombe Mill (Bath, UK, 2010); (e) Archimedes screw turbines at Esholt Sewage Treatment Works (Yorkshire, UK, 2011); (f) Laxey Wheel (Isle of Man, 2010).

Table 6.1 Schemes in the UK by country and capacity.

		England	Northern Ireland	Scotland	Wales[a]	UK
>5 MW	No.	1	-	43	6	50
	Σ kW	5500	-	1 114 876	136 780	1 257 156
2 to 5 MW	No.	1	-	30	3	34
	Σ kW	2600	-	93 744	12 220	108 564
500 kW to 2 MW	No.	5	2	53	7	67
	Σ kW	4969	1315	53 917	4994	65 195
100 to 500 kW	No.	43	6	62	19	130
	Σ kW	10 957	1195	18 028	4508	34 688
15 to 100 kW	No.	43	17	38	16	114
	Σ kW	2065	753	1849	870	5537
<15 kW	No.	77	3	28	21	129
	Σ kW	494	37	171	154	856
All hydro	No.	170	28	254	72	524
	Σ kW	26 585	3300	1 282 585	159 526	1 471 996

[a] Wales also has four schemes whose power is defined in the BHA database as 'variable'.

The particular application with respect to the available head and flow (and to some extent the flow variation) will determine the most efficient turbine type. Pelton and Turgo turbines are best suited to high head/low flow, whereas Propeller and Kaplan turbines are better for high flow/low head applications; Crossflow and Archimedes turbines are more common when both head and flow are low and Francis turbines have the widest range of application, being best at medium to high heads and flows.

A definitive list of all hydropower installations in the UK has been collated from the online databases of the British Hydropower Association (BHA) (2011), the Renewable Energy Foundation (REF) (2011) and the Department of Energy and Climate Change (DECC) (2011). From this it can be seen that there are currently 528 schemes with an installed capacity of 1.47 GW (excluding pumped storage, which are net consumers of electricity), as shown in Table 6.1. As would be expected due to topography, 87% of the capacity is in Scotland, with Wales the second largest (11%), followed by England (1.8%) and Northern Ireland (0.2%).

6.3.2 Examples of Conventional Schemes in the UK Water Industry

The water industry has a long association with hydropower and many water companies own, operate or lease schemes; hydro is a core business for Scottish Water who have seven sites that have been generating for over 40 years. Most of the UK's current capacity was installed in the 1950s (Bartle and Hallows, 2005) and the technology is well established. At dams, compensation schemes operate on the minimum flow that must always be allowed to pass, with other turbines generating when reservoirs are at capacity and spilling. Run-of-river schemes do not use storage but make use of the flows in a river as they occur, abstracting water at weirs.

Example 6.2 Mary Tavy hydropower station (Devon, UK)

South West Water can generate 2670 kW of energy from the two hydropower schemes at Mary Tavy in Devon. Mary Tavy No. 1 (Figure 6.3a) was completed in 1932 and is fed by Wheal Bennetts Reservoir with a capacity of 2.5 million gallons (11 megalitres) via a 700 yard (640 m) long, 36 inch (914 mm) diameter ductile iron pipe. The 230 ft (70 m) head allows the three Francis turbines to generate 720 kW. A second scheme was completed in 1936, Mary Tavy No. 2 (Figure 6.3b), which utilises a 200 year old, 4½ mile long mine leat from Wheal Jewell Reservoir that has a capacity of 16 million gallons (73 megalitres) via a 39 inch (991 mm) ductile iron pipe. Three Pelton wheel turbines use the 560 ft (170 m) head to generate 1950 kW. When the electricity industry was privatised in 1990, National Power inherited Mary Tavy, which it immediately sold to South West Water (Lamb and Lodge, 2000). The reservoirs were purpose-built to supply water for hydropower, for which the Environment Agency determines the flows available for generation.

Figure 6.3 Mary Tavy: (a) one of the Francis turbines of No. 1 and (b) the Pelton turbines of No. 2 (taken in 2011).

Example 6.3 Elan Valley water supply and hydropower scheme (Powys, UK)

A schematic of the reservoir system and installed capacity of the turbines is shown in Figure 6.4. Water is extracted from the River Elan in mid Wales and transferred 118 km to Birmingham via a gravity aqueduct, completed in 1904. The second phase of work on the River Claerwen to increase capacity was delayed by the two wars but was finished in 1952, with engineering advancements allowing one larger dam to be constructed rather than three smaller ones originally envisaged. Water enters the aqueduct via Foel Tower, which sits within the reservoir behind Caban Coch Dam (where water is also extracted for local supply), with its water level guaranteed by the Garreg Ddu Submerged Dam. Water is stored upstream by Craig Goch and Pen-y-Garreg Dams and in times of low flow is supplemented via a tunnel from Dol-y-Mynach Unfinished Dam on the River Claerwen; water here is stored upstream at Claerwen Dam.

There are four stakeholders on the Elan Valley Scheme: Welsh Water, Severn Trent, the Environment Agency and Infinis. The Elan is a tributary of the Wye, so the dams and associated infrastructure are a Welsh Water asset. Severn Trent supplies Birmingham's water and act as agents to look after the dams, pipework and ancillaries up to the inlet valves of the turbine houses. From here responsibility lays with Infinis, the renewable energy generator. The Environment Agency determines how much can be abstracted by the water companies, Severn

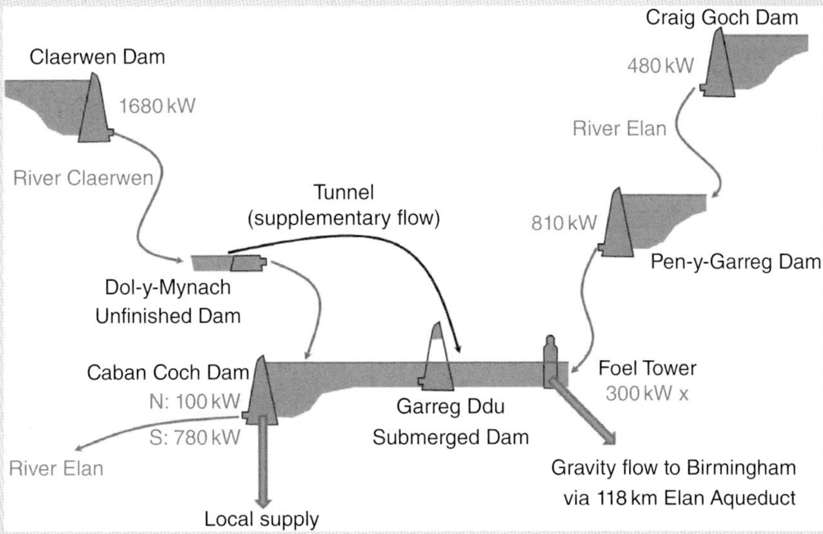

Figure 6.4 The Elan Valley scheme.

Figure 6.5 Elan Valley turbine houses at (a) Caban Coch (North and South) and (b) Claerwen Dam, where the turbines are housed under the floor level roof in the bottom right-hand corner (taken in 2011).

Trent at Foel Tower for supply into Birmingham and Welsh Water for local supplies at Caban Coch; the compensation flow at Caban Coch is also set by the Agency, which is used for generation at Caban Coch North. All other hydro turbine releases at the Elan/Claerwen Dams are managed according to Severn Trent's operational rules of the Elan catchment resource; if the dam is overflowing the turbine may be operated or if there is a release required into the lower reservoirs, this is passed through the hydro turbine (Hughes, 2012).

The first turbines at Elan were installed as part of the original scheme in ornate buildings at Caban Coch North and South in 1904 (Figure 6.5a). These provided direct current electricity for local supply, but were both replaced in 1952. Generation commenced in 1997 at the other four sites, where the turbines are largely hidden (e.g. Figure 6.5b). To coincide with this the Caban Coch South turbine was replaced with one of much larger capacity and in 2011 the Caban Coch North turbine was refurbished. With the exception of the Kaplan turbine at Foel Tower, which was decommissioned in 2010, Francis turbines are employed throughout.

6.4 Micro and Small Hydros

In order to meet the UK government's renewable energy target of 15% by 2020 (House of Lords, 2008), opportunities with hydropower are being explored. According to The UK Renewable Energy Strategy (2009) Command 7686:

> Hydropower is a reliable and generally predictable source of renewable electricity and one of the few that is not intermittent. Although the UK hydro sector is a mature sector, there remain good opportunities to exploit hydropower resources, for micro and small-scale hydro development.

With the government's strategy promoting smaller schemes, it is interesting to note in Table 6.1 the spread according to scheme size; England already has nearly as many <100 kW schemes as the rest of the United Kingdom combined. Demonstrating its support of the government's commitment to hydropower, the Environment Agency published a report (Entec, 2010) suggesting there are a further 4190 'win/win' run-of-river locations in England and Wales that could generate a combined power of 526 MW. These 'win/win' sites were so called because they have existing barriers (weirs) on the rivers where a micro-hydro turbine along with a fish-pass could be installed, thus generating electricity while also improving the ecological status. Historically, settlements would have established around mills associated with the weirs so there may be local demand for any electricity produced, but in reality there will be fewer viable schemes once the coarse nature of the data and the ease with which any electricity can be exported to the grid have been factored. However, there are plenty of schemes currently going ahead, with a recent report for the DECC (Arup, 2011) identifying that the 'main constraints to development are capacity of the environmental regulators to process a high number of applications'.

6.4.1 Water Company Applications with Micro-hydro

A Department of Energy study from 1989 suggested there were 17 MW of exploitable capacity by the water industry (Kirk, 1999), but a later study (BHA, 2010) suggests this will be an underestimate. Water companies have been exploring their potential for installing micro-hydro and small-scale conventional opportunities with compensation flows have been investigated (e.g. South West Water at Meldon Dam), but there are also novel approaches being assessed on flows at the inlet or outlet of treatment works. They have two distinct advantages compared to run-of-river schemes; environmental regulation is not necessary and treatment works are generally grid-connected. Alternatively, if they are far from the national grid or have an intermittent supply, hydro could be used as a local energy source.

Another opportunity being developed by a number of water companies is to install turbines at pressure reducing valves (PRVs). The most cost-effective way of achieving this is with pump-as-turbine technology, which is also the subject of international research for small schemes in isolated areas (Williams, 2010; Nautiyal *et al.*, 2010b). It is a much cheaper alternative than conventional turbines because pumps are mass produced, the motor is used as a generator and standard pipework can be used rather than special penstocks and draft tubes. The application with the water industry is not new; in 1998 a pump-as-turbine was installed at a PRV at Blackacre Water Dosing Plant by North West Water (Williams *et al.*, 1998), which ran successfully for a number of years before the remote site required additional power and was connected to the national grid. However, lack of guide vanes means a standard pump-as-turbine is only suitable for a narrow range of flows and selecting appropriate turbine characteristics has proven to be very difficult. A number of articles have been published by Singh and Nestmann (2010, 2011) concerning the use of Cordier diagrams to predict turbine operation of centrifugal pumps, and analysis with computational fluid dynamics to help determine pump-as-turbine characteristics has also been undertaken (e.g. Nautiyal *et al.*, 2010a), but there still remains the difficulty of selecting the correct pump.

Examples of novel schemes in the water industry include:

- Sutton Poyntz Water Treatment Works (Wessex Water). An 8 kW Ecowave crossflow turbine was installed in 2011 on the inlet of the spring fed works.
- Bratton Flemming (South West Water). A 40 kW Zeropex Difgen pump-as-turbine was installed in 2011 at a PRV. This particular pump-as-turbine has been designed to cope with more variation in flow than standard reverse running pumps and allows pressure and flow management
- Esholt Sewage Treatment Works (Yorkshire Water). Installed in 2009 by Spaans Babcock, two Archimedes screw turbines with a capacity of 180 kW generate electricity using flow from the inlet works.
- Cotton Valley Sewage Treatment Works (Anglian Water). A WKV crossflow turbine installed in 2006 generates 15 kW on the outlet from the works.

6.4.2 Funding

The UK government is committed to promoting renewable energy for a multitude of benefits to society: to reduce the harmful production of greenhouse gases, to help ensure security of the energy supply and to provide opportunities for investment in new industries and new technologies. To this end, subsidies in the form of Renewables Obligation Certificates were introduced in 2002, followed by the Feed-in Tariffs Order (Great Britain, 2010), which came into force on 1 April 2010; the feed-in tariffs reflect the proportionately higher costs of installing smaller turbines. New hydropower schemes of up to 5 MW can now earn their investors money (prices correct for October 2012) from the following sources:

- Offset value for on-site electricity use – typically saving 12 p/kWh used;
- Export tariff – index linked, 4.5 p/kWh exported;
- Feed-in Tariffs (FITs) – depend on installed capacity: 21 p/ kW h for <15 kW, 19.6 p/kW h for 15 kW to 100 kW, 15.5 p/kW h for 100 kW to 500 kW, 12.1 p/kW h for 500 kW to 2 MW and 4.48 p/kW h for 2 MW to 5 MW generated; and
- Renewable Levy Exemption Certificates (LECs) – 0.509 p/kWh generated.

Schemes can choose to receive market value instead of the export tariff and those greater than 50 kW can opt to have Renewables Obligation Certificates (ROCs) in place of Feed-in Tariffs (Office of Gas and Electricity Markets, 2011), which in 2011–2012 for hydropower traded at an average of 4.7 p/kW h generated (e-ROC, 2012). However, DECC recently undertook a scheduled review of the levels of banded support available under the Renewables Obligation and have reduced the band for hydroelectricity from 1 to 0.7 ROCs for new accreditations (DECC, 2012).

The potential benefits of the water industry installing hydropower are twofold. Firstly, there are many opportunities at treatment works and PRVs that could contribute to the government's strategy and, secondly, by generating cost-effective renewable energy the water companies could lower customer bills, while increasing their green credentials.

6.4.3 Issues of Feed-in Tariff Banding

It has been suggested (Crompton, 2010) that the tariffs for turbines with capacities of less than 50 kW and particularly for <15 kW schemes (Elliot, 2011) are too low, meaning many potential schemes are not financially viable due to payback periods in excess of ten years. A greater problem is the issue of banding itself; it is simple to implement, but has an impact on turbine size. This is demonstrated in Table 6.2, which shows that 15 kW to 16 kW, 100 kW to 126 kW, 500 kW to 640 kW and 2000 kW to 5000 kW schemes are extremely unlikely to be installed. Rather than having a turbine that is organically sized for the site conditions, FITs will artificially restrict capacity in order for the investors to receive the greatest return. For example, a location where a 120 kW turbine should be installed would earn £3 thousand more (£61 295–£58 168) if a 100 kW turbine were installed instead. The crossover kW, which delineates the favourable turbine size, is found by multiplying the lower FIT kW banding by the ratio of a higher tariff to a lower tariff (i.e. $100 \times 19.6/15.5 = 126.452$ kW); this assumes a constant capacity factor.

Despite the recent introduction of an extra banding for 100 kW to 500 kW turbines, it would make sense to have more. Even better would be a linear or logarithmic relationship of turbine capacity to the FIT to avoid the impact of being just within an unfavourable band. If bandings have to be adopted, a fairer alternative would be to use electricity generated in MW h/year (e.g. Loening, 2010) rather than installed capacity. Based on the values in Table 6.2 this could be 21.0 p/kW h for the first 47 MW h, 19.6 p/kW h up to

Table 6.2 Current feed-in tariff capacity banding comparison.

Turbine		FIT banding		kW h banding?	
kW	kW h / year[a]	p / kW h[b]	£ / year	£ / year	Δ £
1.000	3127	21.0	657	657	0
15.000	46 910	21.0	9851	9851	0
15.001	46 913	19.6	9195	9852	657
16.071	50 261	19.6	9851	10 508	657
100.000	312 732	19.6	61 295	61 952	657
100.001	312 735	15.5	48 474	61 953	13 479
120.000	375 278	15.5	58 168	71 647	13 479
126.452	395 455	15.5	61 295	74 774	13 479
500.000	1 563 660	15.5	242 367	255 846	13 479
500.001	1 563 663	12.1	189 203	255 846	66 643
640.496	2 003 036	12.1	242 367	309 010	66 643
2000.000	6 254 640	12.1	756 811	823 455	66 643
2000.001	6 254 643	4.48	280 208	823 455	543 247
5401.786	16 893 113	4.48	756 811	1 300 058	543 247
5000.000	15 636 600	4.48	700 520	1 243 766	543 247

[a] Capacity factor of 35.7% from the average performance of existing <5 MW sites between 2005 and 2009 in Arup (2011).

[b] Feed-in Tariffs for December 2012.

Table 6.3 The impact on installed capacity and FIT expenditure if existing schemes had been sized for maximum return from Feed-in Tariffs.

Constructed pre-FIT			If constructed December 2012	
Capacity	No.	Σ kW	kW lost	£FIT extra
15 to 16.1 kW	1	16	1	44
100 to 126 kW	16	1826	226	95 602
500 to 640 kW	11	6175	675	329 385
2000 to 5000 kW	34	108 564	40 564	10 521 345
Total	62	116 581	41 466	10 946 376

313 MW h, 15.5 p/kW h up to 1564 MW h, 12.1 p/kW h up to 6255 MW h and 4.48 p/kW h up to 15 637 MW h. However, if the FIT kW banding scheme were to be replaced with a MW h banding scheme for current turbines, these values would of course result in greater tariffs being paid as the larger schemes would benefit from their initial MW h earning a higher rate.

It is important to consider what positives there may be to downsizing and there is an obvious one: the footprint of a smaller hydropower scheme will be less, so it will have less impact on the environment. It may also operate with a higher load factor, so the actual crossover kW delineating favourable size will be slightly lower than in the discussion above. Another benefit of installing a smaller scheme is less financial outlay, so payback times will be reduced, making some schemes viable that might not have been with organic-sized turbines. However, as long as smaller schemes earn higher subsidies, all of these positives would still apply.

According to Elliot (2011), FIT bandings may be good for installed capacity rather than number of sites, but assuming that if all of the UK's installed turbines falling in the unfavourable bands (15 kW to 16 kW, 100 kW to 126 kW, 500 kW to 640 kW and 2000 kW to 5000 kW) had been reduced to earn greater income, as can be seen from Table 6.3, the UK would lose 41 MW of installed capacity (i.e. more than the combined capacity of England and Northern Ireland). The government would also be paying extra Feed-in Tariffs of £11 million per annum.

With regard to growth in the hydropower sector, the DECC study report (Arup, 2011) suggests that 'the largest number of sites are in the <100 kW range' so the FIT banding will not greatly restrict these schemes, but it also states that 'the biggest contribution to installed capacity is likely to come from the 100–500 kW range', which may well be impacted. FITs banding for other renewables may also be restricting the organic sizing of schemes.

6.4.4 Issues with FITs and Pre-pumped Flows

6.4.4.1 Comparison of ROCs with FITs

More specific legislation of interest to water companies concerns pre-pumped flows. Article 24 of The Renewables Obligation Order (Great Britain, 2009) states that if a station's input electricity exceeds 0.5% of the electricity generated, ROCs are determined based on the gross electricity generated minus the input electricity. The input electricity is defined in paragraph 6(a) as:

> ... the total amount of electricity used by that station for purposes directly related to its operation (including for fuel handling, fuel preparation, maintenance and the pumping of water) whether or not that electricity is generated by the station or used while the station is generating electricity.

This correctly excludes pumped storage schemes from receiving subsidies and accounts for the pumping of cooling water at biomass-fuelled power stations; it also allows water companies to 'net off' any component of flow that is pre-pumped. However, under FITs, OFGEM disqualifies any scheme with pre-pumping from receiving subsidies and does so with an extreme interpretation. For example, a 15 kW turbine on a gravity-fed waterworks in Wales has been deemed ineligible because there are drought pumps installed, despite the fact that they have only operated for approximately 200 hours in the last 12 years; as the capacity is <50 kW the scheme is also unable to opt for ROCs.

6.4.4.2 Energy Recovery

It could be reasoned that as water companies do not pump water (or sewage) for generation but to transfer it for treatment or use, it should not be included as input electricity. Generating electricity from pre-pumped flows may not be renewable, but energy recovery could help with carbon reduction and many opportunities exist at treatment works and pressure-reducing valves. Water UK has so far been unsuccessful in arguing for incentives to be paid (Water UK, 2009), but if policies could be changed to subsidise energy recovery, strict guidelines would need to be in place to ensure that:

- Turbines can only be included if the pumping system cannot be changed for more efficient operation (e.g. pumping over hills can be avoided or the pumping pressure can be reduced).
- Generation is only taking place because the water or sewage requires pumping (e.g. flow is not being pumped in a loop).
- Systems are designed with ethical strategies (e.g. treatment works are not located on tops of hills).

It is possible that subsidies could mean many micro-hydro schemes for energy recovery become financially viable, contributing to the UK government's low-carbon goals. According to the chairman of the water companies' Energy Managers Forum (Burgess, 2010), there are possibly hundreds of megawatts available from energy recovery.

6.5 Other Factors

There are a number of important factors specific to the water industry when exploring opportunities for hydropower. Consideration must be given to pump and system efficiency, operation within OFWAT regulations, the ease with which identified opportunities can become reality and the cost-effectiveness of other renewable energy sources.

6.5.1 Efficiency

Although less glamorous than installing new green energy generation schemes, it is much better for the environment if less energy is used instead. In 2009 the water industry generated around 750 GW h of renewable energy, but used over ten times more for treatment and pumping (Water UK, 2009) and water utilities are among the most significant users of electricity (Harrison, 2004).

Distribution Use of System (DUOS) charges discourages the use of electricity at peak hours by large commercial consumers on weekdays. Transmission Network Use of System (TNUoS) tariffs also apply, which are determined during the three periods of highest transmission system demand, known as Triads. During these Triad periods, electricity use is minimised and generation maximised by the water companies. As demand for electricity often coincides with the diurnal demand for water, a huge amount of effort is given by the industry to achieving efficient operation, from pump refurbishment to whole system analysis; reduction in consumption must continue to be prioritised over generation.

6.5.2 Regulation

Water companies are primarily concerned with supplying potable water rather than energy generation and must operate within the OFWAT regulations, which protect consumers' interests. Hydropower is not supported by OFWAT unless directly related to the appointed business, so where opportunities exist for hydro generation they must be justified for operational reasons (e.g. at Sutton Poyntz Water Treatment Works in Dorset for flow control and Maundown Water Treatment Works in Somerset to combat issues of intermittent supply); any surplus energy generation is a bonus. As they stand, OFWAT regulations could be better aligned with government policy because they are likely to inhibit hydropower development, but there are opportunities that can still be realised and are worthwhile pursuing.

6.5.3 Infrastructure

The difficulty of assessing water company infrastructure can be a hindrance to retrofitting hydropower, as detailed asset surveys are necessary to ensure that it would be cost-effective and convenient. However, there are some schemes that have benefitted from excellent foresight, such as the Llyn Brianne Dam in Powys, whose 4600 kW hydropower scheme was commissioned in 1997 using pipework installed during the dam's construction in the 1970s. On a much smaller scale, but equally valid, construction at Sutton Poyntz Water Treatment Works anticipated the addition of a hydropower scheme at some future stage.

As current legislation and regulation does not always support hydropower schemes, it may be worthwhile for water companies that, where there is a potential, new infrastructure is designed with consideration given to the subsequent installation of a turbine. Future-proofing with respect to the impact of climate change should also be included in design, such as the possibility that compensation flows could be reduced (Hughes, 2011).

6.5.4 Other Renewables

Hydro is a small energy resource compared to other renewables, but it is also small within the water industry, accounting for only 6% of electricity generated (Costyn, 2011). The overwhelming majority (92%) of renewable energy generated by water companies in the United Kingdom uses biogas from sewage digestion (Howe, 2009); the largest UK scheme is Severn Trent's site at Minworth near Birmingham with a capacity of 10.5 MW.

Other techniques that are being explored have included wind, with Yorkshire Water's two 1.3 MW turbines at Loftsome Bridge and one in Hull beginning generation in 2008. In the South, Portsmouth Water installed 50 kW of solar PV panels at six sites in 2012 and Severn Trent has invested in energy crops and a 2 MW biomass plant at Stoke Bardolph, near Nottingham (Dent, 2010).

In terms of energy produced, hydropower can be more effort to install than the other renewables discussed above. However, the established technology and ease of maintenance can still make it an attractive long-term solution for generating green energy and it should therefore not be overlooked.

6.6 Conclusions

The water industry has a proven capability to generate renewable energy from micro-hydro and further installations could help the government meet its 2020 target. However, a number of factors are preventing the industry from realising its full potential and it is, therefore, concluded that:

- If a banding system must be used, FITs should be based on MW h generated to prevent downsizing.
- FITs should have an allowance for input electricity to 'net-off' pre-pumping as with ROCs.
- Energy recovery systems should be subsidised.
- Efficiency measures should be prioritised over the installation of new schemes.
- OFWAT regulations should be better aligned with government policy so as not to inhibit hydropower.
- Water companies should consider the future application of hydropower in the design of infrastructure at locations that have potential.
- Micro-hydro should be considered for electricity generation alongside other renewable techniques.

Acknowledgements

The author is very grateful to the many professionals listed below, who have answered numerous questions in the course of researching this chapter. Any errors, factual or of understanding, remain the author's own.

Peter Beveridge, Scottish Water; Dr Mervyn Bramley OBE, Independent Engineer and Environmentalist; Dr Adam Brookes, Anglian Water; Alan Burgess, South West Water; Andrew Calvert, Yorkshire Water; Claire Chapman, Scottish Water; Duncan Collins, Zeropex; Simon Dennis, Renewables First; Martin Dent, Severn Trent Water; David Harpur, South West Water; Professor Gareth Harrison, Edinburgh University; Noel

Hughes, Severn Trent Water; Clive Hulbert, Wessex Water; John Leonard, Wessex Water; Ross Mitchell, South West Water; Dr Mike Pedley, Dwr Cymru/Welsh Water; Emily Sherwood, Wessex Water; Mike Smith, Infinis; John Willows, Wessex Water.

References

Arup (2011) *Review of the Generation Costs and Deployment Potential of Renewable Electricity Technologies in the UK*. Study Report – Department of Energy and Climate Change. Online, accessed June 2012: http://www.decc.gov.uk/assets/decc/11/consultation/ro-banding/3237-cons-ro-banding-arup-report.pdf.

Bartle, A. and Hallows, G. (2005) Hydroelectric power: present role and future prospects. *Proceedings of the ICE, Civil Engineering*, **156**, 28–31.

British Hydropower Association (BHA) (2010) *England and Wales Hydropower Resource Final Report (Version 13)*. DECC and WAG. Online, accessed June 2012: http://www.british-hydro.org/UK%20Hydro%20Resource/England%20and%20Wales%20Resource%20Study%20Oct%202010.pdf .

British Hydropower Association (BHA) (2011) *List of Installations*. British Hydropower Association. Online, accessed September 2011: http://www.british-hydro.org/installations/installations.html.

Burgess, A. (2010) Email: aburgess@southwestwater.co.uk. Re: Micro hydro in the water industry, 18 August 2010. Email to: J. Millington, jay.millington@uwe.ac.uk.

Costyn, J. (2011) *Feed-in Tariffs: Current Situation and Comprehensive Review*. BHA Hydropower Conference, Cardiff, 29 June 2011. British Hydropower Association. Online, accessed June 2012: http://www.british-hydro.org/downloads/2011/Cardiff/04%20John%20Costyn%20[Compatibility%20Mode].pdf.

Crompton, C. (2010) *The Feed-in Tariff – The Consultation Story and Opportunities for the Hydro Industry*. BHA Annual Conference, Glasgow, 13–14 October 2010. British Hydropower Association. Online, accessed June 2012: http://www.british-hydro.org/Presentations/BHA%20Annual%20Conference%202010/Carl%20Crompton%20[Compatibility%20Mode].pdf.

Dent, M. (2010) *Stoke Bardolph Energy Crop AD (Anaerobic Digestion) Plant*. Severn Trent Water. Online, accessed June 2012: http://www.waterprojectsonline.com/case_studies/2010/Severn_Trent_Stoke_Bardolph_2010.pdf.

Department of Energy and Climate Change (DECC) (2011) *List of Generators*. Online, accessed September 2011: https://restats.decc.gov.uk/app/reporting/decc/monthlyextract.

Department of Energy and Climate Change (DECC) (2012) *Government Response to the Consultation on Proposals for the Levels of Banded Support under the Renewables Obligation for the Period 2013–17 and the Renewables Order 2012*. URN 12D/274. Crown, London. Online, accessed October 2012: http://www.decc.gov.uk/assets/decc/11/consultation/ro-banding/5936-renewables-obligation-consultation-the-government.pdf.

Douglas, J.F., Gasiorek, J.M., Swaffield, J.A. and Jack, L.B. (2005) *Fluid Mechanics*, 5th Edition. Pearson Education, Harlow.

Elliot, C. (2011) *Small Hydro and the FIT*. BHA Hydropower Conference, Cardiff, 29 June 2011. British Hydropower Association. Online, accessed June 2012: http://www.british-hydro.org/downloads/2011/Cardiff/04%20Chris%20Elliott%20[Compatibility%20Mode].pdf.

Entec (2010) *Mapping Hydropower Opportunities and Sensitivities in England and Wales – Technical Report*. Environment Agency, Bristol. Online, accessed June 2012: http://publications.environment-agency.gov.uk/PDF/GEHO0310BRZH-E-E.pdf.

e-ROC (2012) *On-line ROC Auction Service, Prices 2011–12*. Online, accessed June 2012: http://www.eroc.co.uk/trackrecord.htm.

Grafton, Q.R. and Hussey, K. (2011) *Water Resources Planning and Management*. Cambridge University Press. Online, accessed June 2012: http://www.knovel.com/web/portal/browse/display?_EXT_KNOVEL_DISPLAY_bookid=3966.

Great Britain (2009) Electricity, England and Wales. *The Renewables Obligation Order*, 2009 No. 785. The Stationery Office, London. Online, accessed June 2012: http://www.legislation.gov.uk/uksi/2009/785/pdfs/uksi_20090785_en.pdf.

Great Britain (2010) Electricity. *The Feed-in Tariffs Order*, 2010 No. 678. The Stationery Office, London. Online, accessed June 2012: http://www.fitariffs.co.uk/library/regulation/uksi_20100678_en1.pdf.

Hamill, L. (2011) *Understanding Hydraulics*, 3rd Edition. Palgrave Macmillan, Basingstoke.

Harrison, G.P. (2004) *Developing Renewable Energy within the Water Industry*. IDS-Water. Online, accessed May 2012: http://www.see.ed.ac.uk/~gph/publications/IDS_Draft.pdf.

House of Lords (2008) *The EU's Target for Renewable Energy: 20% by 2020*. HL Paper 175-I (2007–08). Online, accessed June 2012: http://www.publications.parliament.uk/pa/ld200708/ldselect/ldeucom/175/175.pdf.

Howe, A. (2009) *Renewable Energy Potential for the Water Industry*. Environment Agency, Bristol. Online, accessed June 2012: http://publications.environment-agency.gov.uk/PDF/SCHO1209BROF-E-E.pdf.

Hughes, C. (2011) *Water Resources Planning in Wales – Potential Applications for Hydropower*. BHA Hydropower Conference, Cardiff, 29 June 2011. British Hydropower Association. Online, accessed June 2012: http://www.british-hydro.org/downloads/2011/Cardiff/03%20Chris%20Hughes%20[Compatibility%20Mode].pdf.

Hughes, N. (2012) Email: Noel.Hughes@severntrent.co.uk. Re: Hydro at Elan Valley, 11 June 2012. Email to: J. Millington, jay.millington@uwe.ac.uk.

Kirk, T. (1999) Small scale hydropower in the UK. *Water and Environment Journal*, **13** (3), CIWEM: 207–212.

Lamb, P. and Lodge, E. (2000) *Christy Brothers in the South West*. South Western Electricity Historical Society. Online, accessed June 2012: http://swehs_archive.swelocker.co.uk/news15su.pdf.

Loening, A. (2010) *An Early FIT?* BHA Annual Conference, Glasgow, 13–14 October 2010. British Hydropower Association. Online, accessed June 2012: http://www.british-hydro.org/Presentations/BHA%20Annual%20Conference%202010/Adrian%20Loening.pdf.

Massey, B. (2006) *Mechanics of Fluids*, 8th Edition. Taylor and Francis, Abingdon.

Nautiyal, H., Varun, P.R. and Kumar, A. (2010a) CFD analysis on pumps working as turbines. Hydro Nepal. *Journal of Water Energy and Environment*, **6**, 35–37.

Nautiyal, H., Varun, P.R. and Kumar, A. (2010b) Reverse running pumps analytical, experimental and computational study: a review. *Renewable and Sustainable Energy Reviews*, **14** (7). Elsevier. Online, accessed June 2012: http://ac.els-cdn.com/S1364032110001176/1-s2.0-S1364032110001176-main.pdf?_tid=a09b16ec0d984d9d08038bcfca5576bc&acdnat=1340192412_16149c0cf8ca1a4e9d9fd5d181fec65c.

Office of Gas and Electricity Markets (OFGEM) (2011) *Renewables Obligation: Guidance for Generators 2011*. Online, accessed June 2012: http://www.OFGEM.gov.uk/Sustainability/Environment/RenewablObl/Documents1/RO%20Generator%20Guidance%20May%202011%20final.pdf.

Renewable Energy Foundation (REF) (2011) *List of Generators*. Online, accessed September 2011: http://www.ref.org.uk/roc-generators/.

Savić, D.A. and Banyard, J.K. (2011) *Water Distribution Systems*. ICE – Thomas Telford Ltd, London. Online, accessed June 2012: http://www.icevirtuallibrary.com/content/book/101163.

Singh, P. and Nestmann, F. (2010) An optimization routine on a prediction and selection model for the turbine operation of centrifugal pumps. *Journal of Experimental Thermal Fluid Science*, **34** (2), 152–164.

Singh, P. and Nestmann, F. (2011) A consolidated model for the turbine operation of centrifugal pumps. *Journal of Engineering for Gas Turbines and Power*, **133** (6), Article Number 063002.

The UK Renewable Energy Strategy (2009) Command 7686. Online, accessed June 2012: http://www.decc.gov.uk/assets/decc/what%20we%20do/uk%20energy%20supply/energy%20mix/renewable%20energy/renewable%20energy%20strategy/1_20090717120647_e_@@_theukrenewableenergystrategy2009.pdf.

Water UK (2009) *Consultation of REF incentives*. Water UK. Online, accessed June 2012: http://www.water.org.uk/home/policy/statements-and-responses/renewable-energy-financial-incentives/refi-final-water-uk-response-oct09.pdf.

Williams, A.A. (2010) *Centrifugal Pumps as Turbines: A Review of Technology and Applications*. In: Hidroenergia 2B.06. Online, accessed June 2012: http://2010.hidroenergia.eu/pdf/2B.06.pdf.

Williams, A.A., Smith, N.P.A., Bird, C. and Howard, M. (1998) Pumps as turbines and induction motors as generators for energy recovery in water supply systems. *Water and Environment Journal*, **12** (3), CIWEM, 175–178.

7

Water Quality and Treatment

J. Bryan Ellis

7.1 Introduction

The total demand for water in the United Kingdom is predicted to increase steadily over the next decade, with the 2020 demand estimated to be some 5–12% (800–1000 million l/day) more than it is at present (Environment Agency (EA), 2009a), driven by a combination of population growth, agricultural needs and commercial factors. Current per capita consumption averages 150 l/head/day varying regionally between 107 and 175 l/head/day, but any reductions that might be expected to follow on from future expansions in metering, water efficiency savings and recycling usage are likely to be more than offset by population growth estimated at ~18% (10 million) over the next 20 years. In addition, an increasing awareness of the urban water footprint impacts of the food and drink service industries is also likely to exacerbate future pressures on domestic water resources. The interlinkages between food policy and security, energy, climate change and population growth will collectively pressurise the development and implementation of future sustainable water resource management strategies and delivery policies.

Much of southern and eastern England is classified as being 'under serious stress' from water abstraction, having an exploitation index of 22% with many catchments having no water available at low flows (Environment Agency, 2008). Over the last decade, nontidal surface waters have contributed between 50 and 60%, with groundwater abstraction contributing a fairly constant 10% of the average total nontidal usage of 38 000 million l/day. Water companies in England and Wales abstract nearly half of the total amount for potable supply but return >70% as treated effluent to enhance river flows. The average annual amount abstracted for the public water supply has not varied very much since 2000, averaging some 16 000 million l/day, with general agricultural and spray irrigation uses comprising significant further demands on abstracted water, especially during dry summers. The urbanised zones of the South East and Midlands regions currently operate below

Water Resources in the Built Environment: Management Issues and Solutions, First Edition.
Edited by Colin A. Booth and Susanne M. Charlesworth.

their target headroom in comparison to southern England, East Anglia and South Wales, which have a 10–20% surplus in their supply–demand balance (Department of Environment, Food and Rural Affairs (DEFRA), 2011).

UK River Basin Management Plans (RBMPs) have identified groundwater and surface water of the urbanised South East and Midlands regions as being 'at risk' or 'probably at risk' of failing the legislative requirements of the EU Water Framework Directive (WFD) as a result of abstraction and diffuse pollution pressures. Much of the Chalk aquifer underlying the Thames basin is also classified as being of 'poor' chemical status. It is these regions that are also most likely to feel the full impact of future climate change, with mean monthly summer flows in 2050 predicted to decline by anything between 10 and 50%. Drought orders have been introduced for two successive years (2010–2012) by six water companies in South and East England following dry winter periods, which have left reservoirs, aquifers and rivers well below their normal levels. However, it is estimated that hosepipe bans only give a 5% 'water supply' saving, equivalent to 100 million l/day in the case of the Thames Water region, compared to very much larger savings that might be gained from metering, leakage control and rainwater harvesting technologies. In addition, it is argued that better public education and awareness of water conservation might also yield a more effective long-term strategy to reduce household water supplies. There are also issues relating to 40 000 private water supplies serving over 1 million rural premises, which are not accounted for in drought contingency planning and which also have concerns regarding water quality, with one in four such supplies exceeding standards for a range of substances (Drinking Water Inspectorate (DWI), 2011).

Whilst it may be argued that post-privatisation improvements in potable supplies have addressed most fundamental national drinking water problems over the last 20 years, there still remain significant local/regional issues relating to metering, leakage and demand supply as well as substantial issues of consumer acceptability. It is clear that UK water resources are currently used intensively and are subject to significant and diverse pressures, ensuring that sustainable future water supplies will continue to present a considerable challenge for the water industry, especially in heavily urbanised areas. This challenge may well require the adoption of a national strategic policy for water abstraction with the development and introduction of new organisational structures and management approaches, as acknowledged in the recent government review of the UK water industry (DEFRA, 2011).

7.2 Water Quality

7.2.1 Potable Supply Quality

It is estimated that UK water companies have invested over £41 billion during the past 20 years to improve the quality of drinking water and to meet an increasingly stringent regulatory regime (Water UK, 2010). The post-privatisation shift towards a proactive 'catchment to tap' approach to the management of water resources has resulted in considerable improvements in the quality and appearance of consumed drinking water. Figure 7.1 confirms the scale of these improvements as measured by the total number of failures in reported water quality standards over the period.

Compliance with the regulatory standards as laid down in the EU Drinking Water Directive (Water Framework Directive (WFD); EC/98/83) has made improvements from 99.00% (in 1990) to 99.96% (in 2010), with very significant improvements for certain individual parameters such as turbidity. However, the raw catchments providing

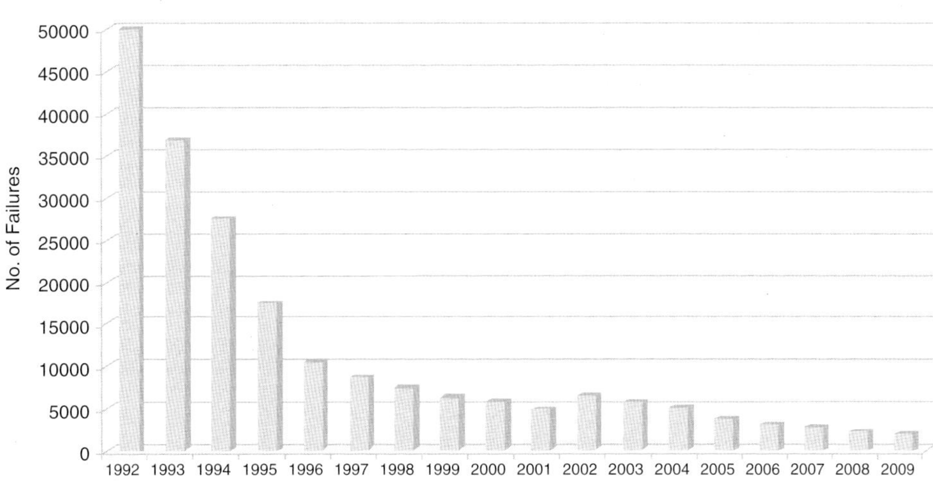

Figure 7.1 UK Drinking Water Compliance Failures 1992–2009 (data sourced from DWI, 2011).

Table 7.1 Capital investment in drinking water quality.

Water quality parameter	AMP5 2010–2015 investment (£ million)	New works expenditure 2010–2015 (%)
Turbidity reduction	5	0.5
Trihalomethane (THM) reduction	37	3.4
Pesticide removal	42	3.8
Nitrate removal	70	6.4
Cryptosporidium risk reduction	89	8.1
Plumbosolvency problems and pipe lead control	170	9.9
Improving acceptability of drinking water: taste, colour, odour.	171	15.6
Water quality monitoring	7	0.6
Meeting environmental obligations (WFD, Biological Action Plans (BAPs), Habitats and Birds Directive, etc.)	97	8.9

Source: OFWAT (2009). Adapted from Table 34.

the source supplies are not always fully under the control of the water companies and these management difficulties are compounded by functional tensions between the regulatory Drinking Water Inspectorate (DWI) and Environment Agency (EA) authorities.

The fundamental causes of these failures can be attributed to a persistent number of raw water agents liable to prejudice consumer health as indicated in the 2010–2015 capital expenditure programmes of water companies in England and Wales for their current Asset Management Plan quinquennial (AMP5) period (Table 7.1). Apart from prime turbidity, bacterial/pathogen/viruses and chemical agents will collectively account for one-third of the AMP5 (2010–2015) water company capital expenditure. A further 16% (£171 million) will be invested in improving consumer acceptability of the potable water supply in terms of taste, colour and odour. The introduction (since 1990) of systematic monitoring protocols and analytical tools has largely been responsible for the identification of quality problems and now also serve as confirmation of the integrity of implemented solutions to deliver compliance. During 2009, >2 million sample tests for 39 different parameters were taken by water companies from consumer taps to assess compliance and only a very small variation exists in the compliance data between different regions of the country.

Whilst this intensive monitoring programme has helped to ensure the quality of potable water delivered to the consumer, it still cannot fully guarantee the quality of water actually consumed from the tap within particular building premises (Jackson *et al.*, 2004). Achieving good customer quality drinking water does not always depend on treatment but on management of the supply chain (May, 2006). In this regard there is no single serviceability indicator relating to water quality in the 400 000 km of the mains distribution supply to fingerprint the system integrity and there is also a need to consider local conditions and modes of deterioration to ensure that the routine daily quality is maintained. There still remains an open question as to whether companies should have responsibility for water quality beyond delivery in their distribution pipe network. At present, water supply quality in building premises, which is subject to variable pipework and plumbing conditions, remains a customer responsibility. Despite these reservations, there has clearly been a step change in compliance, as noted in Figure 7.1, which has largely come about as a result of water companies putting in place legally binding work programmes (or 'undertakings') to design and deliver technically appropriate and fit-for-purpose asset improvement and/or remedial schemes.

7.2.2 Groundwater Source Quality

It is clear from Table 7.1 that nitrates, pesticides and solvents still pose a source quality problem for many water companies, with nearly 150 groundwater sources having been closed between 1975 and 2004 in England and Wales because of water quality issues, costing water companies some £750 million (UK Water Industry Research (WIR), 2004). The 'prevent and limit' and 'reversal of upward trends' objectives of the WFD are significant groundwater markers for drinking water abstraction, and therefore it is important to understand acceptable input limits for even common conservative pollutants. Nitrate is the most widespread of such diffuse pollutants causing particular problems in south, east and central England, with >15% of monitoring sites exceeding the 50 mg/l drinking water Maximum Acceptable Concentration (MAC) Standard. Whilst agricultural runoff undoubtedly comprises the major diffuse source, other important sources include leaking urban sewers and water mains, septic tanks, landfills, stables, atmospheric deposition (e.g. power generation, transport, etc.) as well as urban amenity fertilisers. Potentially up to 60% of raw groundwater sources in England are at serious risk of failing the WFD objectives as a result of high nitrate concentrations, with between 24 and 34% of this total being related to diffuse urban source contributions according to the UK Article 5 river basin characterisation summaries reported to the EU Commission in 2005 (www.defra.gov.uk/wfd).

It is estimated that some 50% of the peak nitrate concentrations recorded in the Thames Basin are subject to substantial delayed effects following diffusion into the Chalk aquifer matrix, which will take some 30–50 years to work through the groundwater system (Harden *et al.*, 2010). The delayed releases from wartime 'dig-up' campaigns and arable subsidies were amplified by the EU Common Agricultural Policy (CAP) ramp-up in the

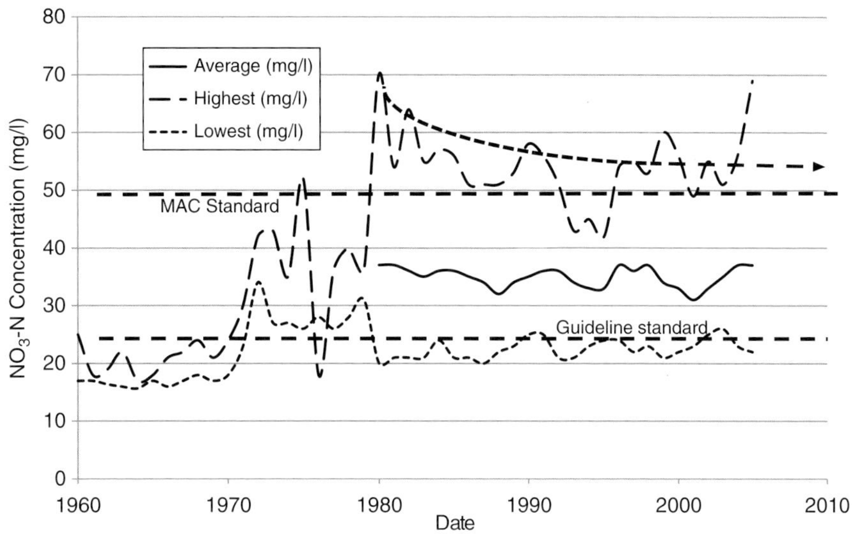

Figure 7.2 Nitrate concentration in the River Thames at Walton, Surrey (1960–2005) (data sourced from DEFRA, 2007).

1960/1970 period, all leading to a substantial uplift of the nitrate decline curve, as seen from Figure 7.2. The net result is that benefits of the introduction in the 1990s of nitrate vulnerable zones (NVZs) cannot be properly assessed for at least another 15–20 years. Wang *et al.* (2012) have taken a more pessimistic view predicting that some 60% of peak nitrate concentrations are still passing through the unsaturated zone in the Thames Basin and that it will take another 50–60 years for them to reach the underlying water table. Such predictions make it very unlikely that the WFD objectives of reversal of upward trends for this diffuse conservative pollutant will be achievable in the envisaged 2027 timescale. This is particularly the case if the UKTAG (2012) recommendation for a single threshold nitrate (NO_3–N) standard of 37.5 mg/l was to be adopted for application in designated drinking water protection zones (WPZs).

Reversal, as well as prevent and limit, approaches to the trends shown in Figure 7.2 will require fundamental changes in land use measures combined with the imposition of strong and effective catchment management strategies. This might include extensive conversion of arable land to forestry as well as on-farm minimal impact practices such as adjustment of fertiliser inputs together with manure management and incorporation of maize silage to reduce nutrient production and leaching. The tiered application by the EA of WPZs and safeguard zones (SGZs) to target and focus mitigating measures and awareness campaigns is intended to highlight and address groundwater 'hotspots', such as those associated with nitrate storage and release from dual-porosity aquifers (Environment Agency, 2011). However, the mechanisms of fissure and matrix flow for such hydrogeological conditions are still poorly understood and remediation measures to reverse matrix pollutant build-up will be very difficult to achieve in any cost-effective manner. There is a need for the development of rapid, high-resolution field tools and techniques to assess the significance of heterogeneity of both properties and process as well as improving the monitoring and modelling of spatial and temporal groundwater flow variability.

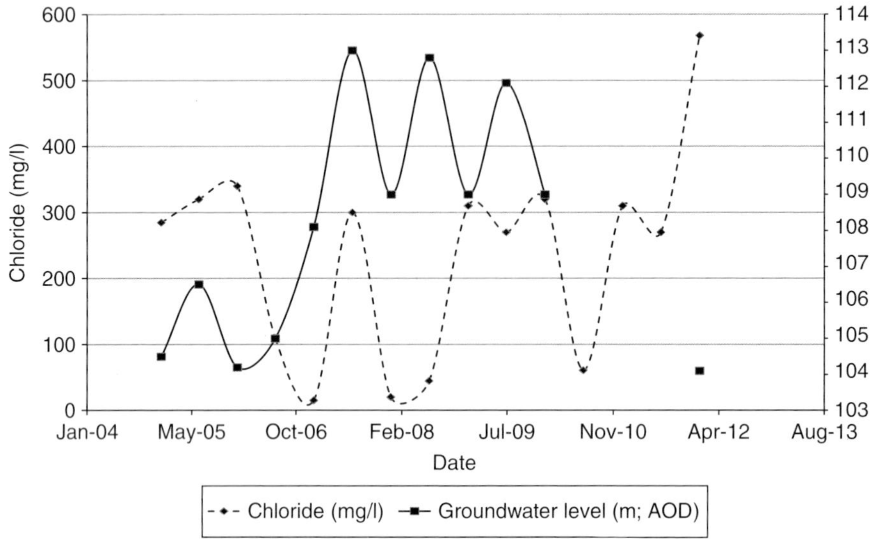

Figure 7.3 Chloride variations with groundwater levels.

Chloride represents another conservative substance that can be of local concern in urban aquifers. Chloride concentrations in motorway and trunk road runoff following winter de-icing application can reach 1500–2000 mg/l (Rivett *et al.*, 2011), but studies in the urbanised M4 and M25 corridors west and north-west of London generally indicate chloride concentrations down-gradient of runoff soakaways to range between background levels of <10 mg/l to just >500 mg/l (Ellis, 2006). Peak concentrations in chloride levels adjacent to the soakaways correlate with low groundwater levels such that, as shown in Figure 7.3, the highest concentration (568 mg/l) recorded in January 2012 follows an exceptionally mild winter, corresponding to an extremely low groundwater level.

The data suggest a time lag of a few months as a consequence of passage through the 40 m unsaturated zone thickness at this M25 site. The lowest concentrations (15–20 mg/l) are only marginally above background, suggesting that chloride passes through the aquifer relatively quickly; concentrations some 1–2 km from the motorway decrease to ~5 mg/l. To replicate values at off-site locations using a probabilistic model such as ConSim (Hall *et al.*, 1998) demands a very low lateral dispersivity, implying that transport in and through the underlying chalk aquifer is primarily by means of fracture/fissure flow with relatively fast travel times (Price, 1994). For conservative ions in a complex aquifer it is particularly important to characterise the flow correctly and thus the application of the MODFLOW model may be more useful than ConSim in this context. However, the fissure flow mechanism at this location may be the result of a substantial driving hydraulic head created by the soakaway receiving a high volume of road drainage. Nevertheless, only at abstraction locations very close (<0.5 km) to the point source are there likely to be any possibilities of water quality standards being breached. Turbulent mixing could occur due to pressure transients in the fracture zones affecting shallow solute distributions. It is also possible that unstable unsaturated zone flow becomes important under such conditions, leading to rapid 'funnel' flow in vertical fracture 'fingers'. This causes rapid water and solute transfer through the unsaturated zone, with other parts of the zone being relatively flow-inactive (Environment Agency, 2009b).

This still leaves questions of what the critical input concentrations and duration of conservative substances from a point (or line) source are, and whether the impact on groundwater only becomes evident once the pollutant diffuses back out from the matrix block to contaminate the groundwater flow. In addition, there are questions relating to the capacity of, and decline equations for, dual-porosity chalk in terms of dilution and dispersion of conservative ions. Risks to groundwater might be reduced by stipulating that no road soakaways should be allowed in source protection zones (SPZs) 1, 2 and 3.

Similar observations of variations in groundwater levels over extended time periods in the Luton urban area show fluctuations to be mirrored by changes in chlorinated solvent levels, with trichloroethylene (TCE) and perchloroethylene (PCE) at combined levels above the 7.5 µg/l standard (Longstaff *et al.*, 1992). High groundwater levels appear to act to dilute the solvents that are presumably emanating from a historic subsurface urban dense nonaqueous phase liquid (DNAPL) source. There seems to be some correlation between the abstraction regime above urban aquifers and solvent concentrations where prolonged periods of pumping tend to reduce contaminant levels (Rivett *et al.*, 2011). This is presumably due to either cleaner groundwater being drawn in from more pristine areas or perhaps to shorter residence times for groundwater in the urban catchment zone. However, once again dilution in the aquifer appears to be a major factor.

Most chlorinated solvents (and pesticides) undergo degradation to daughter products through mechanisms such as reductive dechlorination. This would give rise to exponential declines in concentrations following cessation or prohibition of the point (or diffuse) source inputs, a feature that is commonly observed for solvents and pesticides in groundwater. However, there is little convincing evidence for significant concentration of the daughter products in groundwater and therefore it might be concluded that there is only very limited degradation occurring in the deep aquifer. Methodologically, such contaminant response behaviour poses questions of how the effects of changing groundwater levels and abstraction rates can be 'removed' to obtain baseline declining plots to check the mathematical solutions for the noted correlations. Should the equations generated be based on worst-case scenarios, such as drought and minimal abstraction or the effects of climate change, since peaks and frequency of contaminant peaks often trigger the need for water treatment? It may, however, be better to base the equations on average conditions although water companies are essentially concerned with peak values that are more likely to lead to noncompliance.

It also leaves open the question of appropriate remedial measures if the decline curves still predict breaches of threshold 'good status' beyond the 2027 WFD time limit. Stable conservative pollutants not subject to attenuation in the groundwater environment will be particularly difficult to remediate within the defined WFD timescales up to 2027, especially in dual-porosity aquifers for which there is no easy 'silver bullet' remediation. In the case of nitrate concentrations, a 50–60 year timescale is clearly inappropriate to the WFD 3 × 5 year cycle period during which problems should be fixed. It is clear that there are still many questions to be resolved regarding groundwater–surface water processes and interactions in urban catchments before satisfactory mitigating measures can be implemented (Environment Agency, 2009b). The biggest unknown in flow modelling is the recharge component. Complex, time-varying and low flows cannot be accurately replicated in models such as MODFLOW using 4R coding (Heathcote *et al.*, 2004). This is partly due to the time lag for surface runoff to reach the water table as a result of attenuation and temporary storage by surficial deposits. It may also be partly due to the lack of understanding of varying unsaturated zone flows, particularly in the 'putty-like' characteristics of the chalk. It is also clear that control of point and diffuse source pollution will require catchment-scale measures as well as supplementary education campaigns with farmers, growers and horticultural groups, transport operators and local authorities on the safe storage, use and disposal of chemicals.

7.3 Drinking Water Safety Plans

The concept of a preventative 'hazard and critical control point' (HACCP) approach to drinking water supply management combines elements of risk analysis, quality assurance and multiple-barrier principles. It provides a framework for evaluating treatment and control measures by focusing on process steps that are critical inputs for ensuring water supply quality. The concepts of such Drinking Water Safety Plans (DWSPs) were introduced in World Health Organisation (WHO) guidelines (Bartram *et al.*, 2009). The decision of the UK Drinking Water Inspectorate to require water company AMP5 improvement programmes to be developed within a DWSP approach has given a new focus to risk management of the water supply. The approach goes further than just hazard identification, risk assessment and control, extending to embrace catchment, treatment, distribution and the consumer. It also incorporates operations linked to water supply, including organisational structures and operating procedures, training, communications, monitoring, reporting, incident and emergency procedures. This comprehensive DWSP approach emphasises identifying and controlling risks where they arise and places less reliance on the treatment process stream as the primary control option and where measuring risk was essentially an exercise in numbers and percentage compliance. As such it is considered to offer more sustainable solutions, with the added benefits of reduced energy and carbon footprints.

Under a DWSP approach, monitoring becomes more targeted towards demonstrating that implemented mitigating controls work at optimal levels with risks and incidents being more predictable and identifiable at an earlier stage. A DWSP needs to be developed for each water system with the simplest form consisting of four elements: catchment (including raw source waters), storage and treatment, distribution and customer use. The typical structure for the development of a DWSP is given in Figure 7.4, with the working elements categorised in terms of potential failure, current operational controls and identified interventions, with final risk occurrence probabilities and consequence scores being evaluated for each risk. Such detailed DWSPs give the water companies and regulatory agencies better insights into how the individual suppliers understand, operate and protect their systems (Water Industry Research (WIR), 2009). Given that the Drinking Water Inspectorate (DWI) is an official regulatory body, it has not issued prescriptive generic procedures to water companies for undertaking DWSPs, preferring to advocate guidelines as indicated in Figure 7.4 and leaving the companies to develop their own individual approaches tailored to their particular operational sizes, prevailing technologies, land ownership, consumer base, etc.

Whilst the DWI guidance approach led to early misconceptions (May, 2007), the working principles of a DWSP process were already familiar to most water companies having based their distribution operation and maintenance (DOM) strategies on a similar hazard/risk assessment structure. Thames Water developed their DWSP primarily through conducting optioneering scenarios, a practicability review and cost–benefit analysis to identify 15 optimum solutions offering the highest net investment benefits. Their £6.5 billion plan primarily focuses on ensuring that existing assets will be maintained in good condition,

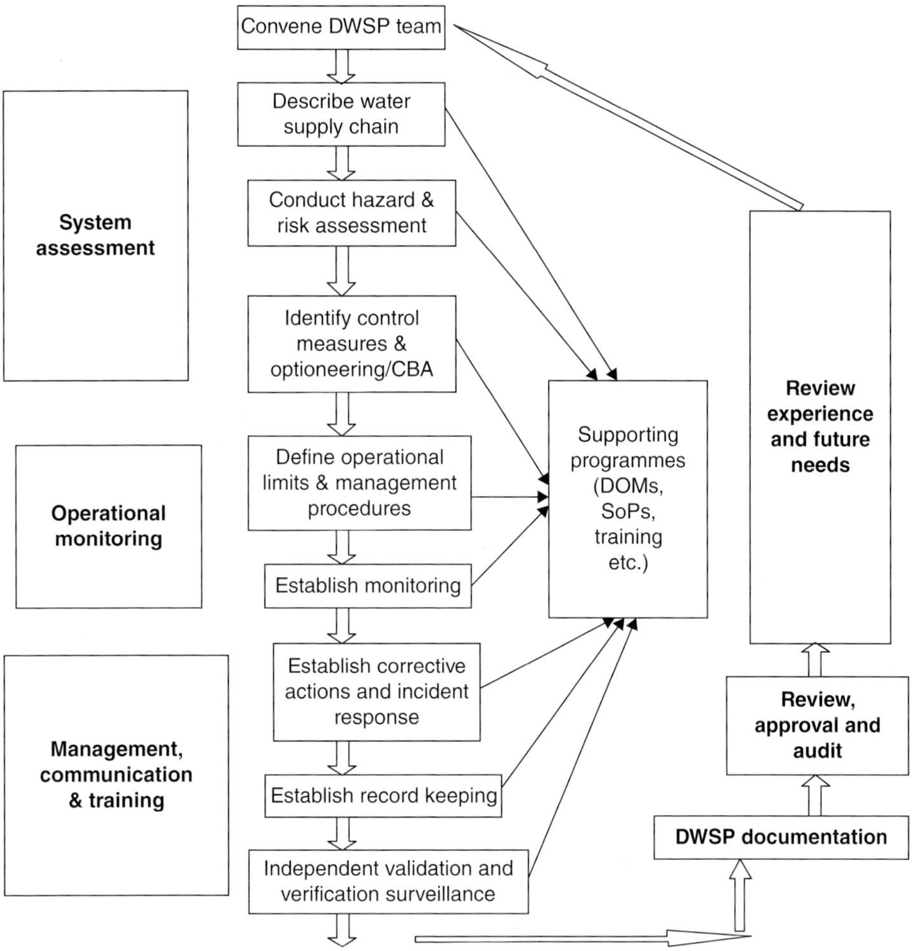

Figure 7.4 The structure and development of a DWSP (adapted from DWI, 2005).

although it also proposes a new reservoir in Oxfordshire to provide up to 10% of the regional water supply when completed in 2021 (Thames Water, 2008). Major expenditure by the company will be targeted at control and reduction of pesticides, nitrates, cryptosporidium and lead pipe replacement. The latter is considered urgent in order to meet the new 10 μg/l drinking water standard to be introduced in 2013. Phosphate dosing has already reduced failure rates to 1.7%, but remaining 'hotspots' will be targeted where ≥20% of random daytime samples are found to exceed the new standard.

However, given the greater emphasis on targeted operational monitoring inherent in the DWSP approach, compliance monitoring will be reduced, making it more difficult to evaluate percentage compliance and to compare relative company performance. Given the more extended timescales for the DWSP approach to deliver benefits, there is also an issue of how the approach will be justified through the cost and environmental appraisal required under the EU Drinking Water Directive. Irrespective of this, the benefits of the

DWSP approach includes better clarity and understanding of the hazard and risk situations causing drinking water problems and it provides a robust transparent system for identifying cost-effective solutions. The DWSP approach will be integrated into the EU Drinking Water Directive as part of its next revision. This will give the DWI a mandatory process rather than an advocacy strategy, which forms the basis for their current policy approach.

7.4 Urban Growth and Water Demand

A key driver in water demand is demographic, with population growth having significant future impacts on household usage rates. However, it is not a straightforward exercise to calculate such future demands given trends for decreasing average household sizes and potential changes in per capita usage rates. The Barker (2004) report suggested that new housing supply in South East England is likely to be some 200 000 above that already contained in regional plans. The South East Regional Planning Committee has also independently indicated, for example, that an average of 29 000 new homes will be built each year over the next 20 years. A sustainability impact study (Office of the Deputy Prime Minister (ODPM), 2005) concluded this would result in water demand increasing by 728 millon l/day by 2016 and 1102 million l/day by 2031 – 6.1% and 9.2% respectively above the current 2011 demand. However, the failure to include water companies and the regulatory Environment Agency as statutory consultees for the water supply in the planning system and in the development of Regional Spatial Strategies (RSSs) and Local Development Frameworks (LDFs) prejudices sustainable resource development and security of supply. The new regional planning arrangements proposed in Policy and Planning Statements PPS11 and PPS12 now recommend that both utilities and regulatory agencies should be involved in the preparation of LDFs and with long-term consultative supply infrastructure planning. In addition, the current quinquennial periodic review(PR) cycle places water companies under considerable financial and logistic uncertainty in terms of long-term planning for future supply development.

The only demographic factor that can be directly addressed by the water industry and regulators is the growth in per capita water usage. However, if domestic usage is to be reduced to more sustainable levels of 110–125 l/head/day, issues of water efficiency and mandatory standards for equipment and fixtures need to be addressed in addition to leakage control. The public are highly unlikely to accept prolonged supply restrictions as long as realistic leakage targets remain a problem. Total leakage in England and Wales for 2010 stood at some 3608 million l/day of a 15 378 million l/day distribution output, which represents an industry average of 23% (OFWAT, 2010). It must be recognised, however, that a significant proportion (up to one-third) of the leakage occurs within the domestic curtilage and as such lies outside the immediate control of the water companies. Irrespective of this, it still aggravates public sensitivity that Thames Water, for example, has opened a costly £250 million desalinisation plant whilst consistently failing to meet leakage control targets, despite the company arguing that they have achieved an 'economic level' of leakage control and that the plant will deliver 140–150 million l/day for 1 million customers in north-east London.

New emerging technology will certainly provide water companies with better solutions for identifying and repairing leakage, including new pipe materials, increased use of remote sensors and valves, as well as improved leak-detection technologies such as acoustic, thermal and noise imaging. However, different pipe materials have different lifetimes and much more work is necessary to understand the operational lifespan of new materials such as polyethylene and epoxy resins. Trunk mains or critical mains, which can have a high impact on water supplies if they fail, should have risk assessment based on known pipeline deterioration factors, with regular inspection using nondestructive techniques. Water efficiency savings and new technology also have an important role in containing and balancing future supply and demand in an environmentally sustainable manner to meet increasing population growth. However, the work of both water companies and government agencies in respect of water efficiency is relatively small scale and piecemeal in nature at present, lacking the strategic catchment approach evident, for example, in Australian policy.

Sydney Water (2010) has retrofitted and subsidised some 300 000 properties with water efficiency devices, achieving an average annual saving of 21million l per household, equivalent to 12% of domestic water usage. The Environment Agency in England and Wales similarly claim that water efficient fittings and appliances could reduce per capita daily consumption for new dwellings to some 105 litres (DEFRA, 2008). In Victoria and New South Wales, all new housing has to meet the Australian 'five-star' rating, which requires that mains water consumption be reduced by 40% compared to the prevailing average for similar sized houses (Australian Bureau of Statistics (ABS), 2010). Whilst such levels and options may not be appropriate in the United Kingdom, the concept of setting challenging water efficiency standards, accompanied by flexibility for developers, must be part of a basis to meet the future water demands of increasing urban growth. It is clear that the water industry must apply a twin-track approach to achieve a balance between resource development and demand management. Behavioural change, new technologies and leakage control need to be accompanied by timely resource development, although in the stressed urban growth regions of South and South East England, it makes economic and logistic sense to emphasise demand management techniques as the front-end driver.

An independent government report (Council for Science and Technology (CST), 2009) has taken the view that the UK water sector has been far too concerned with efficiency performance measures and price controls, paying relatively little attention to innovative solutions and long-term R&D technology planning as well as lacking leading-edge investment and skill capacities. The water sector as a whole only spends an average of 0.3% of its annual turnover on R&D. The report argues that the industry is largely driven by the regulatory framework and there is little evidence of joined-up thinking on coordination between future water quality supply targets and those for energy efficiency and low-carbon solutions, future climate change or consumer engagement.

The government White Paper (DEFRA, 2011) on future water resource development suggests that there is very likely to be closer working relationships between the water companies, regulators and stakeholders, with River Basin Management Plans (RBMPs), Water Resource Management Plans, Drought Plans and Price Reviews being considered in a complementary manner. The structure and process whereby this coordination will be achieved remains to be addressed, but undoubtedly future central government direction will be needed to secure this objective. A national strategic overview of the quality and capacity of water infrastructure is essential to put in place robust and sustainable future service delivery, ensuring resilient long-term supply as well as a customer focus on affordability, competition and demand management.

As indicated in Figure 7.5, there are also forecasts for rapidly increasing energy use in the UK water sector over the next 30–40 years (Ainger *et al.*, 2010). Optimisation of existing technologies will not materially alter this upward trend and new innovative approaches will be required to achieve substantial reductions in long-term energy use. Even accepting that such alternative approaches are adopted, much further reductions in energy use will be required to meet government targets for carbon reduction in the order of 34% by 2020.

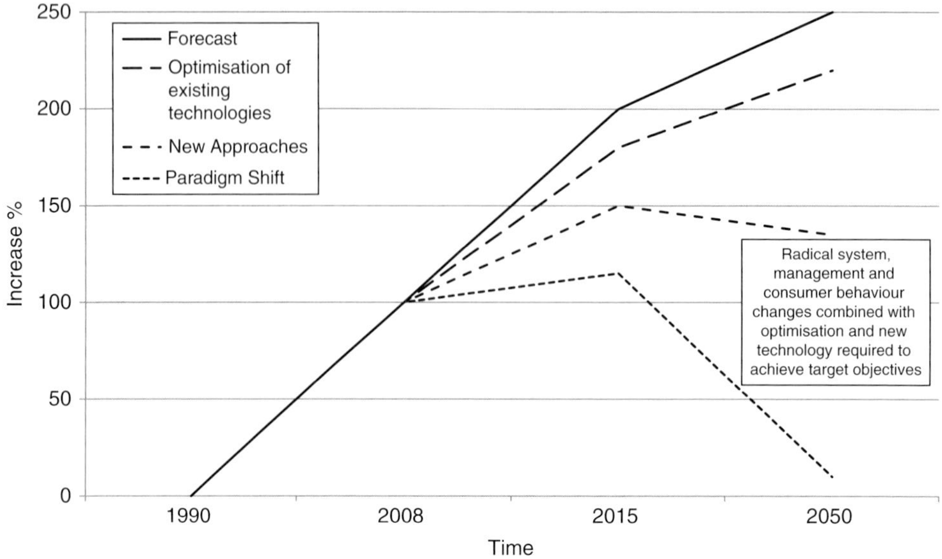

Figure 7.5 Forecasted percentage energy use by the UK water industry (adapted from Ainger *et al.*, 2010).

The water industry as a whole currently accounts for 5 million tonnes of CO_2 emissions, which are only 1% of total UK greenhouse gas emissions, with water supply responsible for 40% of the industry total. However, emissions related to household water use are very much higher (35 million tonnes), accounting for 5.5% of total UK emissions. Metering in those urban areas subject to serious water stress could potentially achieve a savings of 1 million tonnes per year, whilst full customer metering would deliver a 30–40% emission reduction, as well as potentially reducing customer water and energy bills.

The shift to a new paradigm of water management as indicated in Figure 7.5 will demand not only substantial shifts in technology and accompanying optimisation procedures but also major changes in management structures and procedures as well as in consumer behaviour. There may also be an advantage that might be gained to stimulate water efficiency approaches from the establishment of an independent Water Savings Trust or similar Nongovernmental Organisation (NGO) alternative along the lines of the EU Energy Savings Trust. It is clear that increased and more unified company and government promotion of public awareness campaigns for water conservation and the need to behave in a more water efficient manner is needed to drive forward water efficiency initiatives. The introduction of mandatory water labelling for household water use products, analogous to that of the EU energy labelling scheme, would also serve to enhance public awareness.

Revisions to Part G of the UK Building Regulations (NBS, 2010) covering house fittings for new premises might also help establish minimum water efficiency standards, perhaps specified on a region-specific basis to meet future demographic and urban growth patterns. Such awareness-raising approaches must be accompanied by increased metering and Smart real-time billing if long-term reductions in consumption rates are to be achieved. Increased metering also facilitates effective block tariff management; collectively tariff and metering measures could reduce usage by as much as 10–20%. The extension of regulatory responsibilities to water supply systems in buildings remains a contentious issue although there is increasing dialogue between the DWI and the Health and Safety Executive (HSE) on potential frameworks within which Local Authorities (LAs) and their Environmental Health Officers (EHOs) might undertake inspections and share information. The evolution of DWSPs will certainly highlight the critical relationship and gap analysis existing between the water company supply and public health in a much sharper manner.

As acknowledged in the government White Paper (DEFRA, 2011), the development of a national pipeline network for large-scale connectivity and bulk transfer would also provide a strategic solution to the existing and future water demands in the urban areas of South and South East England, although there may be serious ecological issues arising from mixing of source waters. Bulk transfer would require a central government initiative to subsidise the significant costs and to overcome logistical issues as well as providing fiscal incentives for water companies to trade abstraction and transfer rights. Such a national strategic approach would necessitate considerable reform of the management structures and processes currently prevailing in the water industry.

7.5 Conclusions

There have undoubtedly been considerable advancements in drinking water supply quality over the past two decades, with risk assessment and management now comprising the core of all strategic planning and operational delivery. This risk-based approach to drinking water has also been more effective in bringing together a wider stakeholder engagement as required under the WFD. However, there still remain considerable challenges to be resolved as the water industry looks to meet their full Directive obligations. Clearly in the case of many conservative source pollutants, for example, there will be a continued need for a twin-track approach of upgraded treatment and associated water demand control measures as well as improved catchment management. A strategy of continued abstraction (with enhanced treatment), rather than seeking total prevention and trend reversal, may not fulfil the WFD objectives, but it may well be the only feasible option for future control management of diffuse groundwater substances such as nitrates, pesticides, solvents, etc. Challenging scientific questions and issues remain to be resolved before these conservative pollutants can be fully and cost-effectively contained. The question of ensuring sufficient future supplies will also continue to be a contentious issue that might only be resolved through fundamental and radical organisational reform of the UK water industry. Future demand management must also take into consideration how closer legislative and organisational connections can be made between land use planning and water resource planning. The lack of coordination between water companies, regulators and planning policy may be a more significant barrier to future water resource development than any need for improved or innovative treatment technologies as a basis to secure safe, clean drinking water.

References

Ainger, G., Butler, D., Caffor, I., Crawford-Brown, D., Helm, D and Stephenson, T. (2010) *A Low Carbon Water Industry in 2050*. Report SC070010/R3. Evidence Directorate, Resource Efficiency Programme. Environment Agency, Bristol, UK.

Australian Bureau of Statistics (ABS) (2010) *Water and Energy Efficiency Elements of Households in Older and Newer Dwellings*. State and Regional Indicators, Victoria. No. 1367.2, June 2010. ABS, Melbourne, Australia.

Barker, K. (2004) *Review of Housing Supply: Delivering Stability; Serving our Future Housing Needs.* Final Report, March 2004. HM Treasury, London, UK. ISBN 1845320107.

Bartram, J., Corrales, L., Davison, A., Deere, D., Drury, D., Gordon, B., Howard, G., Rhinehold, A. and Stevens, M. (2009) *Water Safety Plan Manual: Step-by-Step Risk Management for Drinking Water Supplies.* World Health Organisation (WHO), Geneva, Switzerland. ISBN 9789241562638.

Council for Science and Technology (CST) (2009) *Improving Innovation in the Water Industry: 21st Century Challenges and Opportunities.*CST, London, UK.

Department of Environment, Food and Rural Affairs (DEFRA) (2007) *Digest of Environmental Statistics. Nitrate Concentrations in Rivers.* DEFRA, London, UK.

Department of Environment, Food and Rural Affairs (DEFRA) (2008) *Future Water: The Government's Water Strategy for England.* Cm. 7319. HMSO, London.

Department of Environment, Food and Rural Affairs (DEFRA) (2011) *Water For Life.* Cm. 8230. The Stationery Office, Norwich, UK. ISBN 9780101823029.

Drinking Water Inspectorate (DWI) (2005) *A Brief Guide to Drinking Water Supply Plans.* DWI, London, UK.

Drinking Water Inspectorate (DWI) (2011) *Annual Report; Drinking Water 2010.* DWI, London, UK.

Ellis, J.B. (2006) A UK and European perspective of sustainable urban drainage systems (SUDS) with particular reference to infiltration systems and groundwater. Sections 2.1–2.8 in *Sustainable Urban Drainage Systems and Groundwater Monitoring.* Proceedings 26th Annual Conference, International Association of Hydrogeologists (Irish Group), April 2006, Tullamore, Ireland. ISNN 1393–1806.

Environment Agency (EA) (2008) *Water Resources in England and Wales: Current State and Future Pressures.* Environment Agency, Bristol, UK.

Environment Agency (EA) (2009a) *Water for People and the Environment. Water Resources Strategy for England and Wales.* Environment Agency, Bristol, UK.

Environment Agency (EA) (2009b) *The Hyporheic Handbook: A Handbook of the Groundwater–Surface Water Interface and Hyporheic Zone for Environment Managers.* Report SC050070. Environment Agency, Bristol, UK. ISBN 9781849111317.

Environment Agency (EA) (2011) *Groundwater Protection; Principles and Practice,* GP3, Parts 3, 4 and 5. Draft for Consultation. Environment Agency, Bristol, UK.

Hall, D.H., Harvey, R.C., Moncaster, S., Plimmer, B. and Rogers, M. (1998) Quantifying the risk posed to groundwater from contaminated land: the ConSim model. In: *Proceedings of the 6th International Conference on Contaminated Soil (ConSoil98),* 17–21 May 1998, Edinburgh, Scotland, pp. 1035–1036. Thomas Telford Ltd, London, UK. ISBN 0717726757.

Harden, N.J.K., Burt, T.P., Worall, F., Whelan, M.J. and Bieneza, M. (2010) Nitrate concentration and fluxes in the River Thames over 140 years. Are increases irreversible? *Hydrological Processes*, **24**, 2657–2602.

Heathcote, J.A., Lewis, R.T. and Soley, R.W.N. (2004) Rainfall routing to runoff and recharge for regional groundwater resource models. *Quarterly Journal of Engineering Geology and Hydrogeology*, **37**, 113–130.

Jackson, P.J., Williams, N.M., Rule, K,L., Davis, L.J., Blake, S., Warburton, S.G. and Ellis, J. (2004) *Quality of Drinking Water in Public Buildings.* Report to Drinking Water Inspectorate, DWI6348 (DWI Reference DWI 70/2/64). WRcNSF Ltd, Marlow, Buckinghamshire, UK.

Longstaff, S.L., Aldous, D.J., Clark, L., Flavin, R.J. and Partington, J. (1992) Contamination of the chalk aquifer by chlorinated solvents: a case study of the Luton and Dunstable area. *Water and Environment Journal*, **6**, 541–550.

May, A. (2006) The benefits of drinking water quality regulation: England and Wales. *Water Science and Technology*, **54** (11/12), 357–393.

May, A. (2007) *An Assessment of the Impact of Regulatory Models for Drinking Water Quality in the UK.* Report May 2007. Drinking Water Inspectorate, London, UK.

NBS (2010) *Building Regulations Part G 2010.* NBS, Newcastle-upon-Tyne, UK. ISBN 9781859463239.

Office of the Deputy Prime Minister (ODPM) (2005) *A Sustainability Impact Study of Additional Housing Scenarios in England*, December 2005. ODPM, London, UK. ISBN 185112828X.

OFWAT (2009) *Future Water and Sewage Charges 2010–15; Final Determinations*. OFWAT, Birmingham, UK. ISBN 1904655521.

OFWAT (2010) *Service and Delivery: Performance of the Water Companies in England and Wales 2008–2009*. OFWAT, Birmingham, UK.

Price, M. (1994) Drainage from roads and airfields to soakaways: groundwater pollution or valuable recharge? *Journal of Chartered Institution of Water and Environmental Management*, **8**, 468–479.

Rivett, M.O., Wealthall, G.P., Dearden, R.A. and McAlary, T.A. (2011) Review of unsaturated zone transport and attenuation of organic compound (VOC) plumes leached from shallow source zones. *Journal Contaminant Hydrology*, **123**, 130–156.

Sydney Water (2010) *Water Efficiency Report*, pp 43. Sydney Water, Parramatta, Sydney, Australia.

Thames Water (2008) *Water Resources Management Plan*, Volume 6, SEA Environmental Report. Cascade Consulting on behalf of Thames Water Utilities Ltd, Reading, UK.

UK Technical Advisory Group (UKTAG) (2012) *Derivation of Nitrate Threshold Values for Assessing Risks to Groundwater Drinking Water Resources*. Guidance Paper, Groundwater Task Team (GWTT). UKTAG, Edinburgh, Scotland.

UK Water Industry Research (UKWIR) (2004) *Implications of Changing Groundwater Quality for Water Resources and the UK Water Industry*. Report 04/WR/09/8. UKWIR, London, UK ,70 pp.

Wang, L., Stuart, M.E., Bloomfield, J.P., Butcher, A.S., Gooddy, D.C., McKenzie, A.A., Lewis, M.A. and Williams, A.T. (2012) Prediction of the arrival of peak nitrate concentrations at the water table at the regional scale in Great Britain. *Hydrological Processes*, **26**, 226–239.

Water Industry Research (WIR) (2009) *Drinking Water Safety Plans: Risk Based Approach*. Report 08/DW/02/48. WIR, London, UK. ISBN 1840574992.

Water UK (2010) *Drinking Water Quality in the UK*. Water UK, London, UK.

8

Desalination

Augustine Ifelebuegu, Susanne M. Charlesworth and Colin A. Booth

8.1 Introduction

Availability and accessibility of fresh water resources will continue to be a major issue in the world economy (United Nations Educational, Scientific and Cultural Organisation (UNESCO), 2012), even though the Earth is a clear watery oasis with ~71% of the surface covered by water. Over 99% of this is found in seas, oceans and the polar icecaps, with less than a percent attributed to land. In total, only ~3% of water is fresh and suitable for drinking with the other 97% being salty and not directly usable (Kalogirou, 2005). As the global demand for water increases due to an increasing population with their expected increased standard of living, plus requirements for water due to agricultural activities and industrialisation, alternative sources of potable water are being explored and developed to meet this escalating demand. Desalination, as a process, offers a technological alternative to transform the inexhaustible supply of sea water into potable fresh water and could assist alleviating future water scarcity. It is the preferred choice over water reclamation, particularly for potable supplies (Dolnicar *et al.*, 2011), due to the level of water purity that can be attained using desalination methods. Also, the process is largely independent of the weather rather than being dependent on the continued operation of associated infrastructure.

The desalination process, called 'desalting' in the early days, is simply the means of removing salt from water to make it suitable for human consumption and industrial usage (European Union (EU), 2008). The concept of desalination has been around from medieval times, through the middle ages to modern times. Greek sailors thousands of years ago boiled seawater to separate out fresh water from salt, just as the Romans used clay filters to achieve the same objectives. As an industrial process, desalinisation began as early as 1912 in Egypt (El-Dessouky, 2007), with production capacity increasing during 1929–1937 due to the beginnings of the oil industry. During and immediately after the Second

Water Resources in the Built Environment: Management Issues and Solutions, First Edition.
Edited by Colin A. Booth and Susanne M. Charlesworth.

World War, desalination became more popular as hundreds of desalination mobile units were made for military use in US units in arid areas and across the globe. By the 1980s desalination became fully commercialised, making it more cost-effective in many countries. By the end of the last millennium many large-scale plants were built or were under construction in the Middle East, Israel, Australia and the United States, where concerted efforts were made to add various desalination plants to their water management portfolios. Today, there are >15 300 desalination plants around the world, with a daily production capacity in excess of 70 million m^3 (Zimmerman *et al.*, 2008). This is projected to nearly double to 130 million m^3/day by 2016. Approximately 47% of the global desalination capacity is located in the Middle East, mainly in Saudi Arabia, United Arab Emirates and Qatar (Hepbasli and Alsuhaibani, 2011). Currently between 50 and 70% of drinking water in Saudi Arabia is desalinated water produced from ~30 different water desalination plants. One of these plants, the Jebel Ali plant in Jubail is currently the largest water desalination plant in the world (Rodriguez, 2011).

Desalination plants vary in size from as small as 20 m^3/day to 1 million m^3/day. There are also domestic-sized desalination plants in use in various countries that purify just a few litres of water per day. These are usually reverse osmosis plants (see Section 8.2 and Peñate and Garcia-Rodriguez, 2012) that purify fresh water sources for domestic or local consumption. Desalination is a water purification process that utilises energy to separate salt from water. The principle of operation involves separating saline water into two streams, namely the brine, which is a high concentration of salt, and the potable water stream. However, this process requires both a high energy demand and other technologies to drive it (Khawaji *et al.*, 2008). Various separation techniques and technologies are employed globally, depending on the level of salinity and a host of other factors.

The concentration of salt in fresh water and sea water ranges from <500 ppm to >50 000 ppm. Water with salt concentrations <500 ppm, 500–30 000 pm and 30 000–50 000 ppm are classified as fresh, brackish and saline water, respectively. The World Health Organisation (WHO) (2003a) suggests that levels of chloride >400 ppm may affect the taste of hot drinks and those levels of sodium >200 ppm may affect the taste of drinking water (WHO, 2003b). Desalination, therefore, is expected to reduce salt concentration in sea water by up to approximately 100 times.

8.2 Desalination Technologies

Desalination technologies are classified under two broad categories (European Union (EU), 2008): distillation processes (Khawaji *et al.*, 2008) and membrane processes (Kim, 2011). Thermal desalination was the main method used before the early 1960s when membrane processes came on stream. Thermal distillation technologies use heat to convert water to steam, which is then condensed to produce distilled water. Membrane desalination technologies uses a thin film of porous material called semi-permeable membranes, which allows water molecules to pass through it but simultaneously minimise or prevent the passage of other ions like salt. Membrane technologies are either pressure-driven, as in reverse osmosis and nanofiltration, or electric potential-driven, as in electrodialysis and electrodialysis reversal.

The major thermal and membrane technologies used globally and their acronyms (EU, 2008) are given in Table 8.1. Thermal technologies require a higher energy input than those utilising membranes, although, as discussed in Section 8.7, renewables are increasingly

Table 8.1 Summary of the main thermal and membrane technologies currently in use.

Phase change (distillation) processes	Single-phase (Membrane) Processes
Multistage flash distillation (MFD)	Reverse osmosis (RO)
Multiple effect distillation (MSF)	Nanofiltration (NF)
Vapour compression evaporation (VCE)	Electrodialysis (ED)
Solar distillation (SD)	Electrodialysis reversal (EDR)

becoming more accepted as an energy source. Tables 8.2, 8.3 and 8.4 give further details of the most widely used approaches taken from Table 8.1.

8.3 Developing Technologies

In addition to the more establish technologies used today, there are emerging methods under active development, with some nearing commercialisation. These new emerging desalination technologies are described in Table 8.4.

8.4 Economics of Desalination

The first issue that negates the use of desalination is the substantial cost associated with it, mainly due to the high energy cost of the processes involved (Carter, 2011). This has prevented the application of desalination technologies in most parts of the globe. However, over the last decade, the cost of desalination has decreased significantly as a result of improvements and advances in thermal and membrane desalination technologies, especially in the case of reverse osmosis (RO) (WateReuse Association, 2012). At the same time, the cost of alternative sources of water has increased due to the rising costs of energy and the increased level of treatment that is required to comply with more stringent water quality standards. Costs depend on a variety of factors, including location of plants, site conditions, technology type, feed water quality, production capacity plant efficiency and options for concentrate and waste disposal. Currently, the cost of desalination has become significantly competitive with alternative treatment types in certain parts of the globe, where desalination offers some advantages, such as in drought-prone regions. According to Reddy and Ghaffour (2007), this is particularly the case for RO technology due to the potential for hybrid systems. The cost of desalination also varies with the technologies applied in the treatment process. Globally, the predominant technologies used are multistage flash (MSF) distillation, multiple effect distillation (MED) and RO. The average unit costs of water produced for these different technologies are presented in Table 8.5.

The current global average cost of water production is ~ $1.14/m^3$ (Thye, 2010). Compared with the costs shown in Table 8.5, desalination is now increasingly competitive with other sources of drinking water. Reduction in the cost of membranes, high-permeability membranes, improved operational efficiency and integration of energy recovery processes in the last few years have made membrane technology more economical to thermal desalination processes and competitive with other alternative and conventional

Table 8.2 Summary of thermal desalination technologies.

Desalination types	Brief Description	Competitive advantage	Limitations
Multistage flash distillation	MSF is the commonest thermal desalination process used globally with 90% of global installed thermal desalination capacity and 46% of global desalination capacity. It involves distilling water through multistage flash chambers numbering from 10 to 25. Feedwater is heated under pressure and passed through the various flash chambers where pressure is successively reduced, causing the water to evaporate rapidly (flashing). The water vapour generated by the flashing process is condensed on heat exchanger tubing that usually runs through the various chambers to produce fresh water. Typical capacity is 4000–880 000 m^3/day; thermal energy consumption is 55–220 kW h/m^3; and electrical energy consumption is 4–6 kW h/m^3.	Possibilities of energy recovery and cogeneration. Proven and reliable. High daily production capacity. Can handle very high salt concentration. Minimal pre-treatment requirement.	Recovery efficiency for sea water is limited to between 10 and 30%. Highest capital expenditure and operational expenditure. Plant corrosion. High technical expertise required.
Multiple effect distillation	MED applies the principle of evaporation and condensation using multiple vessels (multieffect evaporator), typically up to 16, arranged in series with successively lower pressures like in MSF. In the MED process the feed sea water is distributed on the outside of the multieffect evaporators in thin films and condensed in the inside surface . The vapour that is produced in each vessel is used to heat up the feed into the next vessel. Typical capacity is 100–60 000 m^3/day; thermal energy consumption is 40–220 kW h/m^3; and electrical energy consumption is 1.5–2.5 kW h/m^3.	Less energy cost compared to MSF. Better recovery efficiency for sea water compared to MSF.	Recovery efficiency for sea water is limited to 20–35%. Larger footprint.
Vapour compression	VC uses heat from the compression of vapour to evaporate the feed water. There are two main methods of compression, mechanical vapour compression that uses electricity and thermo vapour compression that uses steams from steam jet ejectors. Typical capacity is 5–17 000 m^3/day and electrical energy consumption is 6–12 kW h/m^3.	Higher recovery efficiency for sea water up to 50%. Low energy requirement. Small footprint and portable. Can handle very high salt concentration. Minimal pre-treatment required.	Requires large and expensive steam compressors. Limited to smaller size plants.

Source: Khawaji *et al.* (2008).

Table 8.3 Summary of membrane technologies (see Kim, 2011).

Desalination types	Brief description	Competitive advantage	Limitations
Reverse osmosis	RO process uses pressure to transfer/push water through a semi-permeable membrane at 17–25 bar for brackish water and 55–82 bar for sea water. The RO membranes are designed to be able to retain salts and other ions while allowing water to pass through. The product 'permeate' has around 500 ppm of total dissolved solids. Typical capacity is 0.01–36 000 m^3/day and electrical energy consumption is 2–10 kW h/m^3.	Cheap and quick to build. A wide range of capacity possible. Low energy cost compared to thermal processes. Removes other contaminants.	Pre-treatment required. RO membranes are expensive with an average lifespan of three years. Plant operates at high pressure; hence risk of mechanical failures. Fouling of membrane.
Electrodialysis/electrodialysis reversal	ED and EDR processes use electrochemical separation methods to separate salt from water by using an ion selective membrane and an electrical potential as the driving force to overcome the membrane resistance. Both processes are more suited for the desalination of brackish water as it becomes more expensive with higher salinity, with a maximum feedwater total dissolved solids (TDS) of 12 000 ppm.	Long membrane life of 7–10 years. Very high recovery efficiency up to 95%. Can treat water with a high level of suspended solids.	High capital expenditure and operational expenditure. Cannot remove nonionic contaminants. Not very suitable for sea water desalination. Not affected by scaling. Periodic cleaning of membrane required.

Table 8.4 Emerging desalination technologies (adapted from Carter, 2011; Penate and Garcia-Rodriguez, 2012).

Emerging technology	Brief description
Membrane distillation	This technology involves the use of a thermal process to convert saline water to vapour, which is then passed through a hydrophobic membrane and subsequently condensed as pure water. Changes in partial pressure due to thermal gradient drives the water vapour through the membrane.
Forward osmosis	Forward osmosis uses a semi-permeable membrane similar to reverse osmosis. The basic difference is that instead of using external hydraulic pressure to drive the water through the membrane, it uses the natural pressure gradient that is usually provided by using a higher salinity solution like ammonium carbonate.
Freezing distillation	This involves freezing of the salt water. When water freezes, salts are not involved in the formation of ice crystals so that the salts can be separated and washed off. Much less energy is required in principle for the phase change from liquid to solid compared to the phase change that happens in thermal distillation.
Supercritical desalination	This method involves the generation of supercritical water at elevated temperature and pressure. Salts have very low solubility in supercritical water (for a mixture of liquid and gaseous water at high temperature and pressure) and are removed from water by precipitation.
Capacitive deionisation	Capacitive deionisation uses porous carbon electrodes and low voltage to produce a flow of ions between electrodes producing deionised water.

Table 8.5 Unit cost of desalination technologies (adapted from Ribeiro, 1996; Karagiannis and Soldatos, 2008).

Desalination types	Average cost range ($/m^3)		
	Large-scale plants	Medium-scale plants	Small-scale plants
MSF	0.52–1.7	0.87–2.0	2.7–20
MED	0.52–1.01	0.95–1.5	2.5–15
RO (brackish water)	0.26–0.54	0.78–1.33	5.6–13
RO (sea water)	0.45–0.66	0.48–1.62	0.70–19

Figure 8.1 Aquamaster reverse osmosis plant (30 m^3/h capacity) (Reproduced by permission of Aquamaster Water Treatment Ltd; http://www.businessmagnet.co.uk/company/kineticowatersofteners-128790.htm)

means of water supply. Desalination of brackish water is now usually done using membrane processes. Thermal processes like MSF and MED only compete favourably with RO in the desalination of sea water on the very large scale, especially when energy recovery processes like cogeneration are implemented (Goebel, 2003). Some plants have attempted to drive down the unit cost by hybridisation using a combination of thermal and membrane processes.

8.5 Small and Domestic Scale Desalination Plants

The majority of the small-scale desalination plants commercially available are reverse osmosis plants. They vary in sizes from as small as producing just a 1 litre bottle of water to thousands of litres, which can provide supplies for larger groups from houses up to small communities. A sizeable number of these plants are currently in use, especially in ships and offshore facilities. Increasingly, due to concerns with water quality issues, some houses are installing RO and distillation units for point-of-use purification of their source of drinking water. Figure 8.1 shows an example of a commercially available small RO plant.

There are also small renewable energy powered plants that are available in the market. Some of these are targeted at small, remote communities and islands, where electricity may not be available from the grid and other sources of energy are expensive.

8.6 Environmental Impacts

Water desalination has positive impacts on the environment, such as the recycling and reuse of wastewater. Although desalination is gaining grounds as a source of water supply globally, there are still various environmental impacts of concern, one of which is the impact on marine organisms at the sea water feed intake. These organisms can be entrained with the sea water and carried into the desalination plant or they can be impinged, whereupon they are too large to pass through the intake but are trapped against the screen by the flowing water. This can cause the deaths of large numbers of juvenile-stage aquatic organisms (WateReuse Association, 2011). This impact can be reduced by the use of wedge wire screens to minimise entrainment and whose design includes implementing low velocities at the intake (RBF Consulting, 2009), which decreases the likelihood of organisms being sucked into the intake pipe.

A further barrier to the use of desalination, particularly modern sea water reverse osmosis (SWRO) plants, is that they can generate an average of 1.6 kg CO_2/m^3 of fresh water, with MSF plants individually generating up to 10, MED 3.2 and SWRO 3.6 kg CO_2/m^3 (Bushnak, 2010), although the values do vary depending on their source such that the World Wildlife Fund (WWF) in 2007 quote figures for RO of 1.78, MSF 23.41 and MED 18.05 kg CO_2/m^3 of produced water. Whilst not strictly comparable, the carbon footprint of power generation plants varies between about 0.5 and 0.8 kg CO_2 kW h depending on the fuel used (Bushnak, 2010). Whatever values are used, the carbon footprint for large-scale desalination plants is substantial. The use of nuclear and renewable energy, however, has the potential to reduce the carbon footprint of desalination plants (Kalogirou, 2005; National Centre of Excellence in Desalination, 2011; Voutchkov, 2008) (see Section 8.7).

The main environmental issue resulting from the widespread use of desalination as a source of fresh water is that of brine or waste concentrate disposal (Carter, 2011). Methods include direct sea discharge, which is the cheapest option, particularly if the plant is sited near the coast, but it depends on what chemical treatments have been used in the desalination process as to the possibility of this approach. However, in an Australian study of National Action Plan Regions (URS Australia, 2002), in which the plants were too far from the coast for direct disposal, evaporation ponds, aquifer well injection and surface water body release were utilised, which adds to the cost of disposal significantly. The concentrated brine, which often has elevated temperatures and also contains waste chemicals, must be properly managed as it can significantly impact marine life and the environment. There are no specific standards for impact assessments, only guidelines drawn up by the United Nations Environment Programme (Lattemann and Höpner, 2008), although the implementation of these has been proposed in the past (Hoepner, 1999). To date, environmental impact assessments have not been integrated into management policies, although technology and management options to reduce impacts are available and implemented in most countries. There are opportunities for harvesting the salt contained in the brine for stock feed, medical and chemical uses (URS, Australia, 2002), with Ahmed *et al.* (2003) suggesting that products from such a process (including gypsum, sodium chloride and calcium chloride, amongst others) can potentially provide a not inconsiderable income from what was a waste product. Careful irrigation of salt-tolerant crops (such as olives, almonds and pistachios) is also possible (URS, Australia, 2002). Salinity gradient solar ponds have been studied for their potential to collect and store incident solar radiation (Lu *et al.*, 2001; Velmurugan and Srithar, 2008) by utilising zero discharge desalination to concentrate brine reject streams. Lu *et al.* (2001) found that this use of saline ponds was both a reliable and environmentally friendly source of the heating and cooling energy required during thermal desalination processes.

8.7 Renewable Energy Sources and Desalination

The use of renewable energy sources (RES) to power desalination plants is also growing and increasingly becoming main stream (IEA-ETSAP and IRENA, 2012). Solar, wind, geothermal and nuclear energy powered desalination plants are being widely explored. There are increasing market potentials for RES powered desalination plants due to the increasing demand of desalinated water by energy-importing countries like China and India. There are now commercially available small-scale RO plants powered by solar energy in the form of photovoltaic cells (see Section 8.5).

The renewable energy potentials, which are available in both the Middle East and other arid regions, make the potential of RES in powering desalination plants attractive (Al-Karaghouli *et al.*, 2009). It is estimated that Saudi Arabia uses energy for thermal desalination equivalent to ~1.5 million barrels of oil per day. There is, therefore, the need for an alternative source of energy and, based on this, the first large-scale renewable desalination plant is currently being constructed by an IBM joint venture with King Abdulaziz City for Science and Technology (KACST), which will consist of a desalination plant powered by solar energy in the City of Al-Khafji. The plant will be powered by ultra-high concentrator photovoltaic (UHCPV) technology, with an estimated daily production capacity of 30 000 m^3 of drinking water (Dawoud and Mulla, 2012).

8.8 The Future of Desalination and Sustainable Water Supplies

Although the cost of sea water desalination has decreased significantly over the last few decades it is still higher than conventional sources of drinking water supply. However, desalination will play an increasingly significant role in the future supply of potable water. Existing fresh water sources and other alternatives, like water conservation, recycling and reuse, will not be sufficient to meet the rapidly increasing fresh water demand, especially in drought-affected and arid regions of the world. Unlike most fresh water sources, sea water is readily available and not affected by climatic and environmental conditions like drought or the vagaries of the weather. Hence, sea water desalination will continue to offer a large potential for an abundant and regular supply of fresh clean water and remain a significant component of the global portfolio of water supply options.

There is much ongoing effort in the industry to make desalination more affordable and sustainable. These efforts are geared towards improving efficiency and reducing energy

Table 8.6 Present and future forecast for seawater RO desalination (see Ribeiro, 1996; Karagiannis and Soldatos, 2008).

	2012	Projection for 2030
Cost of water (US$/m^3)	0.45–0.8	0.15–0.4
Construction cost (US$/m^3)	120–3500	500–100
Power utilisation (kWh/m^3)	2–3	1–2
Plant recovery (%)	45–55	55–70
Membrane productivity (m^3/day/RO membrane)	25–50	95–125

consumption through the integration of energy recovery devices (ERDs), which are usually centrifugal or positive displacement devices (Stover, 2007). Some of the ERDs now in use for energy recovery include: the Francis turbine, Pelton Impulse turbines, pressure exchangers and hydraulic turbochargers. Also in use for energy recovery are Isobaric systems. With these devices the energy requirements for driving high-pressure pumps is considerably reduced and hence, the overall cost of desalination is also reduced (Fritzmann *et al.*, 2007).

Sea water reverse osmosis is growing faster than other technologies and more plants of various sizes are being built around the world. The technology is expected to improve considerably over the next two decades with the development of larger-diameter and high-injection membranes, chlorine and fouling resistant membranes and advances in nanotechnology (Humplik *et al.*, 2011; Kim *et al.*, 2010). The current and projected unit cost, construction cost, energy usage, recovery and productivity of sea water RO desalination are shown in Table 8.6.

8.9 Case Study: The Thames Water Desalination Plant

According to the Environmental Agency (EA) (no date), much of the Thames Water area of the United Kingdm is classed as 'seriously water stressed', with a score of 40 where seriously stressed is >34 based on criteria such as: current water demand, forecast growth in water demand, forecasts of population growth in the area, availability of water resources currently and the forecast of availability of water resources. In extended periods of dry weather, inhabitants of the nearby places of Swindon and Oxford are at particular risk of water restrictions being enforced.

In 2010, therefore, the Thames Water Desalination Plant opened in Beckton, East London, and is the first plant of its kind to be built in the United Kingdom (Greenlee *et al.*, 2009). It is an RO plant sourcing water from the Thames Estuary during the last three hours of the ebb tide, and is designed to serve up to 900 000 customers, producing ~150 000 m^3/day of potable water for the public water supply (Lopez-Gunn and Llamas, 2008) and will be in operation during times of drought or when existing supplies need support when necessary. The plant runs on 100% renewable energy from biodiesel made from used cooking oil and the waste saline solution is mixed with wastewater effluent from the sewage treatment works outfall.

8.10 Conclusions

Water scarcity across the world is a serious and escalating issue (Elimelech and Phillip, 2011) and it is anticipated that the impacts of global climate change will only exacerbate this problem. Water desalination technology has become a significant player in the portfolio of drinking water supply options for meeting global demand, and in particular large desalination plants, for example those based on concentrating solar power, have been constructed in the Arabian Gulf (Trieb, 2007). The cost of desalination is falling considerably and is expected to continue to fall as a result of advances in energy use and recovery, improvements in membrane technology and the increasing success of RES-powered desalination plants (Elimelech and Phillip, 2011). The global water desalination plant capacity is expected to continue to grow to nearly double digit rates over the next decade. No longer will there 'not be a drop to drink'.

References

Ahmed, M., Arakel, A., Hoey, D., Thumarukudy, M.R., Goosen, M.F.A., Al-Haddabi, M. and Al-Belushi, A. (2003) Feasibility of salt production from inland RO desalination plant reject brine: a case study. *Desalination*, **158**, 109–117.

Al-Karaghouli, A.A., Renne, D. and Kazmerski, L.L. (2009) Solar and wind opportunities for water desalination in the Arab regions. *Renewable and Sustainable Energy Reviews*, **13**, 2397–2407.

Bushnak, A.A. (2010) Desalination. In: El-Ashry M., N. Saab and B. Zeitoon (eds), *Arab Environment: Water – Sustainable Management of a Scarce Resource*. Report of the Arab Forum for Environment and Development. Available at: http://jkbharris.files.wordpress.com/2010/09/afed-sustainable-management-of-a-scarce-resouce1.pdf.

Carter, N.T. (2011) *Desalination: Technologies, Use, and Congressional Issues*. Congressional Research Service. 14 pp.

Dawoud, M.A. and Al Mulla, M.M. (2012) Environmental impacts of seawater desalination: Arabian Gulf Case Study. *International Journal of Environment and Sustainability*, **1** (3), 22–37.

Dolnicar, S., Hurlimann, A. and Grun, B. (2011) What affects public acceptance of recycled and desalinated water? *Water Research*, **45**, 933–943.

El-Dessouky, H. (2007) Sea water desalination: importance, need, methods and historical developments. *Journal of the Pakistan Materials Society*, **1** (1), 34–35.

Elimelech, M. and Phillip, W.A. (2011) The future of seawater desalination: energy, technology, and the environment. *Science*, **333**, 712–717.

Environment Agency (EA) (no date) *Areas of Water Stress: Final Classification*, 9 pp. Available at: http://cdn.environment–agency.gov.uk/geho1207bnoc–e–e.pdf.

European Union (EU) (2008) *ADIRA Handbook – A Guide to Desalination System Concepts*. Euro-Mediterranean Regional Programme for Water Management (MEDA), 94 pp. ISBN 9789755613116.

Fritzmann, C., Löwenberg, J., Wintgens, T. and Melin T. (2007) State-of-the-art of reverse osmosis desalination. *Desalination*, **216**, 1–76.

Greenlee, L.F., Lawlerb, D.F., Freemana, B.D., Marrotc, B. and Moulin P. (2009) Reverse osmosis desalination: water sources, technology, and today's challenges. *Water Research*, **43**, 2317–2348.

Goebel, O. (2003) *Co-Generation of Power and Water – Selection of Desalination Process*. Lahmeyer International GmbH, Germany.

Hepbasli, A. and Alsuhaibani, Z. (2011) A key review on present status and future directions of solar energy studies and applications in Saudi Arabia. *Renewable and Sustainable Energy Reviews*, **15**, 5021–5050.

Hoepner, T. (1999) A procedure for environmental impact assessments (EIA) for seawater desalination plants. *Desalination*, **124**, 1–12.

Humplik, T., Lee, J., O'Hern, S.C., Fellman, B.A., Baig, M.A., Hassan, S.F., Atieh, M.A., Rahman, F., Laoui, T., Karnik, R. and Wang, E.N. (2011) Nanostructured materials for water desalination. *Nanotechnology*, **22** (29), 292001–292019.

IEA-ETSAP and IRENA (2012) *Water Desalination Using Renewable Energy: Technology Brief*, 28 pp. Available at: http://www.irena.org/DocumentDownloads/Publications/Water_Desalination_Using_Renewable_Energy_-_Technology_Brief.pdf.

Kalogirou, S.A. (2005) Seawater desalination using renewable energy sources. *Progress in Energy and Combustion Science*, **31**, 242–281.

Karagiannis, I.C. and Soldatos, P.G. (2008) Water desalination cost literature: review and assessment. *Desalination*, **223**, 448–456.

Khawaji, A.D., Kutubkhanah, I.K. and Wie, J.-M. (2008). Advances in seawater desalination technologies. *Desalination*, **221**, 47–69.

Kim, D.H. (2011). A review of desalting process techniques and economic analysis of the recovery of salts from retentates. *Desalination*, **270**, 1–8.

Kim, S.J., Ko, S.H., Kang, K.H. and Han, J. (2010). Direct seawater desalination by ion concentration polarization. *Nature Nanotechnology*, **5**, 297–301.

Lattemann, S. and Höpner, T. (2008) Environmental impact and impact assessment of seawater desalination. *Desalination*, **220**, 1–15.

Lopez-Gunn, E. and Llamas, M.R. (2008) Re-thinking water scarcity: Can science and technology solve the global water crisis? *Natural Resources Forum*, **32**, 228–238.

Lu, H., Walton, J.C., and Swift, A.H.P. (2001) Desalination coupled with salinity-gradient solar ponds. *Desalination*, **136**, 13–23.

National Centre of Excellence in Desalination (2011) *Australian Desalination Research Roadmap*, 114 pp. Available at: http://desalination.edu.au/wp-content/uploads/2010/02/NCEDA-Research-Roadmap-2011.pdf.

Peñate, B. and Garcia-Rodriguez, L. (2012) Current trends and future prospects in the design of seawater reverse osmosis desalination technology. *Desalination*, **284**, 1–8.

RBF Consulting (2009) *Camp Pendleton: Seawater Desalination Project Feasibility Study*. Prepared for San Diego County Water Authority, 330 pp. Available at: http://www.sdcwa.org/sites/default/files/files/water-management/desal/vol1_report_desal-study_Dec09.pdf.

Reddy, K.V. and Ghaffour, N. (2007) Overview of the cost of desalinated water and costing methodologies. *Desalination*, **205**, 340–353.

Ribeiro, J. (1996) *Desalination Technology – Survey and Prospects*. Institute for Prospective Technological Studies, European Commission.

Rodríguez, L.L. (2011) Desalination: towards a sustainable source of drinking water. *EnviroNews*, **20**, 5–10.

Stover, R.L. (2007) Seawater reverse osmosis with isobaric energy recovery devices. *Desalination*, **203**, 168–175.

Thye, J.F. (2010) *Desalination: Can It Be Greenhouse Gas Free and Cost Competitive?* Unpublished Masters thesis, Yale School of Forestry and Environmental Studies, United States, 66 pp.

Trieb, F. (2007) *AQUA–CSP, Concentrating Solar Power for Seawater Desalination*. DLR Institute of Technical Thermodynamics, Section System Analysis and Technology Assessment, Germany, 279 pp. Available at: http://www.dlr.de/tt/en/Portaldata/41/Resources/dokumente/institut/system/projects/aqua-csp/AQUA-CSP-Full-Report-Final.pdf.

United Nations Educational, Scientific and Cultural Organisation (UNESCO) (2012) *Managing Water under Uncertainty and Risk*. The United Nations World Water Development Report 4. Available at: http://unesdoc.unesco.org/images/0021/002171/217175E.pdf.

URS Australia (2002) *Economic and Technical Assessment of Desalination Technologies in Australia: with particular reference to national and action priority regions*, 80 pp. Available at: http://www.environment.gov.au/water/publications/urban/pubs/desalination-full-report.pdf.

Velmurugan, V. and Srithar, K. (2008) Prospects and scopes of solar pond: A detailed review. *Renewable and Sustainable Energy Reviews*, **12**, 2253–2263.

Voutchkov, N.S. (2008) Planning for carbon-neutral desalination in Carlsbad, California. *Environmental Engineer*, **6**, 2–11. Available at: http://www.aaee.net/DownloadCenter/EEJournalV06P2.pdf.

WateReuse Association (2012) *Seawater Desalination Costs*, 19 pp. Available at: http://www.watereuse.org/sites/default/files/u8/WateReuse_Desal_Cost_White_Paper.pdf.

WateReuse Association (2011) *Desalination Plant Intakes Impingement and Entrainment Impacts and Solutions*, 22 pp. Available at: http://www.watereuse.org/sites/default/files/u8/IE_White_Paper.pdf.

World Health Organisation (WHO) (2003a) *Chloride in Drinking-Water: Background Document for Development of WHO Guidelines for Drinking-Water Quality*. WHO/SDE/WSH/03.04/03. Available at: http://www.who.int/water_sanitation_health/dwq/chloride.pdf.

World Health Organisation (WHO) (2003b) *Sodium in Drinking-Water: Background Document for Development of WHO Guidelines for Drinking-Water Quality*. WHO/SDE/WSH/03.04/15. Available at: http://www.who.int/water_sanitation_health/dwq/chemicals/sodium.pdf.

Zimmerman, J.B., Mihelcic, J.R. and Smith, J. (2008) Global stressors on water quality and quantity. *Environmental Science and Technology*, **42**, 4247–4254.

9

Delivering and Designing for Potable Water in Buildings

Phil Harris

9.1 Introduction

The dependence of developed countries on treated water is incalculable and much work is needed to conserve and reduce reliance on this supply of water. Water conservation measures are increasingly being introduced into the built environment. They are becoming standard considerations when designing new buildings and are now an integral part of Planning Legislation and Building Regulations.

In the United Kingdom water is collected in various ways to supply buildings. For instance, water from building roofs, paved areas and shallow wells are not deemed suitable for consumption unless there are no alternatives due to the location of the buildings and their proximity to a mains supply. Such collected water would need to be treated to ensure that it is safe to drink. Other methods of collecting water are from boreholes dug into aquifers and surface water sources such as abstraction from large rivers and lakes.

In the Middle Ages water was drawn directly from streams and wells or even artificial canals, ditches, rain barrels or cisterns with buckets or by installing hand pumps (Magnusson, 2001; Newman, 2001). This daily chore was easier should the town be sited close to a suitable water supply. This practice of taking water directly from its source by hand is still common in many developing countries where a sophisticated water infrastructure has not been developed.

Globally, the poor face constant water shortages; it is estimated that up to 1 billion people do not have access to clean water (see http://www.ppiaf.org/page/sectors/water-and-sanitation). The World Health Organisation has reported that there are 1.6 million deaths per year attributable to diseases due to the lack of access to safe drinking water as well as basic sanitation (see: http://www.who.int/water_sanitation_health/mdg1/en/index.html). It is essential to provide access to clean water and improved sanitation facilities in order for economic development to occur. The challenge of providing water rises as global water demand

Water Resources in the Built Environment: Management Issues and Solutions, First Edition.
Edited by Colin A. Booth and Susanne M. Charlesworth.

increases along with world population and economic growth. This is without the impacts of climate change, which is likely to affect the distribution of water globally. It is estimated that by 2030 demand for water will rise by 40% globally and by up to 50% in developing countries (United Nations World Water Development Report (UNWWDR), 2012).

It is particularly the case in some African countries where access to a safe water supply remains problematic, with efforts to increase access to piped water being less than successful (WaterAid, 2011). Small-scale providers, whose supplies have previously been seen as temporary, unsafe and expensive, are now increasingly being used as a practical way of supplying infrastructure services to low-income households and dispersed populations in rural regions.

In developed countries availability of fresh clean water in buildings is taken for granted. Complex sophisticated water infrastructure networks have been developed that move water, often over great distances, from catchment areas to point-of-use, filtering, cleaning and treating the water in the process.

The remainder of this chapter concentrates on water supply practice in the United Kingdom.

9.2 Regulating Water Supply

The Building Regulations Approved Document Part G (2010) covers the UK requirements for water supply to buildings (see: http://www.planningportal.gov.uk/uploads/br/BR_PDF_AD_G_2010.pdf). The introduction of the revised Regulations were delayed twice due to government intervention (the sheer volume of work involved in revising Approved Document Part L) and then by falling foul of European legislation. The document covers sanitation, hot water safety and cold water efficiency. For the purposes of this chapter, only areas related to water supply have been included. The revised document has increased from 14 to 43 pages and now has six sections compared to three previously. The main change to Part G concerned with water supply is the requirement for all new dwellings to achieve a water efficiency standard of 125 litres of wholesome water per person per day (Hassell, 2010). 'Wholesome water' in this context being: water supplied to a building by a statutory water undertaker or a licensed water supplier through an installation complying with the requirements of the Water Supply (Water Fittings) Regulations 1999 (SI1999/1148 as amended) (OFWAT, 2011). Other updates include the option of using rainwater or greywater (see Chapters 12 and 13 in this volume) in buildings for certain purposes such as toilet flushing. If solar thermal systems are to be used to heat domestic hot water, they require automatic protection against Legionella bacteria (Makin, no date). This is generally achieved by raised temperatures; the death of Legionella bacteria generally starts at 50 °C, but by 60 °C the majority are killed within a few minutes. This hot water will then need to be mixed with cold before being allowed to be run off at the hot tap in the building, the resultant temperature depending on the use to which the water is put – for example 41 °C for showers and hand basins (Thermodynamic Mixing Valve Association (TMVA), 2000).

9.2.1 Rainwater and Greywater

Approved Document Part G1 covers the provision of water of 'suitable quality' to any sanitary appliance fitted with a flushing device, whilst requiring a supply of 'wholesome water' to showers, baths, bidets, washbasins, sinks located in an area where food is prepared and in any place where drinking water is drawn off. The document classifies alternatives to wholesome water as that from wells, springs, boreholes or water courses; harvested rainwater; reclaimed greywater; and reclaimed industrial process water. Rainwater and greywater can be used for WC and urinal flushing, washing machines and irrigation, provided an appropriate risk assessment has been carried out. The risk assessment should ensure that any rainwater or greywater system does not cause waste, misuse, undue consumption or contamination of wholesome water.

9.2.2 Water efficiency and Regulation 17K

Regulation 17K refers to the maximum allowable amount of wholesome water per person per day in dwellings being 125 litres – this is the first time the Building Regulations have included concern for water efficiency. This volume of water is significant when consideration is given to the fact that the average UK consumption is currently 150 litres/person/day, and therefore this regulation strives to reduce average water consumption. To show compliance, WCs, baths and taps need to be carefully chosen and the flushing volume or flow rates of all the appliances must be used to calculate total water consumption in litres/person/day and the results then presented to Building Control.

9.3 Water Supply to Domestic Low-Rise Buildings

The majority of water for use in buildings is stored in reservoirs or impounding reservoirs (valleys that have been dammed to trap the water), which are also often used to generate hydroelectric power. Sand filters are used to remove dirt and impurities, and the water is sterilised using chlorination. Water from boreholes is normally naturally filtered and is free from debris when extracted, but still must be sterilised. Treated water is then pumped and stored in high-level reservoirs or water towers to feed the demand via gravity.

The water is then distributed to buildings via an underground pipework distribution grid, which enables sections of the grid to be isolated for maintenance and repair without wholesale disruption of supply. Isolation valves are positioned at, or near to, the mains pipe, which connects to the communicating supply pipe into the building. Another isolating valve is positioned inside the property boundary for the owner's use. The length of communication pipe up to the boundary remains the property of the water authority; from the boundary to the building is the property owners and is often called a service pipe.

There are two distinct systems of supplying water into buildings: (i) direct and (ii) indirect.

9.3.1 Direct Systems

These require a high-pressure mains water supply. All sanitary appliances in the building are fed directly from the water main; a storage cistern is only required for the hot water supply if indirect hot water is being installed. With direct systems, fresh drinking water is available at all taps, less pipework is necessary and smaller storage vessels are required. The main problem is that if the water main has low pressure, then taps on higher floors tend to have very little flow, especially when two or more taps are open. With this system the storage cistern supplying the hot water cylinder only needs to be of 115 litres capacity

(Hall and Greeno, 2007). Maintenance valves need to be fitted in order to isolate each section of the pipework. The cistern would be omitted where hot water is supplied from an unvented hot water system or mains fed into a water heater or combination boiler, which is now often the case. Figure 9.1 shows the infrastructure involved in utilising a direct system of water supply.

Due to the fact that every outlet is supplied from the main there is a possibility that back siphonage will take place, particularly when there is a high demand on the water main. Negative pressure can draw water back into the main from a submerged inlet (Institute of Plumbing (IOP), 2002). An example would be when a shower head connected to a tap is left in the bath, which could result in contaminated water entering the water supply.

9.3.2 Indirect Systems

In an indirect system, cold water is normally stored in a cistern in the roof, which usually has a minimum capacity of 230 litres (Water Regulations Advisory Scheme (WRAS), 2005). The main water pipe supplies a drinking water outlet, which is usually situated in the kitchen sink of a domestic dwelling (see Figure 9.1). Any other cold water needs are supplied from this cistern, which also feeds the hot water cylinder, if present. The fact that the water is stored means that there is always a water supply even if the main is cut off for repair. Indirect systems require more pipework than direct ones, the heavy cistern in the roof may cause structural issues and other problems may include potential freezing of fittings in the winter. Further information relating to the water supply to buildings can be found in The Water Supply (Water Fitting) Regulations 1999.

9.3.3 Water Storage in Domestic Low-Rise Buildings

Drinking water storage cisterns and lids, including all those used for domestic purposes, should not impact on taste, colour, odour or toxicity to the water or promote bacterial growth. BS EN806 specifies that the water cisterns for domestic purposes must be of the protected type (Garrett, 2008), assuming that drinking water could be used at more than one outlet. Cisterns are generally manufactured from plastic or glass-reinforced plastic (GRP), but mild steel is also still available although their use should be restricted to feed the expansion tanks of heating systems. Cisterns may be located in cupboards, usually above the hot water storage cylinder since this reduces the probability of them freezing. However, if a gravity-fed shower is installed it will perform badly; generally a distance of 1 m is required between the underside of a cold water storage tank and the head of the shower – this distance is known as the static head.

Automatic flow control devices or float valves are fitted in order to maintain an appropriate volume of water in the cistern (see Figure 9.2). As water is drawn off, the float valve opens and allows fresh water to fill the cistern to a certain level and at this point shuts off the float valve, stopping the water flow. There are a variety of float valves on the market including Portsmouth and Croydon valves (Hall and Greeno, 2007), which include pistons that move horizontally or vertically respectively. However, Portsmouth valves are now mostly obsolete and are generally only found in older properties.

Water supply pipework can be copper (BS 2871, Tables X, Y and Z) with external bore diameters of 15, 22, 28, 35, 42 and 54 mm, stainless steel, ABS (acrilonitrile–butadiene–styrene) or PVC (polyvinyl chloride–U (Masterman and Boyce, 1984)). Building services installations generally comprise of Tables X and Y copper, where Table X is the semi-hard

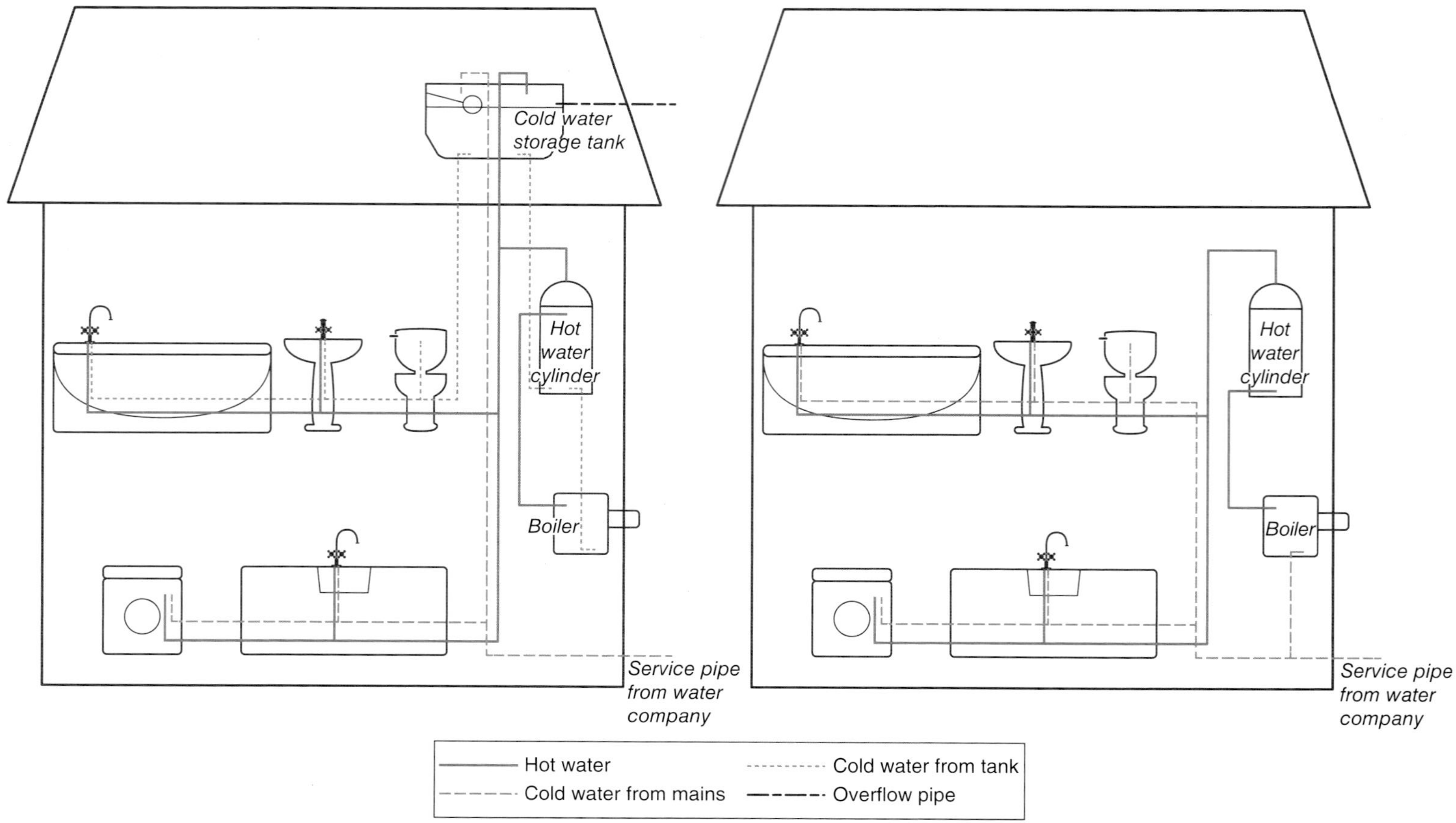

Figure 9.1 Direct and indirect domestic cold water supply.

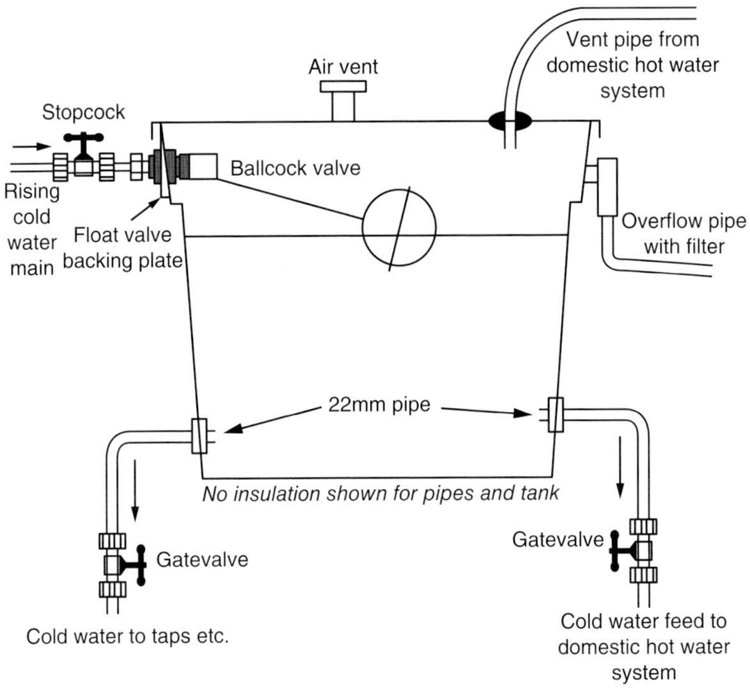

Figure 9.2 Cold water storage cistern.

Table 9.1 General cistern requirements.

Materials should be suitable for maintaining potable water quality and must not corrode or deform in use.
All cisterns and pipes should be insulated against the effects of frost and heat.
They should be situated away from heat sources.
Access should be provided for inspection and maintenance internally and externally.
Those supplying drinking water must comply with the requirements of the Water Regulations 1999.
Lids must be rigid, close fitting and securely fixed and should fit closely around any vented pipe.
They should be supported on a firm level base capable of taking the load of the cistern when full. If the cistern is located in a timber roof space, the load should be distributed over at least three roof joists using suitable bearers and boards.
Those situated in larger buildings with higher water demand will require plant rooms at roof level or within the structure, which should be ventilated and insulated and have a suitable thermostatically controlled heating facility to reduce the risk of freezing.
British Standards that cisterns must comply with include: BS 7181 for water storage cisterns up to 500 litre in domestic buildings; BS EN13280 for thermoplastic cisterns up to 500 litres; BS 417–2 for galvanised low carbon steel cisterns; Float valves must conform to BS 1212–2 and 3.

copper and Table Y is soft copper, sometimes available in coils and can be small diameter (microbore). Table Z is hard copper and is suitable for fitting in premises where long straight pipe runs can be facilitated.

Table 9.1 lists some general requirements for cisterns, emphasising their impact on the quality of water stored in them and also the necessity for it to be of potable quality.

9.4 Water Supply to Medium and High-Rise Buildings (or Those with Insufficient Mains Pressure)

The Guaranteed Standards Scheme (GSS) Regulations (OFWAT, 2010), which establishes minimum standards of service that each company must provide to its consumers, states that water companies will maintain a minimum pressure of water in the communication pipe serving premises supplied with water with seven metres of static head (equivalent to 70 kPa) (Waste and Resources Action Programme (W&RAP), 2010). Mains pressure during peak demand is 300 kPa; this pressure will supply water inside a building up to a height of approximately 30 m. Controlling pressure can save the direct cost of the water, but also the cost of, for example, heating water unnecessarily (W&RAP, 2010). The following formula allows calculation of the pressure required to supply a building with mains water:

$$\rho g h = Pa$$

where

ρ = density of water = 10^3 kg/m^3
g = gravity = 9.81 m/s
h = height (m)

Therefore, 1 m of head requires 10 kPa (or 1 bar) of pressure. If the building has a draw-off point at 30 metres above ground level there would not be enough residual pressure at this point to give an adequate supply of water, so to allow for sufficient pressure to supply water it is usual to deduct a 6 m head from the mains pressure available to allow adequate pressure at the highest draw-off point. This is clearly a rule of thumb and in reality internal water pipe sizing and storage requirements would be calculated by a Building Services Engineer.

If a high-rise building is above the water level in the service reservoir, or the water tower that supplies it, a centrifugal pump must be used to deliver the water to the upper floors. There are a number of ways this can be achieved.

9.4.1 Direct Pumping from the Mains Water Supply

It may be possible to pump water to the highest points of a building directly from the mains. This will depend upon the mains water pressure available. If the mains pressure in the area is low the Water Company will probably not allow this as there is a danger that the pump activity will cause negative pressure (partial vacuum) to develop in the main, risking back-siphonage, which can cause contamination of the supply (see Section 9.3.1). The water is then stored in header tanks at various levels in the building to be charged with water as necessary (see Figure 9.3).

9.4.2 Indirect Pumping from the Mains Water Supply

Often the mains pressure is insufficient to allow direct pumping from the main. In this case a 'break tank' is introduced at a low level in the building, such that it fills under mains pressure (see Figure 9.4). The tank must be noncorrosive, ensure the water does not become

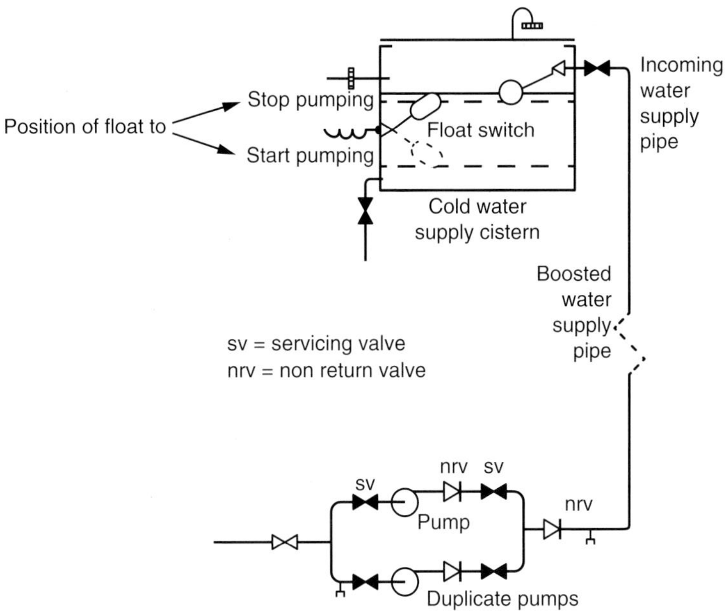

Figure 9.3 Direct pumping from the mains.

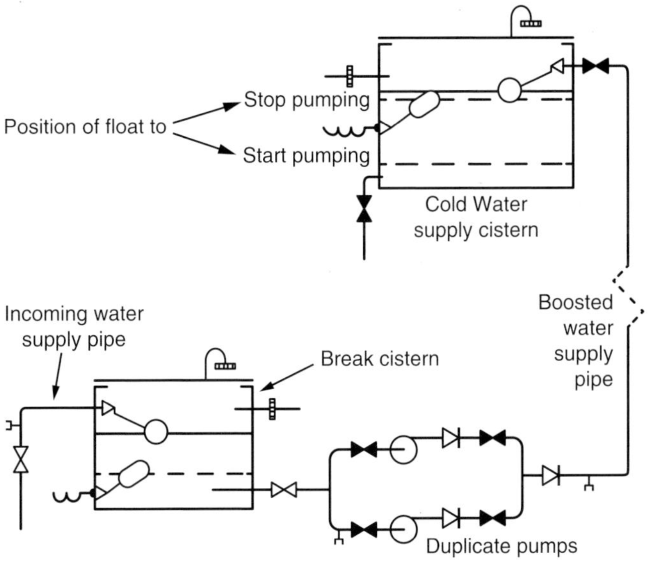

Figure 9.4 Indirect pumping from the mains.

contaminated and have sufficient capacity to allow the pumps to feed off this reservoir for a minimum of 15 minutes continuous usage. A low-level float switch should be installed to turn off the pump in case the mains water supply is cut off. Lower-level floors would be supplied by mains water as normal, but nondrinking water would be supplied to the

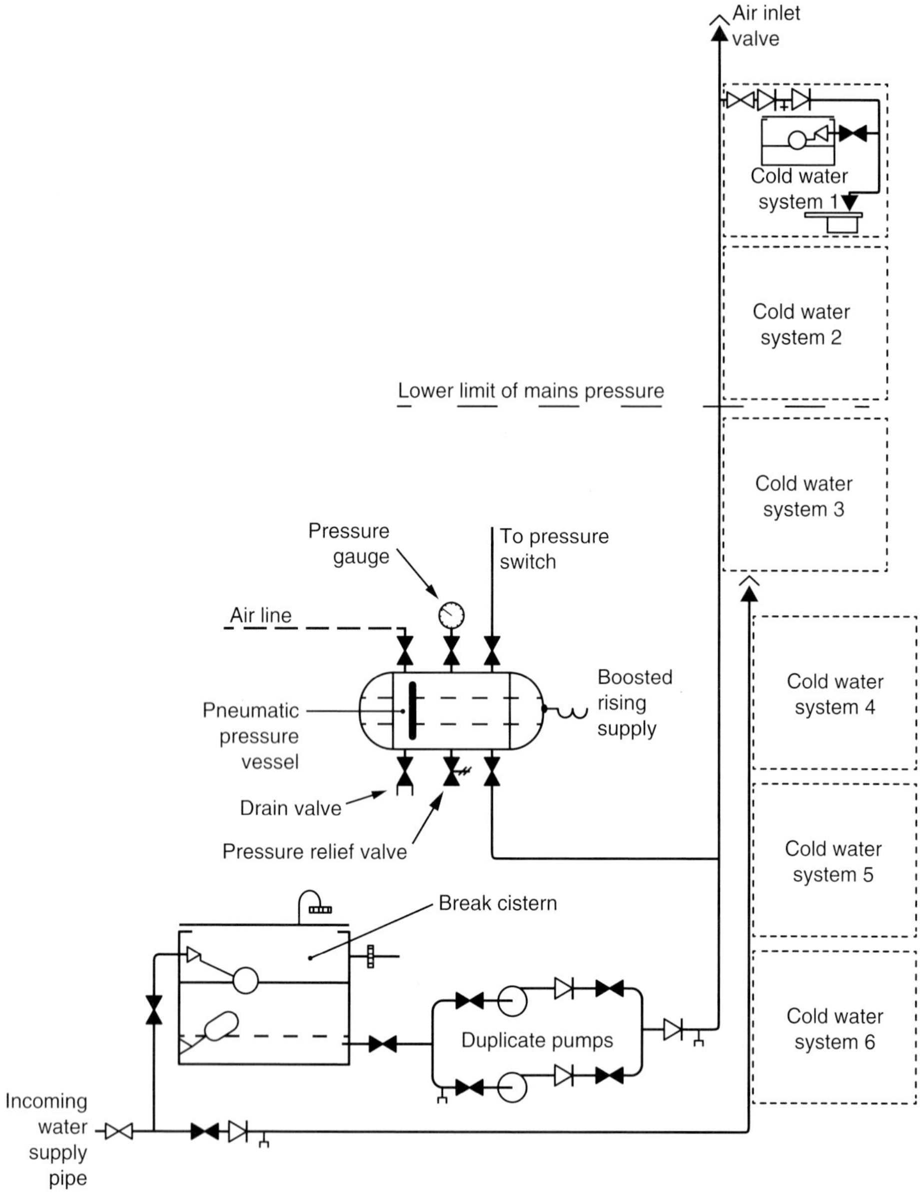

Figure 9.5 Pneumatic cylinder/pressure vessel.

higher-level floors by a header cistern positioned either at the top of the building or at various heights within the building if it is very tall. Drinking water, which should not be stored in large quantities to keep it palatable, can be supplied by installing a small header vessel next to the header cistern. This is filled by the pumps at the same time as the header cistern and feeds dwellings with fresh palatable drinking water.

9.4.3 Automatic Pneumatic Cylinder (Most Commonly Used in Practice)

A preferred alternative to drinking water header tanks is the use of a pressure or pneumatic cylinder; this is favoured as all controls can be housed in a lower-level plant room (see Figure 9.5). The cylinder is partially charged with water and an air compressor is used to pressurise a cushion of air, thus putting the water in the vessel under pressure. Float valves, or pressure sensors, are used to restart the pump and/or air compressor when the water level and/or pressure drops to a given value due to draw-off from drinking water taps or absorption of air during off-peak times.

9.5 Pipe Sizing and Flow Rate Design in Buildings

The design of a building's water storage, distribution and supply system is the responsibility of a specialist Building Services Engineer. For example, in England, design criteria are stipulated by the Chartered Institution of Building Services Engineers (CIBSE) (2004) and regulated by Approved Document G of the Building Regulations, 2010. There are a number of different methodologies that can be used to estimate water demand, required flow rates and water storage requirements; currently specialist software programmes are used in almost all cases.

9.5.1 Flow Rates via Tables and Charts

The loading, or demand, unit value (Table 9.2) and corresponding loading unit chart (Figure 9.6) are commonly used to determine flow rates for pre-pressurised systems (IOP, 2002). Alternatively, a recommended flow rate per fitting chart (Table 9.3) can also be used, just for reference or for times when continuous usage is expected. However, even as the quantity of fittings increases (e.g. three toilets with full fittings such as basin, tap, WC and shower) it is still not usual for them all to be utilised at the same time. Therefore, the loading unit chart (Figure 9.6) provides a recommended flow rate, which considers the case of probable simultaneous usage. These have been ascertained by statistical surveys of intermittent use of various fittings. After working out the number of loading units from the number and type of appliances identified on the plans of the building, Figure 9.6 can be used to read off the required flow rate of water to meet the demand of the appliances. A worked example is given below:

Appliance	Loading unit per appliance	Total loading units
10 × wash hand basins	1.5 × 10	15
10 × WCs	2 × 10	20
2 large sinks	2 × 3	6
		Overall loading unit = 41

Therefore, plotting a total loading unit of 41 on to Figure 9.6 shows that a design flow rate of 0.68 l/s would be required.

Table 9.2 Loading unit per fitting (based on recommendations by Anglian Water: http://www.anglianwater.co.uk/_assets/media/DS_wastewater_charges_2012_to_2013.pdf).

Appliance	Building type	Loading units
Flushing cistern attached to WC (single or dual flush)	Domestic dwelling Office School Industrial	2
Wash basin	Domestic dwelling Office School Industrial	1.5 1.5 or 3 (for use in rapid succession) 3 (for use in rapid succession) 3 (for use in rapid succession)
Bath	Domestic dwelling	10
Sink	Domestic dwelling	3–5
Shower	Domestic dwelling	3
Bidet	Domestic dwelling	1.5

Table 9.3 Recommended flow rate per fitting (l/s) (based on recommendations by Anglian Water: http://www.anglianwater.co.uk/_assets/media/DS_wastewater_charges_2012_to_2013.pdf).

Type of fitting	Flow rate
Flushing cistern attached to WC	0.1
Flush valve attached to WC	1.5
Discharge per flush should not exceed a total volume of 9 litres per WC	
Tap on wash basin	0.15
Bidet	0.15
25 mm bath tap	0.6
Shower head	0.2
12 mm sink tap	0.2
20 mm sink tap	0.3
25 mm sink tap	0.6
Flushing cistern attached to urinal (per bowl)	0.004
Discharge per flush should not exceed a total volume of 4.5 litres per bowl urinal	

Water Companies (e.g. Anglian Water: http://www.anglianwater.co.uk/_assets/media/DS_wastewater_charges_2012_to_2013.pdf and Wessex Water: www.wessexwater.co.uk/WorkArea//DownloadAsset.aspx?id=8175) use the total loading units of the fixtures and fittings to calculate infrastructure charges to be levied under the Water Industry Act 1991 when building or developing a property. These charges cover the costs incurred in order to improve the existing network to accommodate the extra capacity required for new developments. For example, a new build development of 10 communal flats has a total loading unit of 183; the relevant multiplier (RM) is found by dividing the number of flats + the standard loading unit per dwelling (taken as 24) by the total loading unit, that is

$$RM = 183/(10 \times 24) = 0.7625$$

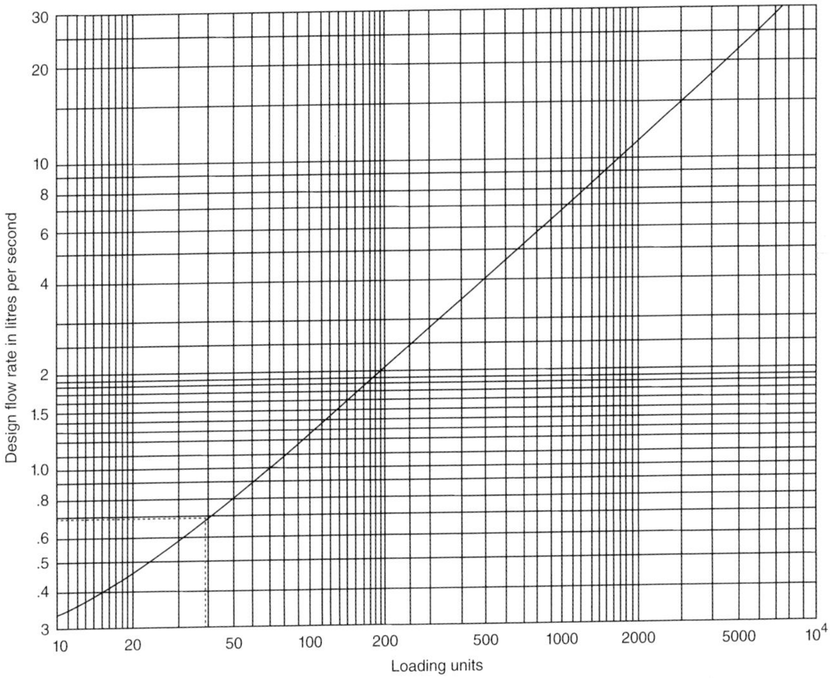

Figure 9.6 Loading unit chart (dashed lines, see Section 9.5.1).

The standard charge levied is set annually by the Water Services Regulation Authority, OFWAT, and applies to water and sewerage for each building that is served with its own standard-sized domestic water supply connection – for 2012–2013, this has been set at £328 for water and the same for sewerage. In the case above of the 10 flats, the charge for water infrastructure for the whole development is therefore

$$£328 \times 10 \times 0.7625 = £2501$$

9.5.2 Pipe Sizing

Correct pipe sizes will ensure adequate flow rates for appliances and avoid problems. For example:

Oversized pipework	Undersized pipework
Additional and unnecessary installation costs Delays in obtaining water at draw-off points Increased heat losses from hot water pipes	Inadequate delivery from outlets Variation and fluctuation in pressure at outlets (e.g. showers and other mixers) Increase in noise levels

For small, simple installations, pipes are often sized based on the experience of the plumber and convention. There are a number of methods that can be used to determine the correct and most economical diameter of pipe to deliver the required amount of water based on the flow rate and static head of pressure. Two simple methods are by calculation and by using a nomogram.

9.5.2.1 By Calculation

A popular method is to use the Thomas box formula (Hall and Greeno, 2007; IOP 2002):

$$q = \sqrt{\frac{d^5 x H}{25 x L x 10^5}} \qquad d = \sqrt[5]{\frac{q^2 x 25 x L x 10^5}{H}}$$

where
q = flow rate through pipe (l/s)
d = internal diameter of pipe (mm)
H = head (pressure) of water (m)
L = length of pipe (m) allowing for bends, tees, etc.

For example, where q = 1 l/s, H = 3 m and L = 20 m, then

$$d = \sqrt[5]{\frac{1^2 x 25 x 20 x 10^5}{3}}$$
$$= 27.83\text{mm}$$

The closest commercially available piping with this diameter in steel has a 32 mm bore and for copper has an outside diameter of 35 mm.

9.5.2.2 Nomogram

Nomograms are charts (Figure 9.7) which are used to obtain the pipe diameter based on a known flow rate in litres per second and static head loss per metre (Majumdar, 2001). There are different nomograms for different types of pipe material – copper, plastic, steel, etc. The head loss, effective length (actual length + additional lengths based upon pressure losses due to friction as the water passes through fittings) and flow rate coordinates are plotted on the nomogram and the line joining these points are extrapolated to find the pipe diameter.

9.6 Pipework Maintenance Issues

Water is classed as either hard or soft (Table 9.4) and is a measure of the content of calcium and magnesium salts dissolved in the water. Low concentrations of these salts are found in soft water, high concentrations in hard water. Total hardness can be calculated by measuring the total concentration of calcium and magnesium salts in the water and using the following formula:

[Calcium (mg/l) × (2.497) + magnesium (mg/l) × (4.188)] × 0.4 = total hardness as a function of calcium concentration mg/l (see Table 9.4).
See: http://www.anglianwater.co.uk/household/water-quality/facts/hardness/.

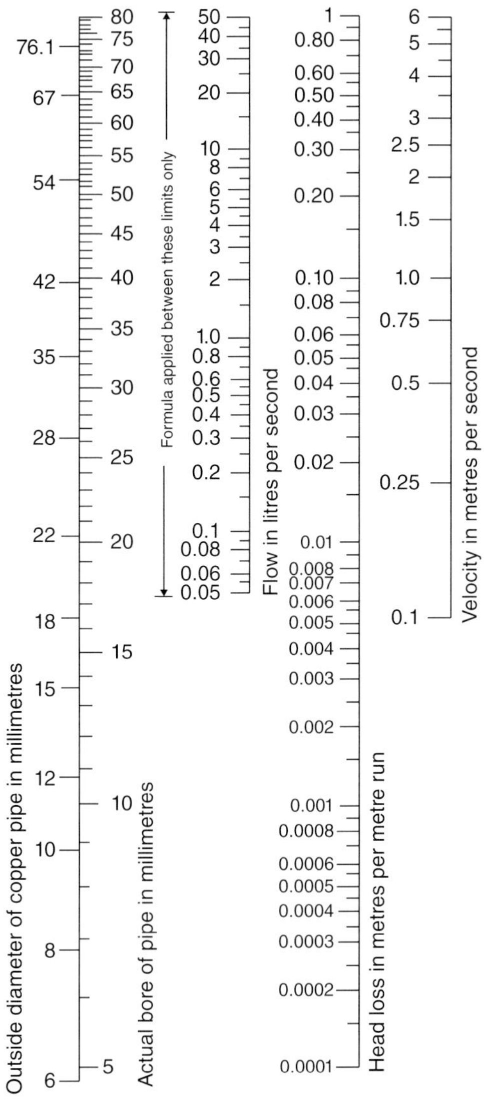

Figure 9.7 Pipe sizing nomogram.

Table 9.4 Water classification based on total calcium and magnesium salt concentration.

Designation	ppm (or mg/litre)
Soft	0–50
Moderately soft	50–100
Slightly hard	100–150
Moderately hard	150–200
Hard	200–300
Very hard	>300

Hardness or softness depends on the underlying lithology of the particular location (Drinking Water Inspectorate (DWI), 2011). The solubility of the salts decrease with an increase in temperature, so when water is heated the salts deposit themselves on surrounding surfaces, leading to scale build-up on kettle elements or the furring-up of pipes in direct hot water heating systems; this is termed 'temporary hardness'.

There are two main processes for removing salts causing temporary hardness: firstly by boiling, which is very expensive and time consuming, but efficient, or secondly by adding small quantities of limewater or cream of lime to the water, known as Clark's process. Carbon dioxide and bicarbonate react with the calcium causing the salts to precipitate out of the water. The salt crystals then either settle at the bottom of the tank and/or get filtered out as the water is processed further. Noncarbonate salts such as calcium chloride, calcium sulphate and magnesium chloride cause permanent hardness and cannot be removed by boiling. These salts can only be removed by chemical reaction.

A more controversial method marketed as a nonchemical alternative to water softeners is that of magnetic water treatment (also known as anti-scale magnetic treatment or AMT). Manufacturers of magnetic water treatment devices claim that powerful magnetic fields can affect the structure of water molecules or the properties of solutes passing through the magnetic field, thus eliminating the need for chemical softening agents. Scientific evidence for their efficacy is contradictory, with Krauter *et al.* (1996) stating that 'No beneficial effect was found when using the magnetic device', but Kobe *et al.* (2001) found 'no doubt that magnetic water treatment works'.

To prevent boilers and pipework from being blocked with lime scale, phosphates may be added to inhibit formation of the scale.

9.7 Future Issues

It has been predicted that over the next 20 years demand for water is likely to increase substantially as a result of population growth coupled with increased development, on the one hand, and increased consumption per capita, on the other. Predicted climate change is forecast to make it drier during the summer months and slightly wetter during the winter months in the United Kingdom. The overall picture is one of increasing demand for an ever-reducing resource. As a result the UK government has outlined its strategy to improve the environmental performance of the built environment, which includes water efficiency (Department of Environment, Food and Rural Affairs (DEFRA), 2008). There are four policy strands where legislation has been used to control the use of water in the built environment.

9.7.1 New Homes

Between now and 2050, it is likely that one-third of the housing stock will have been built (Barker, 2004). A reduction in water consumption in new homes has been addressed through a review of Part G of the Building Regulations, which came into effect in April 2010, and the introduction of the Code for Sustainable Homes (DCLG, 2009). The government has also given powers through the planning system to go beyond current policy to give water efficiency targets in excess of the Building Regulations, provided local need can be demonstrated (DEFRA, 2008). It is also now compulsory to install a water meter in new homes.

9.7.2 Existing Housing Stock

In support of the sustainability agenda in the refurbishment/improvement of existing buildings a variety of initiatives has been introduced including Energy Performance Certificates, statutory duties on water companies to promote water efficiency and the regulation of individual fittings and appliances. The revised Part G of the Building Regulations applies to all domestic refurbishment projects where there is a material change of use.

9.7.3 Water Using Fixtures and Fittings

Individual water using fixtures and fittings are regulated though the Water Supply (Water Fittings) Regulations 1999 and aim to reduce water consumption in buildings.

9.7.4 Nondomestic Buildings

Water consumption in nondomestic buildings is currently only addressed through the Water Supply (Water Fittings) Regulations 1999, which regulates individual fittings and fixtures. The main driver for water efficiency in private and public nondomestic buildings will be cost savings and various corporate social and environmental responsibility drivers. The government's Enhanced Capital Allowance Scheme enables businesses to claim 100% of the first year capital allowances on investments in technologies and products that encourage sustainable water use.

These policy strands support the government's vision to reduce per capita water consumption from a national average of approximately 150 litres per person per day in 2012 to a more sustainable level of 125 litres per person per day by 2030 (DCLG, 2007).

References

Barker, K. (2004) *Delivering Stability: Securing our Future Housing Needs*. HM Treasury Publication.

Building Regulations Approved Document Part G (2010) UK Government.

CIBSE (2004) *Guide G: Public Health Engineering*.

DCLG (2007) *Water Efficiency in New Buildings: A Joint DEFRA and DCLG Policy Statement*. Communities and Local Government Publications, Wetherby.

DCLG (2009) *The Code for Sustainable Homes: Setting the Standard in Sustainability for New Homes*. Communities and Local Government Publications, Wetherby.

Department of Environment, Food and Rural Affairs (DEFRA) (2008) *Future Water: The Government's Water Strategy for England*. UK Government.

Drinking Water Inspectorate (DWI) (2011) *Hard Water*. Available at: http://dwi.defra.gov.uk/consumers/advice-leaflets/hardness.pdf.

Garrett, R.H. (2008) *Hot and Cold Water Supply*,Third Edition. Wiley-Blackwell.

Hall, F. and Greeno, R. (2007) *Building Services Handbook*. Butterworth Heinemann.

Hassell, C. (2010) Changes to Part G Building Regulations: will they deliver 'water smart' homes? *Green Building*, Summer, 53–56.

Institute of Plumbing (IOP) (2002) *Plumbing Engineering Services Design Guide*. IOP, 248 pp.

Kobe, S., Dražic, G., McGuiness, P.J. and Stražišar, J. (2001) The influence of the magnetic field on the crystallisation form of calcium carbonate and the testing of a magnetic water-treatment device. *Journal of Magnetism and Magnetic Materials*, **236**, 71–76.

Krauter, P.W., Harrar, J.E., Orloff, S.P. and.Bahowick, S.M (1996) *Test of a Magnetic Device for the Amelioration of Scale Formation at Treatment Facility D*. Lawrence Livermore Laboratory, Contract: W-7405-ENG-48.

Magnusson, R.J. (2001) *Water Technology in the Middle Ages: Cities, Monasteries, and Waterworks*. John Hopkins University Press, 256 pp.

Majumdar, S.R. (2001) *Pnuematic Systems: Principles and Maintenance*. Tata McGraw-Hill Education, 282 pp.

Makin, T. (no date) *Legionella Bacteria and Solar Pre-heating of Water for Domestic Purposes*. Water Regulations Advisory Scheme. Available at: http://www.wras.co.uk/PDF_Files/Preheated_Water_Report.pdf.

Masterman, A.H. and Boyce, R.M. (1984) *Plumbing and Mechanical Services*. Nelson Thornes Limited, 248 pp.

Newman, P.B. (2001) *Daily Life in the Middle Ages*. McFarland, 291 pp.

OFWAT (2011) *Water Supply Licensing: Guidance on Eligibility*. Available at: www.ofwat.gov.uk.

OFWAT (2010) *Service and Delivery – Performance of the Water Companies in England and Wales 2009–10: Water Today, Water Tomorrow*. Available at: www.ofwat.gov.uk.

Thermostatic Mixing Valve Association (TMVA) (2000) *Recommended Code of Practice for Safe Water Temperatures*, 15 pp.

The Water Supply (Water Fittings) Regulations (1999) HMSO.

United Nations World Water Development Report (UNWWDR) (2012) *Managing Water under Uncertainty and Risk. Facts and Figures from United Nations World Water Development Report 4*. UNESCO-WWAP.

Waste and Resources Action Programme (W&RAP) (2010) *Reducing Water Use – Pressure, Pipework and Hoses*. Available at: www.wrap.org.uk/rippleffect.

WaterAid (2011) *Off-Track, Off-Target: Why Investment in Water, Sanitation and Hygiene Is not Reaching Those Who Need It Most*.

Water Regulations Advisory Scheme (WRAS) (2005) *Cold Water Storage Cisterns – Design Recommendations for Mains Supply Inlets*, No. 9-04-04, Issue 2.

Section 4 Water Conservation Strategies

10

Water Neutrality – An Overview

Victoria Ashton

10.1 Introduction

Observations of spatial variance and demands on water have led to the development of knowledge around the concept of water footprints of countries, cities, businesses and individuals. Water foot-printing methodology is likely to provide the technical basis on which a credible concept for improving the visibility and management of water resources can be based. One such approach is 'water neutrality'.

Water neutrality is an important but relatively new concept for managing water resources in the context of a new development and the associated demand for water. The basic concept of water neutrality is that the demand for water should be the same after a new development is built as it was before. This is achieved by making both new and existing homes and buildings in the area more water efficient.

Water neutrality could be achieved in a combination of ways. New developments could be made superefficient, but will still require water to fulfil essential needs. This water can be 'offset' by retrofitting existing buildings within the area with more efficient devices and appliances, expanding metering and introducing innovative tariffs for water use, which reward moderate water users. As well as the household sector, ways to use water more efficiently with nondomestic users should be developed. Water companies should also reduce demand by improving management of leakage where it is cost-effective. These are all techniques applied within the current water resource management planning process. Water neutrality may give additional focus to these activities and a coherence to demand management activities that balances the coherence of supply-side options.

Taking a strict interpretation, no individual or entity that uses water can ever be entirely water neutral, as water use cannot be reduced to zero. It is possible, however, that the term used in a consistent and transparent manner could drive positive action on water issues and will, therefore, have potential, similar to that of carbon neutrality (Department of

Water Resources in the Built Environment: Management Issues and Solutions, First Edition.
Edited by Colin A. Booth and Susanne M. Charlesworth.

Energy and Climate Change (DECC), 2009). There are similarities between the water neutral concept and the carbon neutral concept. It is possible to take lessons from 'carbon', but only as far as the similarities go, because water has its own very specific characteristics – like its geographically confined nature and the fact that most water is a renewable and not a fossil resource – to which the carbon lessons will not fully apply.

10.1.1 Background – UK Context

Water is a precious resource that is under increased pressure from climate change and population growth. By 2050, the UK population could increase by as much as 20 million (Minister of State for Housing, 2007) and that fact, coupled with climate change potentially reducing the amount of water available, presents a very real pressure on water resources. Population growth and a change in demographics have resulted in a housing crisis in the United Kingdom, due to a deficit of decent affordable homes. The solution is to build new homes that are the right size, the right price and in the right location (Communities and Local Government, 2007). The addition of homes to an already populated area will result in increased pressure on the environment and natural resources. Further development must, therefore, be sustainable, and water neutrality can help to achieve this in areas where water resources are under pressure and significant development is planned.

Water neutrality is an ambitious concept and aims to ensure that there is enough water to support new development without requiring additional water resources. The concept of 'water neutrality' looks at whether demand for water by new housing and commercial development could or should be offset in the existing community by making existing homes and buildings in the area more water efficient. In effect, 'water neutrality' is a more robust policy to emphasise and implement demand management strategies. This is to ensure that total water use after a development has been built does not exceed the water use before the development.

In the United Kingdom, the development of the concept of water neutrality is particularly pertinent to the Thames Gateway in South East England. Over 165 000 new homes have been proposed in this area between 2001 and 2016 (Communities and Local Government, 2006). Concern has been expressed, however, that the level of housing growth proposed may not be sustainable, due to the water resource situation in the area. It is thought that the sustainability of water resources in the context of this housing growth and associated increased demand is significant and could be an issue. As a consequence, a collaborative piece of research was conducted to look at whether water neutrality could be achieved in the Thames Gateway (Department for Environment, Food and Rural Affairs, Environment Agency and Communities and Local Government, 2007).

10.2 Defining Water Neutrality

There is no universally accepted definition of water neutrality, and the term probably has its origins in the equivalent term 'carbon neutral'. 'Water neutrality' is used in manufacturing (e.g. Coca Cola) and has now been used as a concept in water resource management (e.g. in the Thames Gateway).

The starting point for the definition of water neutrality is:

> For every new development, total water use in the region after the development must be equal to or less than total water use in the region before the new development.
>
> (Therival *et al.*, undated)

The terms in the definition need clarification for further guidance on how to achieve water neutrality. Taking the key terms in order they are defined as follows:

New development. This could be a single property through to a major new community (such as the Thames Gateway), and will include developments such as eco-towns. It is perhaps better to use the term 'significant new development' as this suggests a development on a large scale. It is necessary, however, to then define what is classified as a 'significant development'.

Total water use. This term refers to total demand by all licensed abstractors, for all sectors, including the public water supply (domestic and non-domestic sectors and leakage), industrial abstractors and agriculture. This value needs to be derived for 'before' and 'after' for comparison and to ensure that neutrality has been met. Guidance regarding the methodology of the derivation of this figure will need to take account of variations in total water use that would naturally fluctuate over time, even with no development (such as leakage, manufacturing or agriculture changes, which may increase or decrease total water use). One option would be to consider changing the wording to talk about offsetting the 'increase in predicted total water use due to the new development'.

Region. This is the defined area within which water neutrality is to be attempted. In the Thames Gateway study, for example, this was defined as the area covered by the Thames Gateway Plan. Alternative definitions of the spatial area for the region include local authority boundaries, river catchment boundaries or river abstraction management areas. The final definition of the region should be agreed between the local groups and interest groups based on local requirements, including the need to consider the range of development sites, water catchment and abstraction issues, water supply issues and political boundaries. The definition of the region may also cross several boundaries.

Sustainability or longevity of the water neutral status is not currently included in the definition, but is very important. It is essential that the actions undertaken to achieve water neutrality should be sustained in the long term. The overarching objective of water neutrality is to achieve long-term water savings. This would require long-term monitoring to be put into place to ensure that neutrality is maintained. The ethos behind water neutrality should be to achieve it through interventions that are sustainable, in terms of the longevity of water savings. The issue of sustainability of water neutrality still needs to be addressed and could be defined in the target-setting process for a particular area, and could include energy and carbon targets as well.

10.2.1 Recommended Revised Definition of Water Neutrality

The basic definition has now been refined to:

> For every new significant development, the predicted increase in total water demand in the region due to the development should be offset by reducing demand in the existing community, where practical to do so, and these water savings must be sustained over time.

The definition is supported by the following points:

- 'Predicted increase in total water demand ... due to the development' is used because water neutrality needs to be designed at the planning stage. The aspiration should be that 'actual' total water use in the region after the development should be less than or equal to total water use before development.
- The definition of the size of new development will be dictated by local factors and needs to consider phased developments in the region. There may need to be a test for 'significance' that might be determined through carrying out the water cycle study, which should identify tensions between growth proposals and environmental requirements.
- The definition of the region should be agreed between local groups based on local requirements, including the need to consider the range of development sites, water catchment and abstraction issues, water supply issues, and political boundaries. Suitable definitions for the region might include local authority boundaries, river catchment boundaries or river abstraction management areas.
- The term 'where practical to do so' has been included as this allows scope for local constraints (which may include cost, infrastructure or natural constraints) to modify targets for water neutrality, allowing them to be set below the 100% water neutral level.

10.3 Strategies for Water Neutrality Implementation

The process of achieving water neutrality starts at the planning stage of a new development and the suggested overall process is illustrated in Figure 10.1.

10.3.1 Deciding on Water Neutrality

The decision to pursue a strategy of water neutrality in any particular location involving the development of new housing, businesses and public building is driven by the heightened level of water stress in that location. This stress acts as a constraint on new development and further supply exploitation of the water resource will result in damaging the environment.

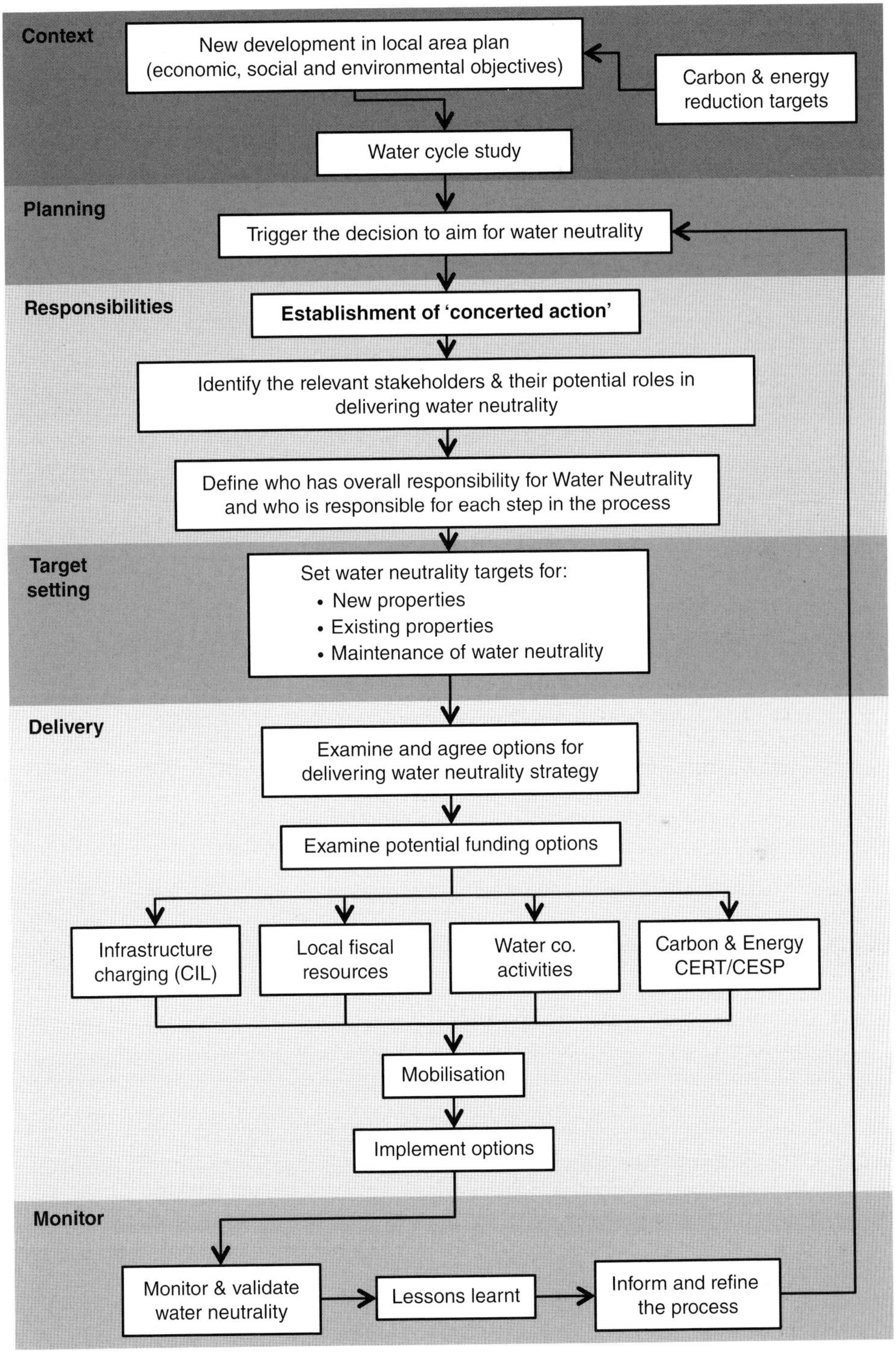

Figure 10.1 Process for achieving and maintaining water neutrality.

Other motivations for striving for water neutrality might include:

- Its contribution to achieving local authority targets on adaptation and mitigation activities to climate change.
- Its contribution to supporting the Regional Development Agency and local authority policies on carbon reduction, energy efficiency and sustainable development.
- Its contribution to local authorities' wider policies for future environmentally sustainable development.

When looking at intervention measures for water neutrality, a number of important issues needs to be considered, which includes:

- The significance of the development, which might be the scale or types of properties involved. This will also need to take into account phased developments in an area, where collectively they might become significant.
- The supply–demand balance (SDB) – whether there is a deficit within the normal water resources management (WRM) planning period, the availability of supply-side options, amount of headroom in the SDB, level of water stress and local catchment abstraction management strategy (CAMS).
- The current level of demand management activity (which may define the capacity for further activity).
- Current meter penetration or use of sophisticated tariffs, which again may define the capacity for further metering or tariff options.
- Sociodemographics of the water resource zones (WRZs): mix and age of properties, social classification groups and so on. All may influence the behavioural responses to demand management.

Before implementing water neutrality it is necessary to have engaged with and brought together the main groups likely to be involved in the planning and delivery of water neutrality. It is imperative to have a concerted action between all stakeholders and interested groups, and this will increase the likelihood of water neutrality being a success. There is a comprehensive set of guidance for water neutrality and the planning process (Environment Agency, 2009).

10.3.2 Options for Implementing Water Neutrality

Water neutrality could be achieved in a combination of ways:

- Making new developments more water-efficient.
- 'Offsetting' new demand by retrofitting existing homes with water-efficient devices.
- Encouraging existing commercial premises to use less water.
- Implementing metering and tariffs to encourage the wise use of water.
- Education and awareness-raising amongst individuals.

A range of devices and strategies for buildings and homes are available; in this chapter, these are termed 'interventions'. Interventions can be designed for incorporation in new buildings or for retrofitting into existing ones, in the domestic or commercial sector. Interventions range from dual-flush toilet retrofits, efficient showerheads and garden water butts to large-scale rainwater recycling systems and extensive publicity campaigns to promote the wise use of water. Potential interventions are listed in Table 10.1 and discussed in Chapter 12.

Table 10.1 A range of water saving devices and strategies for buildings and homes.

Measure	Example
Education and promotional campaigns	– Encourage community establishments (e.g. schools, hospitals) to carry out self-audits on their water use. – Deliver water conservation message to schools and provide educational visual material for schools.
Water-efficient measures for toilets	– Cistern displacement devices, such as a Hippo – Retrofit dual-flush devices – Retrofit interruptible flush devices – Replacement dual-flush toilets – Replacement low-flush toilets
Water-efficient measures for taps	– Tap inserts, such as aerators – Low-flow restrictors – Push taps – Infrared taps
Water-efficient measures for showers and baths	– Low-flow showerheads – Aerated showerheads – Low-flow restrictors – Shower timers – Reduced volume baths (e.g. 60 litres) – Bath measures.
Rainwater harvesting and water reuse	– Large-scale rainwater harvesting – Small-scale rainwater harvesting with water butt – Grey water recycling
Water-efficient measures addressing outdoor use	– Hosepipe flow restrictors – Hosepipe siphons – Hose guns (trigger hoses) – Drip irrigation systems – Mulches and composting
Commercial properties	– Commercial water audits – Rainwater recycling – Grey water recycling – Optimising processes – Provide water efficiency information to all newly metered businesses
Metering	– Promote water companies free meter option. – Compulsory metering (in water-stressed areas). – Smart metering (to engage the customer with their consumption). – Provide interactive websites that allow customers to estimate the savings associated with metering (environmental and financial). – Innovative tariffs (seasonal, peak, rising block). – Customer supply pipe leakage – supply pipe repair and replacement.
Other	– Household water audits, including DIY or with help of plumber. – Seek-and-fix internal leaks and/or dripping taps. – Water-efficient white goods, including washing machines and dishwashers. – Ask customers/public to spot and report leaks (Leakspotters).

Aspiring towards water neutrality is an ideal opportunity to install a 'basket of measures' to maximise water efficiency. A combination of products or actions can yield greater water savings than a single device or intervention. In addition, this may result in economies of scale for delivering measures. For instance, a campaign or delivery programme of water-efficient devices might include a number of different messages or devices or be targeted at specific household or commercial groups. However, some combinations of measures may not be cost-effective, particularly where end users are in more scattered groupings. Decisions on interventions would need to be made in the context of local circumstances. Interventions could be installed into new and existing homes, commercial and public buildings to deliver water savings for water neutrality.

Manufacturers are continuously working to improve the performance, desirability and price of efficient water-using products. Water neutrality would be an ideal opportunity to install the 'best in class', making water neutrality a gold standard of water efficiency and, thus, using the best and most efficient products available. In order for water neutrality to be realised, this would require different water efficiency interventions to be undertaken at their full potential. To achieve water neutrality, the boundaries of interventions in terms of their implementations must be stretched. The devices and the strategy adopted to implement them must be designed to operate at the highest level.

Uptake of such devices will vary depending on cost, ease of installation and manner in which these are promoted. These factors will have a potentially important, but as yet unquantifiable, impact on predicted yields and, therefore, the effectiveness of each measure can be variable across different communities. The uptake rate will also depend on a range of factors and not just cost. The effectiveness and success of a water-efficiency measure and, therefore, its desirability for implementation to help to achieve water neutrality cannot be considered by the yield it may give alone. Delivery mechanisms must include practical and behavioural options. Many interventions are designed to reduce demand when operated in a particular way and, ultimately, rely on the user being aware of and engaged with their water consumption. Savings of some interventions are undermined if

Table 10.2 Possible interventions to deliver water neutrality.

Intervention	Applicable to:		Type of demand reduced:		Reduces energy in the home
	Households	Commercial	Average	Peak	
Toilets/WCs	√	√	√		
Tap use	√	√	√		√
Showers/baths	√		√		√
White goods	√		√		√
Rainwater harvesting	√	√		√	
Greywater reuse	√	√	√		
Water butts	√			√	
Hose trigger guns	√			√	
Drip irrigation	√			√	
Mulches	√	√		√	
Industrial process		√	√		√
Water audits	√	√	√	√	√
Metering	√		√		√
Tariffs	√			√	
Education, publicity and behaviour change	√	√	√	√	

the message of 'wise water use' is not acted on. Publicity and water efficiency advice is, thus, an important intervention itself.

There will be opportunities and benefits of joining up different sectors. There is a link with water and energy as a reduction in the volume of heated water used will save both water and energy, and therefore the carbon footprint and carbon emissions. There is an opportunity to combine water efficiency initiatives with energy efficiency and some devices will be able to deliver savings for both. This is an important synergy to recognise and support in the future, and will have implications for water neutrality in the wider context. Energy savings can also be made within a water company's treatment process and network. If demand is suppressed, less water will need treating and pumping around the infrastructure. Although real, these savings are relatively small compared to energy savings associated with heated water in the home. When household and water company emissions are considered together, around 90% of emissions in the water system can be attributed to 'water in the home' (Environment Agency, 2008). Table 10.2 summarises the interventions and identifies whether the intervention is aimed at peak or average demand, and if it potentially has an impact on energy.

10.4 Funding Mechanisms

When discussing the concept and delivery of water neutrality, it is important to investigate the fiscal implications and how this concept will be financed. There is currently no overarching funding mechanism for water neutrality, and it is unlikely that water neutrality could be met solely through current water company funding arrangements. A range of funding options is possible, but their feasibility in delivering the appropriate level of funding is uncertain. Some funding mechanisms are directly related to raising funds from the planning system by way of infrastructure charges, and others are in the form of fiscal incentives to property owners. These include the possible expansion of competition in the sector and adoption of funding mechanisms being trialled in the energy sector to lower carbon emissions.

The funding mechanisms potentially available to achieve water neutrality fall under the following categories:

- Infrastructure-related funding (generally from developer payments).
- Fiscal incentives at a national or local level to influence buying decisions of households and businesses.
- Water company activities, either directly funded by the five-year price review or as a consequence of competition and individual company strategies.
- Joint funding through energy efficiency schemes (and possibility to integrate with the heat and energy saving strategy).

It may be possible to use international case studies as examples of how to fund water efficiency measures on a large scale. Some successful international case study examples of fiscal incentives or support for water efficiency are as follows:

Australia. As part of the $13 billion national water plan, Water for the Future, the Government is delivering the $250 million National Rainwater and Greywater Initiative to help people use water wisely. This provides grants of up to $10 000 to surfing businesses to install a rainwater tank or undertake a larger water-saving project. For households, rebates of up to $500 are provided to install rainwater tanks or greywater systems (Australian Government, 2008).

Toronto, Canada. The City's Water Efficiency Plan (WEP) supports the identification of actions that will reduce energy consumption and CO_2 emissions. Key elements of the Toronto WEP are: implementation of the WEP will cost approximately $74 million, which compares favourably to the cost of providing an equivalent capacity through the expansion of infrastructure, at an estimated cost of $220 million. It has been estimated that through the implementation period, 90 000 tonnes of CO_2 emissions will have been avoided and, once fully implemented, annual CO_2 emissions of 14 000 tonnes will be avoided. The WEP seeks to achieve reductions in wastewater flows by 123 million litres per day and peak day demands (associated largely with summer lawn watering) of 266 million litres per day. The WEP is based on financial incentives to support implementation of water conservation measures within the following sectors: municipal, single family residential, multiunit residential and industrial, commercial and institutional. All measures are based on voluntary participation and include aggressive targets.

The Toilet Replacement Programme is the measure with the greatest potential water savings in the WEP, requiring the replacement of over 700 000 toilets to achieve savings of around 100 million l/day, and is a priority within the WEP representing approximately $42.5 million in incentives (57% of the implementation budget) and continues to be a focus of the WEP's public education campaign (Toronto City Council, 2003).

Ottawa, Canada. The City Facilities Retrofit Programme consisting of new capital funds is allocated to Real Property Assessment Management (RPAM) for the purpose of undertaking water-efficient retrofits at city facilities in the amount of $100 000 per year, 2007–2009 inclusive. In addition, the City Council makes available a $75 rebate to residents who purchase and install a city approved six-litre water-efficient toilet. They also make available to residents, free of charge, water efficiency kits (one per household) consisting of a water-efficient showerhead, one or more tap aerators and a lavatory displacement bag. For large water users, the city has a 'High Volume User' subsidy programme consisting of two elements: subsidies for pre-approved retrofits (e.g. toilet replacements) and subsidies for the retrofit of process equipment to become more water efficient (e.g. water chillers). Both programmes have a maximum limit of $10 000 per application (Ottawa City Council, April 2007).

10.4.1 Key Findings and Recommendations on Funding Mechanisms

Water neutrality provides a strategic 'meeting point' for mechanisms for water efficiency to be concentrated in a particular place of new development in water-stressed areas. A water neutrality strategy provides the opportunity for those with funding roles and responsibilities to work together to maximise the combined effect of their actions to meet water neutrality. Water Company funding of water efficiency measures does not in itself constitute funding for water neutrality.

Water neutrality is delivered at regional and local levels and the responsibility for funding processes must lie within the local organisations. There will be links between funding processes. For instance, grants from local authorities could in the future be raised from infrastructure agreements with developers. Currently, in the UK, most of the funding options discussed do not exist. The main funding resource for the delivery of water efficiency measures resides with water company funded schemes and the discretionary buying behaviour of property owners and landlords.

The UK Government has many properties within its remit, such as schools, libraries, hospitals, government buildings and other public institutions. Funding for water efficiency in these institutions should be straightforward, by stipulating mandatory clauses in the procurement policies of all these institutions. For water neutrality, any public institutions would be among the first to be targeted for water efficiency retrofitting, funded by the government property owner. All public institutions in the area of water neutrality should have a mandatory requirement to install water-efficient devices and have a water efficiency commitment to match their energy efficiency and carbon reduction commitments.

A policy shift away from regarding water companies as the default funders of measures required for water neutrality is needed. This should be done by evaluating measures that contribute to water neutrality as broader mitigation measures required to accommodate new developments. Integrated within mechanisms for funding energy efficiency, the Government should establish a coordinated package of fiscal incentives for water efficiency measures for property owners and landlords, and business premises.

10.5 Conclusions

The definition of water neutrality has been refined to: 'for every new development, the predicted increase in total water demand in the region due to the development should be offset by reducing demand in the existing community'. This definition should be used to ensure a consistent understanding of the purpose of water neutrality.

It is technically feasible to achieve water neutrality. However, the way in which water-efficient measures are delivered needs to be improved. Water neutrality can be achieved through the mobilisation of such measures, backed up by adequate funding, in the specific water neutrality area. Water neutrality cannot be solely achieved by water companies and these cannot be seen as the default funders of measures to meet water neutrality. The strategy for water neutrality must sit within the planning system if its delivery is to be effective. This will identify the need, bring partners together and focus on the community.

To achieve water neutrality it will be necessary to deliver technical and behavioural water efficiency interventions in a coordinated and targeted way to maximise take-up rates and water savings. Achieving water neutrality should be a 'concerted action' and is more likely to be achieved when delivered in partnership with the local community, local authority, water companies, the regulators and developers. This should seek to involve the community at every level, from schools to parents to businesses, to homeowners to social and private tenants, and importantly be led by the local authority/government.

Achieving water neutrality will rely on a series of mitigation actions or interventions. These include water conservation in domestic, commercial and public properties; metering and alternative tariffs; activities that enforce water conservation standards and regulations; water efficiency labelling, encouraging local retailers to include point of sale information on water conservation; and training of local plumbers to offer water conservation advice.

The right mix of information and interventions should be targeted at specific groups. From the evidence and modelling of scenarios, metering the existing housing stock offers the greatest water saving per household; consequently, water neutrality requires metering to be maximised. This increase in metering is not a stand-alone activity and must be part of a community action programme. Alternative tariffs to encourage water saving could be applied to water neutrality areas in the future, but UK evidence for their effectiveness is limited and the impact on water affordability needs to be considered. Trials of different tariffs are underway in a number of water companies and more are planned in the next water company Asset Management Programme (AMP) planning period and price review. Evidence from these trials should indicate the impact of alternative tariffs in offsetting for water neutrality.

There is currently no overarching funding mechanism to deliver water neutrality. Funding for water efficiency is piecemeal and disjointed. Water neutrality will require substantial funding with available mechanisms working together. Innovative infrastructure tariffs are being developed by some local authorities. A range of fiscal incentives from reward rebates on local taxation, reduction or exemptions on VAT (value added tax) for water-efficient products, subsidised retrofitting of water-efficient fittings and devices, and fiscal benefits to plumbers accredited on water efficiency delivery schemes would all contribute to a situation where customers gain financially from supporting water neutrality. It is crucial to back up education and campaigns for existing householders with incentives to help meet water neutrality.

There is a need to gather more evidence on complex solutions, such as integrated suites of interventions, rainwater and greywater reuse and alternative tariffs. This lack of evidence should not delay implementation of solutions as long as the water, heat, energy and carbon savings are monitored and evaluated.

Monitoring and feedback on the performance of measures are a vital step in the water neutrality process. These will verify that water neutrality has been achieved, assess the longevity of water savings and contribute to the evidence base to inform future strategies.

References

Australian Government (2008) *National Rainwater and Grey-water Initiative*. Department of Sustainability, Environment, Water, Population and Communities. Available at: http://www.environment.gov.au/water/policy-programs/nrgi/index.html.

Communities and Local Government (2006) Available at: http://www.communities.gov.uk/publications/thamesgateway/thamesgateway2.

Communities and Local Government (2007) *Homes for the Future: More Affordable, More Sustainable*.

Department of Energy and Climate Change (DECC) (2009) *Guidance on Carbon Neutrality*. September 2009. Available at: http://www.decc.gov.uk/assets/decc/What%20we%20do/A%20low%20carbon%20UK/carbonneutrality/1_20090930090921_e_@@_carbonneutralityguidance.pdf .

Department for Environment, Food and Rural Affairs, Environment Agency and Communities and Local Government (2007) *Towards Water Neutrality in the Thames Gateway, Summary Report*. Science Report: SC060100/SR3.

Environment Agency (2008) *Greenhouse Gas Emissions of Water Supply and Demand Management Options*. Science Report: SC070010, July 2008.

Environment Agency (2009) *Water Cycle Study Guidelines*.

Minister of State for Housing (2007) Statement to the House of Commons on 23 July 2007.

Ottawa City Council (2007) *Real Property Acquisition Policy*, April 2007. Available at: http://www.ottawa.ca/en/city_hall/policiesadministration/policies/real_property_acquisition/index.html.

Therival, R., Drury, C. and Hepburn, R., (no date) *Achieving Water Neutrality in the South East Region: A Discussion Paper*.

Toronto City Council (2003) *Water Efficiency Plan*. Available at: http://www.toronto.ca/watereff/plan.htm.

11

Building Regulations for Water Conservation

Sean Churchill, Colin A. Booth and Susanne M. Charlesworth

11.1 Introduction

Most developed countries will have a set of building standards, which are primarily intended to control the safety of people in and around buildings. These standards are applied as the building is constructed to ensure that it is safe for use throughout its lifetime. Some countries, such as England and Wales, will have a set of national standards that are adopted right across the country so that designers and builders will be aware of the requirements wherever they undertake building work. In other countries, however, such as the United States, standards will be set locally with each different state or region having its own set of building codes. In either case the intention of these building standards are to ensure that new buildings are structurally sound throughout their lifetime, as well as ensuring life safety, particularly in terms of means of escape in case of fire.

Although building standards and regulations were adopted initially to ensure the physical safety and well-being of the people using new buildings, over time this legislation has been adapted in some countries to control less physical aspects of the performance of the buildings themselves. Two areas where mandatory building standards are commonly used to control the performance of buildings are carbon emissions (or heat loss) and water efficiency. It could be argued that these less tangible aspects of a building's performance do not need to be controlled by legislation as government-led education could help developers realise that they have a moral and environmental responsibility to ensure that the buildings that they construct are sustainable. Unfortunately in some countries, such as England and Wales, the government has felt that the moral obligation will not be a significant enough factor to encourage developers to build sustainably and have introduced legislation to

Water Resources in the Built Environment: Management Issues and Solutions, First Edition.
Edited by Colin A. Booth and Susanne M. Charlesworth.

compel them to do so. This need for legislation over education may be traced primarily to the costs of sustainable construction over other more traditional building methods, mainly due to the technology required to ensure that a building can be used in an environmentally sustainable manner (O'Connor, 2009).

In this chapter an explanation is given as to why there is a necessity to conserve water in buildings and, more particularly, why the house building sector is a prime target when the government looks for areas that can use water more efficiently. It will also detail why legislation is felt to be the best way to achieve sustainable construction, as opposed to education, and the potential effects of this legislation. It should be noted at this point that, although the article primarily deals with the current situation in England and Wales and, therefore, draws much of its research from this geographical area, the principles and problems discussed apply on a global scale and the conclusions can accord with the regulations and construction sector across the globe.

11.2 What are the Building Regulations?

The Building Regulations (2010) is the name given to a set of legislation that is derived from the Building Act 1984. The Building Regulations (2010) are comprised of many Statutory Instruments, which ensure that the policies set out in the Building Act are undertaken and enforced. The government describes the purpose of the Building Regulations as being to set objective and fair building standards for most new buildings and alterations to existing buildings. Note that the Building Regulations (2010) only apply in England and Wales, as Scotland has its own set of Building Standards.

To enable designers and builders to achieve compliance with the requirements of the Building Regulations (2010), the government also produces an accompanying set of advisory 'Approved Documents'. The Approved Documents are lettered from A to P and give suggestions on how the required standards can be met. This is important to note as the Building Regulations are not prescriptive, that is they do not state that projects have to be designed in a certain way but rather that they only have to meet a particular target. How that target is met is down to the discretion of the designer and subject to approval of the relevant Building Control body. The Building Regulations and their associated Approved Documents cover many facets of the construction process, most notably structure, fire safety, conservation of fuel and power, access and water efficiency.

The Building Regulations (2010) apply to almost all new build and extension projects in both the domestic and commercial sectors, with the exception of some smaller domestic schemes, such as small conservatories and porches, which are exempt from the requirements of the regulations. The requirements of the regulations on any given project will be enforced by a Building Control body, either through the Local Authority or a private Approved Inspector, who will check the design for compliance at the plan stage and carry out regular site visits as the build progresses. Although the Building Control body will have powers to take enforcement action on any noncompliant work, it would be wrong to see these organisations simply as a police force for the Building Regulations as their specialist knowledge and advice can be sought and used at the design stage to identify potential problems before they arise on site. Most Building Control bodies will also be happy to give advice on the best possible solution to any issues with the regulations both at plan stage and during the build process.

11.2.1 The Review of Approved Document G

The Building Regulations are regularly reviewed to ensure that their standards are maintained as building practices, technologies and trends change. In April 2010 Part G of the Building Regulations (2010), which was previously entitled *Hygiene* was updated and re-titled *Sanitation, Hot Water Safety and Water Efficiency*. Previously Approved Document G had been a relatively minor part of the requirements of the Building Regulations (2010) and dealt mainly with the need to provide suitable sanitary conveniences and washing facilities in relevant buildings. As the title of the revised document suggests, however, this was reviewed in 2010 and the new document now deals with a requirement for water efficiency.

As part of the review of Approved Document G, for the first time Regulation 17K was inserted into the Building Regulations (2010) to limit the usage of water in new dwellings through Building Control bodies. Note that Regulation 17K has since been moved to Regulation 36 in a further revision of the Building Regulations (2010), although the actual requirement remains the same. Regulation 36 is worded as follows:

> 36.–(1) The potential consumption of wholesome water by persons occupying a dwelling to which this regulation applies must not exceed 125 litres per person per day, calculated in accordance with the methodology set out in the document 'The Water Efficiency Calculator for New Dwellings', published in September 2009 by the Department of Communities and Local Government (DCLG).

Note that although the Building Regulations (2010) are not prescriptive, in this case they do specify a particular set of guidance for the designer to work to in the form of *The Water Efficiency Calculator for New Dwellings*, which we shall examine further in due course.

11.3 Background to the Changes in Approved Document G

Climate change has been a source of much discussion at government level and in the media for many years now. Reports such as the government commissioned Stern Review of 2006 (HM Treasury, 2006) estimate that current levels of greenhouse gas emissions will lead to a rise in global temperatures of 2–5 °C some time within the next 20 to 50 years. The consequences of climate change are many and varied but perhaps chief amongst them is the trend for less predictable weather patterns with many hotter summers than those currently experienced, which are likely to then lead to water shortages.

Climate change is undoubtedly a global issue but this then leads on to local problems, especially where water supply is concerned. In fact, a governmental report on water use, *Future Water* (Department for Environment, Food and Rural Affairs (DEFRA), 2008), states that water use is already exceeding abstraction rates in some areas. It is worth noting at this juncture that overabstraction does not merely lead to water shortages as such activity also has a negative impact on the eco-systems in and around the rivers and the water sources that these supplies are taken from (Elliott, 2009a). Simple mathematics shows that if water usage is exceeding supply action needs to be taken to reduce that level of usage.

Not everyone agrees that reducing water usage will completely solve the problem. In fact, some academics, such as Rajger (2006), suggest that changing global weather conditions mean that water shortages are 'inevitable', regardless of other factors. This is a view supported by the influential Intergovernmental Panel on Climate Change (IPCC), which released a report (IPCC, 2007) stating that the threat of the climate system was 'unequivocal'. A more specific recent report released by the Department of Communities and Local Government on water efficiency in new buildings (DEFRA, 2007) states that if carbon dioxide emissions were reduced immediately, there would still be significant climate change over the next 40 years as a result of past human activity resulting in a less predictable climate and water supply. This is not to say that restricting water usage is a pointless exercise; it means the opposite is true because, if we are to believe that water shortages are 'inevitable', then the sensible use of existing supplies becomes even more important.

Climate change is not the only reason for potential water shortages as population growth may also play a significant role. It is at this point that the reasons for the changes to the Building Regulations (2010) become apparent. The *Future Water* report mentioned previously (DEFRA, 2008) estimates that due to an ageing and increasing population, the number of households in England and Wales increased by around 30% in the last three decades of the 20th century. DEFRA (2007) contends that house building at this time does not meet the increase in population size and, therefore, the government must undertake a significant building programme to provide affordable housing. If the Building Regulations can be used to reduce water usage in new dwellings then this can play a part in an overall strategy of water saving. The Government commissioned the 'Barker Review' (HM Treasury, 2004), which calculated that as far back as 2002, up to 138 000 additional homes were being built in England per year. If this needs to further increase to keep pace with the growth in population then we begin to get an indication of the additional strain that will be placed upon an already scarce resource.

Even without additional housing, water use in England and Wales is already relatively excessive when compared with our European neighbours, with the average water use per person per day currently at 150 litres (Heywood, 2008). This is a rather high figure when there are usage levels of 127 litres in Germany and 125 litres in the Netherlands (Elliott, 2009a). These figures show that England and Wales need to make major advances in water efficiency to reach the levels seen on the Continent. This is illustrated even more clearly when more specific figures are used and we see an extremely high figure of 164 litres per person per day quoted for the South of England (Elliott, 2009a).

11.3.1 Water Use in Existing Dwellings

When considering how to reduce water usage in new dwellings it is important to look at how water is used in existing dwellings. From this it can then be identified where water use can be cut out altogether or even replaced by the use of nonpotable water. On this point, the 2008 report, *Future Water: The Government's Water Strategy for England* (DEFRA, 2008), concludes that of the potable water currently supplied to homes in England and Wales just 7% is used for drinking and cooking. To reinforce this claim, Burkhard *et al.* (2000) calculate that of all water fit for human consumption supplied to homes in England each year, 33% is used to flush toilets and 21% used in washing machines. Overall, at least 54% of the water supplied to homes in England and Wales each year is used in places where alternative water sources could be applied.

Measures have already been taken to reduce the amount of water used in dwellings, with the majority of new homes connected to water meters. This is a move that has had a positive impact on water use as, according to Elliott (2009a), metered houses on average use at least 15% less water per annum than nonmetered houses. This figure can rise to as high as 40% in houses occupied by just one or two people. Consequently, the Environment Agency have called for all homes to be fitted with compulsory water meters within the next six years (Elliott, 2009b). This alone will not be enough, however, as the main aim of the government's *Future Water* report (DEFRA, 2008) is that by 2030 the average water usage for a home in England decreases to 130 litres per person per day. The Environment Agency believes that water meters have a major role in meeting this target (Elliott, 2009b) but further measures, including the use of legislation such as the Building Regulations, will be required as part of a wider strategy.

11.3.2 Previous Government Action

The government has already taken some action to limit water usage in new dwellings through compulsory requirements such as *The Code for Sustatinable Homes*. A revised edition of *The Code for Sustainable Homes* was introduced by the government in 2008 (Department of Communities and Local Government (DCLG), 2008a), which gives mandatory standards of compliance for five levels of sustainable house construction, ranging from 120 litres to 80 litres per person per day. From 2008 the government required all new social housing to be constructed to at least level three of the code, with a potable water consumption of 105 litres per person per day. The water calculator used under *The Code for Sustainable Homes* is that which later lent itself to use under the Revised Building Regulations (2010).

It is worth asking briefly at this point why the government have had to take legislative action to limit water usage in new dwellings rather than any changes being undertaken voluntarily by housing developers. At the risk of oversimplifying the issue, the straightforward answer is because of cost. Sustainability does come at a price; O'Connor (2009) reports that the Home Builders Federation estimates that meeting Government standards adds around £7000 to the cost of a new home. Some of this cost is regained by the homebuyer in cheaper bills but Blackwood (2008) believes that this additional cost to the construction, coupled with the recent downturn in the housing market, has a negative effect on the willingness of house builders to buy-in to theories of sustainability. If the financial payback of water saving technologies is reaped by the homeowner over a long period then the only way for the developer to make this pay is by adding the cost of installation to the price of their properties. They may not see this as a sensible move in a time of financial hardship as prospective buyers may not be enticed by long-term savings for a short-term outlay.

11.4 Changes to Approved Document G and the Water Calculator for New Dwellings

The previous edition of Approved Document G, which was titled simply *Hygiene*, was the oldest of the suite of Approved Documents, having been introduced in 1992. This was a document of relatively minor importance, which effectively did little more than insist that the occupants of a dwelling be supplied with wholesome water. In 2009, the Government announced major revisions to Approved Document G to bring it in line with the other main Approved Documents, all of which have been updated since 2006. Under the proposed revisions to Approved Document G a whole-building performance standard will be mandatory, which states that any new dwelling should be designed to give a maximum water usage of 125 litres per person per day.

In the consultation on these amendments, the Department for Communities and Local Government stated its belief that as the oldest of the approved documents of the Building Regulations, Approved Document G should be updated to reflect current standards and legislation (DCLG, 2008b). The intention seems to be, through the Building Regulations, to slowly bring new housing developments in line with guidance such as *The Code for Sustainable Homes*.

Although the consultation document for Approved Document G (DCLG, 2008b) claims that the government is aware of the unsustainable water demand in many parts of the country, it is not made clear how the figure of 125 litres has been arrived at. Consequently, other governmental departments, such as the Market Transformation Programme (MTP), have questioned the wisdom of the changes. In a follow-up report to the Approved Document G consultation, which dealt with water consumption in both new and existing homes, the MTP expressed concern that the revisions to the Building Regulations would not be effective as there are uncertainties on their basis in actual water use and consumer behaviour (MTP, 2008a). Gilg and Bark (2006) had already raised concerns over the use of such approaches to promoting environmental action as they do not take into account the 'lived experiences' of the people they are supposed to influence.

11.4.1 Effect of the Restrictions of Water Usage

If we use the figures stated in the Barker Review (HM Treasury, 2004), which estimate that 138 000 new dwellings are constructed each year in England, then we see that a huge saving of water used per day can be made by simply reducing the amount each person uses each day by just 25 litres. If the new houses were to continue to run at 150 litres per person per day then in total almost 21 million litres of water would be used in comparison to the 17 million used in total by the houses running at 125 litres per person per day. In this theoretical situation, where each dwelling has single-person occupancy, 3 450 000 litres of water would be saved per day in England alone. Over a year the water demand from new dwellings in this scenario would be reduced by over 1.25 billion litres in total.

Theoretical figures for single-person occupancy of all new dwellings built in one year do not fully convey the potential for water saving if the imposition of the new limit of 125 litres per person per day is successfully implemented, although actual occupancy levels are likely to be much higher than this. In actual fact, the most recent UK census, carried out in 2001, shows that the average occupancy of a house in England is 2.36.

If we recalculate the potential water savings based on these census figures then we see that the limit of 125 litres becomes even more effective. In this scenario, in any given year new dwellings continuing to use 150 litres of water person per day would use a total of almost 49 million litres of water a day compared to those running at 125 litres per person per day, which would use nearly 41 million litres. The difference per day between the two would be around 8 million litres per day, which translates as a 16.6% saving where houses use 125 litres per person per day. Under the revised Building Regulations, there is the potential to save almost 3 billion litres of water per year.

11.4.2 How the Water Calculator for New Dwellings Works

The Water Calculator for New Dwellings is a straightforward tool. It lists the most common water-using fixtures and appliances in any dwelling and attempts to multiply their stated water usage by a 'use factor', which indicates how frequently they are used. These figures are then totalled and multiplied again by a 'normalisation factor', which attempts to adjust the overall figure to take account of people's everyday behaviour. As stated previously, at no point in any of the available literature is any explanation given as to how these set figures have been arrived at and as a result their validity has already been thrown into question. Due to the lack of empirical information available we are unable to prove or disprove this theory initially, but when the use factors are examined in terms of which fixtures or appliances they identify as being the most frequently used, they appear to make sense. Table 11.1 shows an example of the Water Calculator for New Dwellings before any

Table 11.1 The water calculator for new dwellings.

Installation type	Unit of measure	Capacity/flow rate	Use factor	Fixed use (litres/person/day)	Litres/person/day
WC (single flush)	Flush volume (litres)		4.42	0.00	
WC (dual flush)	Full flush volume (litres)		1.46	0.00	
	Part flush volume (litres)		2.96	0.00	
WCs (multiple fittings)	Average effective flushing volume (litres)		4.42	0.00	
Taps (excluding kitchen/utility room taps)	Flow rate (litres/minute)		1.58	1.58	
Bath (where shower also present)	Capacity to overflow (litres)		0.11	0.00	
Shower (where bath also present)	Flow rate (litres/minute)		4.37	0.00	
Bath only	Capacity to overflow (litres)		0.50	0.00	
Shower only	Flow rate (litres/minute)		5.60	0.00	
Kitchen/utility room sink taps	Flow rate (litres/minute)		0.44	10.36	
Washing machine	Litres/kg dry load		2.10	0.00	
Dishwasher	Litres/place setting		3.60	0.00	
Waste disposal unit	Litres/use	0	3.08	0.00	
Water softener	Litres/person/day		1.00	0.00	
			Total use (litres/person/day)		0
Contribution from greywater (litres/person/day)					0
Contribution from rainwater (litres/person/day)					0
Normalisation factor					0.91
Total water consumption (Code for Sustainable Homes) (litres/person/day)					0
External water use					5.0
Total consumption (Building Regulation 17K) (litres/person/day)					5.0

variable figures have been entered. The figures you see in the table are fixed figures, which will apply to any project.

Table 11.1 shows that filling in the calculator itself is a simple process of entering various flow rates, capacities and flushing volumes, which can be obtained from the manufacturer's literature to establish the amount of water that could potentially be used per person per day for each fixture or appliance. It is noteworthy that the calculator contains certain fixed figures that cannot be altered. Some, such as the figures for taps, can be added to whereas others, such as external water use, are fixed.

Appliances that may be added to the dwelling later or are easily changed, in this case washing machines and dishwashers, have fixed figures that must be entered if the actual figure is unknown. These figures must be entered even if there is no intention to install a washing machine or dishwasher at first. This is in order to take into account any future human activity, such as if the ownership of the property was to change hands and the new owners install or change these appliances. In this case the calculator has still made some allowance for this change.

In the event that the figures are entered into the Water Calculator for New Dwellings and a figure is returned that is above the 125 litres per person per day then some re-design will be required. This can be done by selecting fixtures or appliances with a lower level of water usage or, for the first time under the Building Regulations (2010), allowance is made for the designer to offset the amount of potable water used by returning rainwater or greywater back into the system. This nonpotable water can then be used to perhaps flush toilets or wash clothes as mentioned previously. Obviously, if required, a combination of these two options can be used with more efficient appliances and fitting being complemented by the addition of rainwater or greywater systems. When deciding on the best solution to this particular problem it may be wise to first establish where the majority of potable water within the dwelling is being consumed.

Although examination of the Water Efficiency Calculator for New Dwellings does not generally lend itself to statistical analysis until it has been fully completed, the predetermined figures that are entered into the calculator may be analysed in order to demonstrate which fittings and appliances are used most frequently throughout the day. The chart in Figure 11.1 converts the 'use factor' of each fixture or appliance and displays it as a percentage of the total usage each day. Note that this does not relate to how much water each fitting or appliance uses, just how frequently it is used each day. It is hoped that this will identify which of these is used the most frequently so that high-usage fittings or appliances can be targeted specifically when trying to reduce water usage.

Figure 11.1 suggests that showers and toilets are the most frequently used fittings in a new dwelling, with showers accounting for between 12.3% and 15.7% of overall daily usage and toilets in some form accounting for up to 12.4%. This would suggest that as they are used so regularly, these are the fittings that should be considered first when identifying areas where water usage can be reduced. The dishwasher also accounts for a major proportion of daily usage, of 10.1%, but this is not perhaps as relevant as the figures for showers and toilets as not every home will have a dishwasher. If there is a difficulty in meeting the 125 litre target then these figures may act as a guide to the designer when deciding where water-efficient fittings will be most effective.

If it is still clear after this statistical analysis and adjustment to the figures that water-efficient fittings alone will not be sufficient to meet the target set out in Regulation 36, then the designer may revisit the idea of using rainwater and greywater recycling.

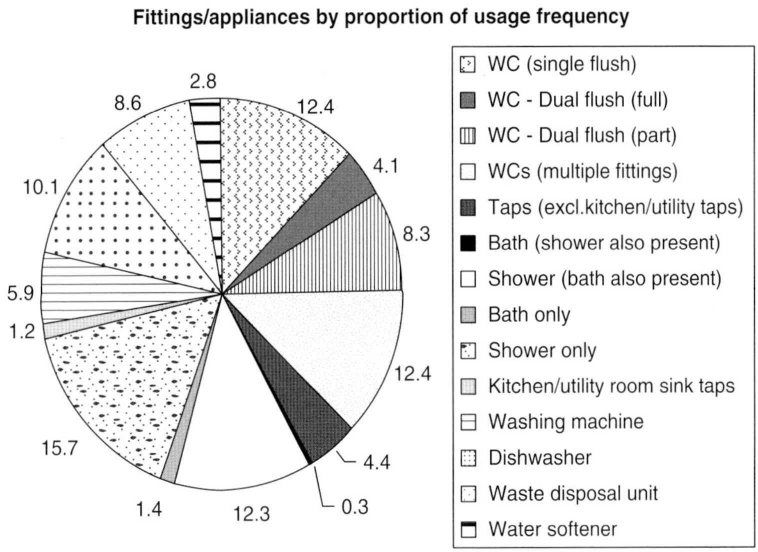

Figure 11.1 Fittings rates by proportion of total water usage.

11.5 Rainwater and Greywater Recycling

The inclusion of water recycling in Approved Document G is a major alteration to the Building Regulations (2010) and Mussett (2009) believes that, as well as attempting to minimise water use, the government now has a clear policy of promoting the reuse of rainwater and greywater. According to Furumai (2008), a move towards fuller understanding of the potential of recycling domestic water is almost inevitable as concern about the sustainability of urban water use grows.

11.5.1 Rainwater Recycling

Rainwater is generally the more straightforward of the two to recycle but its use is currently relatively poorly understood in this country. Again, England and Wales are some way behind their European neighbours here, with countries such as Germany currently having an uptake of 50 000 rainwater recycling plants in new homes per year (Nolde, 2007). Such an uptake suggests that some people feel that there are significant benefits to the use of these systems, but still Hatt *et al.* (2006) believe no long-term survey of rainwater reuse systems has been carried out on a scale wide enough to give us a clear picture of their worth.

Due to a lack of empirical evidence, the potential benefits of rainwater reuse systems on a small domestic scale in England and Wales are still in some doubt. Mustow *et al.* (1997) estimate that rainwater systems can have an initial start-up cost of between £2000 and £3000, which gives a payback period of approximately 30 years. This is a significant proportion of the £7000 quoted by O'Connor (2009) as the figure for making a new dwelling meet overall energy efficiency standards. This initial outlay is even riskier when taking into consideration the beliefs of some commentators that there is still a need for further developments to provide truly useful systems for rainwater collection and reuse (Hatt *et al.*, 2006)

There are figures to indicate that storm water reuse can have significant benefits. The research of Sims (1998) shows that the use of rainwater and greywater in combination can reduce the amount of drinking water used by up to 39%. Fewkes (1996) backs up Burkhard *et al.* (2000) in the belief that the toilet is the greatest consumer of potable water, accounting for 30% of the overall amount used, so the benefits of such systems can be seen in some context.

One of the few examples of rainwater reuse on a large scale in England is in Northampton, where a development of 120 houses has seen a reduction in the consumption of mains water of 40% when compared to older dwellings with no such systems (Rajger, 2006). This backs up the belief of Dixon *et al.* (1999) that small-volume domestic water reuse systems can be successfully applied to the urban housing environment. What this example does not tell us is on what scale rainwater is collected. Economies of scale suggest that many houses sharing one larger recycling system see the greatest benefits, but how feasible would it be to install these on all housing developments? Due to issues of space and to avoid contamination, much of the water used in these systems is taken from the roofs of dwellings (Villareal and Dixon, 2004). The MTP in their investigation into rainwater reuse do not believe this to be sufficient due to both the limited catchment area and the limited space for storage (MTP, 2008b). There does not seem to be enough evidence to support either theory at this time.

For further details on rainwater harvesting (see Chapter 12), the reader is also referred to the literature of the Construction Industry Research and Information Association (Leggett, 2001a, 2001b) and the Environment Agency (2003).

11.5.2 Greywater Recycling

Greywater refers to waste water that has been used in the home, through appliances such as washing machines and showers, but can be treated and reused to flush toilets. Minimal literature exists to show the use of greywater systems domestically, due to their physical size and complexity. Naisby (1997) believes that greywater recycling is not viable in the domestic market due to the initial set-up cost when taken in relation to the relatively low cost of water. The difficulty of maintaining these systems is also considered to render them prohibitive to the domestic market (Wise and Swaffield, 2002). It seems that further guidance on their use from the DCLG is imperative if they are to become widespread. Until these systems are fully implemented it is difficult to speculate on how they will be received as there is currently limited evidence on which to base an opinion. A certain level of buy-in from the end-user is imperative, due to the way in which it will affect their everyday life (Heywood, 2008) and because of the unfamiliar level of physical maintenance that such systems require to function properly (Naisby, 1997).

For specific details on greywater technologies (see Chapter 13), the reader is also referred to the literature of the Construction Industry Research and Information Association (Leggett, 2001a, 2001b) and the Environment Agency (2008).

11.6 Case Study: Calculating Water Usage

Perhaps the most effective way to understand how the Water Calculator works in practice is to determine how difficult it is for new dwellings in England to be designed with a water consumption of 125 litres per person per day and to apply the calculator to the values

Table 11.2 Flow rates of fittings and appliances in a three-bed town house.

THREE BED TOWN HOUSE - APPROX 5 YEARS OLD		
Location	**Fitting**	**Capacity/flow rate**
GROUND FLOOR		
Kitchen	Washing machine	8 litres per kg of dry load
	Mixer tap	6 litres per minute
Cloakroom	WC – dual flush	6/4 litres per flush
	Sink taps	5 litres per minute
FIRST FLOOR		
Bathroom	WC – dual flush	6/4 litres per flush
	Sink taps	5 litres per minute
	Bath	215 litres to overflow
SECOND FLOOR		
Bathroom	WC – dual flush	6/4 litres per flush
	Sink taps	5 litres per minute
	Shower	12 litres per minute
STANDARD FITTINGS TO BE ADDED		
	Dishwasher	1.25 litres per place setting
	External water use	5 litres

gained from an existing dwelling. This case study should not only demonstrate how the calculator works but also how difficult it can be to achieve the required level of water consumption.

11.6.1 The Dwelling and Water Consumption

The dwelling in question is five years old and the water usage figures were derived from investigation into the particular fittings and appliances currently installed. The fittings and appliances that consume water in this example dwelling are listed in Table 11.2. Their flow rates or capacity have been taken from the relevant manufacturer's literature.

The values from Table 11.2 were entered into the Water Efficiency Calculator for New Dwellings to give the actual water consumption per person per day for this dwelling. Note that as there is no dishwasher provided, a standard value of 1.25 litres has to be entered into the calculator. Standard types of fittings (e.g. taps and WCs) were used throughout the property so the water consumption figure was the same for each. Note that if multiple types of fitting are being used, such as different taps to each sink with varying water consumption, then a subcalculator within the Water Efficiency Calculator for new dwellings can be employed to give a single average water consumption figure for all the different types of a particular fitting. The chart in Figure 11.2 shows the amount of water consumed by each fitting in this typical three-bed town house when the figures above are entered into the water calculator. Figure 11.2 shows that in total the three-bed town house currently uses 145.47 litres of water per person per day. When the water calculator's 'normalisation factor' is applied to this then the final figure becomes 132.8 litres per person per day.

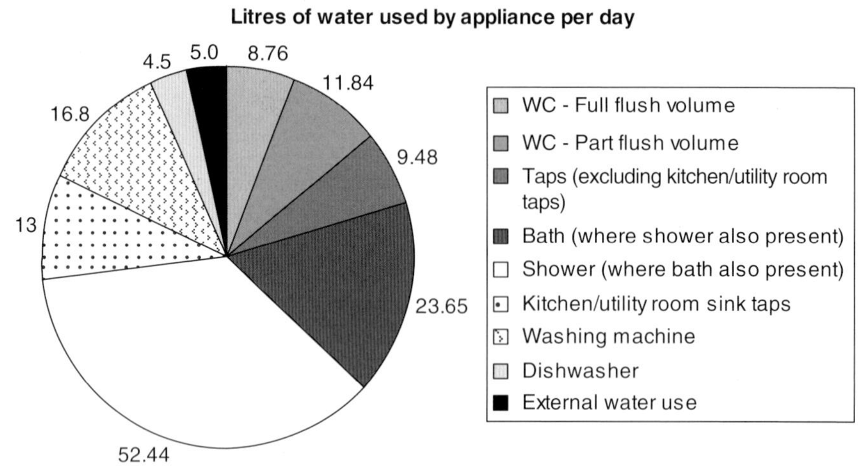

Figure 11.2 Litres of water usage per day by appliance.

11.6.2 Adapting the Design to Meet the Requirements of the Building Regulations

Overall, even though government figures suggest that the average house uses 150 litres of water per person per day, the calculation for this particular case study suggests that a typical three-bed new-build town house is using 132.8 litres of water per person per day, just 7.8 litres above the new target of 125 litres per person per day. The shower and bath, with respective totals per person per day of 52.44 litres and 23.65 litres, are easily the greatest consumers of water, but as the shower in particular has not only a high level of water usage in terms of litres per minute but also has the highest use factor, this would seem to be the most obvious area in which to save a mere 7.8 litres of water. To this end the calculation was amended to include a readily available 'water saver' type of shower, which uses 9.5 litres of water per minute. This is still a greater flow rate than many retrofitted electric showers and is actually the maximum permitted flow rate in the United States. When this type of shower was entered into the calculator it returned a figure of 122.9 litres of water used per person per day, which is compliant with the requirement of Regulation 36. This shows how easily the target of 125 litres per person per day may be achieved without the need for nonpotable water sources.

The fact that it would be generally unnecessary to use rainwater and greywater recycling in a new dwelling (the fittings and appliances in the case study were at the high end of water usage for such devices and compliance was still easily achieved) could possibly be considered to be something of a surprise. There is a belief in some quarters that for a sustainable water supply to be achieved new sources of domestic water supply, such as greywater and rainwater recycling, must be used (Wood, 2003), but the revisions to the Building Regulations do not presently seem to encourage such innovations. Similarly, Mussett (2009) states that these new Building Regulations are specifically intended to promote the reuse of rainwater and greywater but the case studies suggest otherwise.

Regardless of concerns over its potential effectiveness, this legislative attempt to control water usage in new dwellings is not limited exclusively to England and Wales. Since the introduction of the Buildings Energy Performance Directive by the European Parliament in 2003, which was intended to promote improvements in the energy performance of buildings through cost-effective measures, many European member states have moved to reinforce their standards and introduce performance-based building codes for the use of energy and water. Such legislation is not restricted to European nations though, as even countries such as the Gulf States, which have experienced a major construction boom in recent times, have still felt it necessary to apply prescriptive standards controlling water use (Pan and Garmston, 2012). Like England and Wales, in international terms it seems to be too early to comment on the effectiveness of legislation on water conservation in the construction sector as these controls are a relatively recent phenomenon across the globe. Solving the issue of a dwindling water supply is certainly a long-term issue, it seems that the success or otherwise of legislative building codes in tackling the problem may only become obvious in the long term.

11.7 Other Household Water Conservation Measures

The simplest changes in household water use could have an enormous influence on future water demand. Therefore, many examples of domestic water saving devices typically focus on those areas of greatest use, such as toilets, taps, showers, reuse and outdoor aspects (Table 11.3). Some of the most noteworthy examples of domestic water savings include: changing a toilet from an old (9 litre) to a new type (4.5 litre) can halve the water used per flush; fitting an aerated showerhead can reduce the flow rate by 28% (~3 litres per minute); fixing a tap that is leaking only two drops per second can save 9500 litres per year;

Table 11.3 Examples of domestic water-saving devices (adapted from Waterwise, 2008).

Water savings use	Example devices
Toilets	Cistern displacement devices
	Retrofit dual-flush devices
	Retrofit interruptible devices
	Replacement dual-flush toilets
	Replacement low-flush toilets
Taps	Tap insert aerators
	Low-flow restrictors
	Push taps
	Infrared taps
Showers	Low-flow showerheads
	Aerated showerheads
	Low-flow restrictors
	Shower timers
	Bath measures
Reuse	Small-scale rainwater harvesting water butts
	Large-scale rainwater harvesting systems
	Greywater recycling
Outdoors	Hosepipe flow restrictors
	Hosepipe siphons
	Hose guns
	Drip irrigation systems

and updating to a modern water-efficient dishwasher that uses only 12–18 litres of water to wash 12 place settings is notably lower than the equivalent 40 litres of hot water used to wash the same crockery by hand (Environment Agency, 2007). Further achievements will improve the water efficiency in existing housing stock through refurbishment and end-of-life replacement of fittings that are innovative and improve performance.

It has been compulsory to install water meters in all new UK homes since 1997 and, as a consequence, it is now widely acknowledged that metering reduces water usage because it raises awareness of the financial incentive of using less. Moreover, it is generally regarded by many that paying for the quantity of water consumed is the fairest way to pay (similar to other utility services). At present, ~30% of homes have water meters but it is expected that it will take many years before this will hugely alter patterns of water demand. Therefore, the Environment Agency would like the majority of homes (80%) in seriously stressed areas of England and Wales to be metered by 2015 (Thompson *et al.*, 2007), which will also allow more complex tariff structures to be formulated. This accords with the recommendations of the housing review of Barker (2004) for changes in the water charging system to encourage customers to use water wisely, but also highlights that it should be affordable for all.

For further details on water efficiency, the reader is referred to the literature of the Department of Communities and Local Government (2006, 2007), the Environment Agency (2007) and the National House Building Council Foundation (Griggs and Burns, 2009), plus the visionary report of Thompson *et al.* (2007), which outlines the feasibility of approaches to achieve water neutrality (see Chapter 10). For further details on water-saving devices, the reader is referred to the literature of the Environment Agency (2003, 2007), the Water Regulations Advisory Scheme (2005) and Waterwise (2008).

11.8 Conclusions

We have seen that, due to several factors, most notably climate change and a lack of education on how water can be used more sensibly, the government decided that it was necessary to amend the Building Regulations in order to limit, through legislation, the amount of water used in new dwellings. The effectiveness of this legislation is still in question because the case study shows that the target is still relatively easy to achieve.

It seems that the Water Calculator for New Dwellings is fundamentally flawed and that the dwellings that are designed using it will actually use much more water than the figures suggest. Having said this, even if the calculator were totally accurate it seems that the new regulations are not stringent enough to produce any major savings in water consumption. In terms of water efficiency it can only be hoped that these regulations are a starting point from which the industry can get used to the idea of the Building Regulations being used to control domestic water consumption and that the requirements will be tightened over a period of time. The government have a timetable by which all of the major parts of the Building Regulations are amended every three years and perhaps Approved Document G will be included in this, with the current measures merely being an introduction to the changes and more stringent measures being introduced over a period of time in line with *The Code for Sustainable Homes*.

It is clear, however, that this is not a problem that is specific to England and Wales. Water shortages are finally becoming recognised as a major issue both in Europe and around the developed world. Legislation generally seems to have been the way in which Governments have chosen to tackle this problem, such as the European Building Codes, which have arisen as a result of the Buildings Energy Performance Directive, but there is still an argument to say that education is as important as legislation. Due to the short timeframe currently available to assess the proposed solutions to a long-term problem, it is difficult to speculate on the success of the legislative route, but it seems that if England and Wales are used as a barometer then the Building Regulations that have been implemented will need to be far more stringent if they are to be effective.

References

Barker, K. (2004) *Delivering Stability: Securing our Future Housing Needs*. HM Treasury Publication.

Blackwood, L. (2008) Homebuyers pay too much for energy bills. *The Times* [online], 20 June [Accessed 30 November 2009]. Available at: http://property.timesonline.co.uk/tol/life_and_style/property/article4170836.ece.

Burkhard, R., Deletic, A. and Craig, A. (2000) Techniques for water and wastewater management: a review of techniques and their integration in planning. *Urban Water*, **2**, 197–221.

Department of Communities and Local Government (2006) *Water Efficiency in New Buildings: A Consultation Document*. Communities and Local Government Publications, Wetherby.

Department of Communities and Local Government (2007) *Water Efficiency in New Buildings: A Joint DEFRA and DCLG Policy Statement*. Communities and Local Government Publications, Wetherby.

Department of Communities and Local Government (2008a) *The Code for Sustainable Homes: Setting the Standard for Sustainability in New Homes*. Communities and Local Government Publications, Wetherby.

Department of Communities and Local Government (2008b) *The Building Act 1984, The Building Regulations 2000. Proposals for amending Part G (Hygiene) of the Building Regulations and Approved Document G Consultation*. Communities and Local Government Publications, Wetherby.

Department for Environment, Food and Rural Affairs (DEFRA) (2007) *Water Efficiency in New Buildings – A Joint DEFRA and Communities and Local Government Statement*. Communities and Local Government Publications, Wetherby.

Department for Environment, Food and Rural Affairs (DEFRA) (2008) *Future Water: The Government's Water Strategy for England*. The Stationary Office, London.

Dixon, A., Butler, D. and Fewkes, A. (1999) Water saving potential of domestic water reuse systems using greywater and rainwater in combination. *Water Science and Technology*, **39**, 25–32.

Elliot, V. (2009a) Call for compulsory water meters within six years. *The Times* [online], 30 March [Accessed 2 December 2009]. Available at: http://www.timesonline.co.uk/tol/news/environment/article5998849.ece.

Elliot, V. (2009b) Government to promote household water saving in £1 million campaign. *The Times* [online], 24 September [Accessed 2 December 2009]. Available at: http://www.timesonline.co.uk/tol/news/environment/article6847290.ece.

Environment Agency (2003) *Harvesting Rainwater for Domestic Uses: An Information Guide*. Environment Agency Publications, Bristol.

Environment Agency (2007) *Conserving Water in Buildings*. Environment Agency Publications, Bristol.

Environment Agency (2008) *Greywater: An Information Guide*. Environment Agency Publications, Bristol.

Fewkes, A. (1996) The reuse of rainwater for WC flushing. *Building and Environment*, **34**, 765–772.

Furumai, H. (2008) Rainwater and reclaimed wastewater for sustainable urban water use. *Physics and Chemistry of the Earth*, **33**, 340–346.

Gilg, A. and Bark, S. (2006) Behavioural attitudes to water saving. Evidence from a study of environmental actions. *Ecological Economics*, **57**, 400–414.

Griggs, J. and Burns, J. (2009) *Water Efficiency in New Homes: An Introductory Guide for Housebuilders*. BRE Press, National House Building Council Foundation, Amersham.

Hatt, B.E., Deletic, A. and Fletcher, T.D. (2006) Integrated treatment and recycling of stormwater: a review of Australian practice. *Journal of Environmental Management*, **79**, 102–113.

Heywood, J. (2008) The housing market and eco-friendly homes. *The Times* [online], 25 April [Accessed 29 November 2009]. Available at: http://property.timesonline.co.uk/tol/life_and_style/property/buying_and_selling/article3806386.ece.

HM Treasury (2004) *The Barker Review – Final Report*. HM Treasury, London.

HM Treasury (2006) *Stern Review on the Economics of Climate Change*. HM Treasury, London.

Intergovernmental Panel on Climate Change (2007) *Climate Change 2007: Synthesis Report*. United Nations.

Leggett, D., Brown, R., Stanfield, G., Brewer, D. and Holliday, E. (2001a) *Rainwater and Greywater Use in Buildings: Best Practice Guidance*. Construction Industry Research and Information Association, London.

Leggett, D., Brown, R., Stanfield, G., Brewer, D. and Holliday, E. (2001b) *Rainwater and Greywater Use in Buildings: Decision-Making for Water Conservation*. Construction Industry Research and Information Association, London.

Market Transformation Programme (2008a) *Water Consumption in New and Existing Homes*. The Stationary Office, London.

Market Transformation Programme (2008b) *Alternative Sources of Water: Greywater and Rainwater Reuse. Innovation Briefing Note*. The Stationary Office, London.

Mussett, J. (2009) Ingraining green. *Public Service Review: Local Government and the Regions*, **14**, 90–92.

Mustow, S., Grey, R., Smerdon, T., Pinney, C. and Wagget, R. (1997) *Water Conservation: Implications of Using Recycled Greywater and Stored Rainwater in the UK*. BSRIA, Bracknell.

Naisby, C. (1997) *Greywater Recycling and Rainwater Harvesting: A Viable Means of Domestic Water Conservation?* MSc Thesis, University of Leeds.

Nolde, E. (2007) Possibilities of rainwater utilisation in densely populated areas including precipitation runoffs from traffic surfaces. *Desalinisation*, **215**, 1–11.

O'Connor, R. (2009) Builders fear rising prices of 'green' homes. *The Times* [online], 1 September [Accessed 2 December 2009]. Available at: http://www.timesonline.co.uk/tol/money/property_and_mortgages/article6816356.ece.

Pan, W. and Garmston, H. (2012) Compliance with building energy regualtions for new-build dwellings. *Energy*, **48**.

Rajger, G. (2006) Revamped rainwater recycling. *Refocus*, **7**, 68–69.

Sims, B. (1998) Liquid asset – services for housing greywater recycling. *Building Services Journal*, February 1998, 30.

Thompson, J., Langley, E., Ballantyne, J. and Perry, C. (2007) *Towards Water Neutrality in the Thames Gateway*. Environment Agency Publications, Bristol.

Villareal, E.L. and Dixon, A. (2004) Analysis of a rainwater collection for domestic water supply in Ringdansen, Norrkoping, Sweden. *Building and Environment*, **40**, 1174–1184.

Water Regulations Advisory Scheme (2005) *Conservation of Water: An IGN for Architects, Designers, Installers and Occupiers of Premises*. WRAS Information and Guidance Note, Issue 2.

Waterwise (2008) *Evidence Base for Large-Scale Water Efficiency in Homes*. Waterwise Publications, London.

Wise, A.F.E. and Swaffield, J.A. (2002) *Water Conservation: Water, Sanitary and Waste Services for Buildings*, 5th Edition. Butterworth–Heinemann, Oxford, pp.189–206.

Wood, W.W. (2003) Water sustainability: Science or science fiction? Perspective from one scientist. *Developments in Water Science*, **50**, 45–51.

12

Rainwater Harvesting – Reaping a Free and Plentiful Supply of Water

Susanne M. Charlesworth, Colin A. Booth, Frank Warwick, Craig Lashford and Omolara O. Lade

12.1 Introduction

Harvesting rain is not a new technique; 4500 years ago the Ur people of present-day Iraq irrigated their land using harvested rain (African Development Bank, no date). Rainwater harvesting (RwH) remained important until the introduction of mains water in the 19th century and in some dry countries it is still a vital part of their water supply (Gerolin, 2009). Mbilinyi *et al.* (2005) suggested, therefore, that much could be learnt from indigenous peoples of the world, such as those living in the Kilimanjaro region of Tanzania, who have relied on rainwater for many generations. Lessons learnt complete the circle by encouraging other peoples to take up RwH. For instance, the Maasai Kajiado women of Sub–Saharan Africa have addressed increased variability of rainfall by constructing cement water storage tanks to collect rainwater from the sheet iron roofs on their houses (Nanzala, 2008).

Recent acceptance of problems with water shortages (United Nations Environment Programme (UNEP), 2009), flooding (Pitt, 2008) and the potential for modification of precipitation patterns, due to climate change, have reawakened interest in RwH as a means of reducing reliance on mains water and providing a certain amount of flood peak attenuation. In developing countries, ~884 million people do not have access to a safe source of drinking water (World Health Organisation (WHO) and United Nations Children's Fund (UNICEF), 2010), so RwH has the potential to provide an alternative source (Meera and Mansoor Ahammed, 2006). According to the UK Environment Agency (EA) (2010), each person in the UK uses ~150 litres of potable water per day (l/p/day), some of which is used for activities not requiring high-quality supplies. For example, toilet flushing at ~26% (39 l/p/day) of the total water used could be replaced with harvested rain, as could the 12% used to wash clothes and 7% for outdoor activities (such as car washing and watering the garden).

Water Resources in the Built Environment: Management Issues and Solutions, First Edition.
Edited by Colin A. Booth and Susanne M. Charlesworth.

12.2 What is Rainwater Harvesting?

Rainwater harvesting is the capture of water that would otherwise be lost. RwH collects rain falling on buildings, roads or pathways that would normally infiltrate into the soil, evaporate or be directed to the receiving watercourse via the storm sewerage network (EA, 2009, 2010). Rainwater differs from greywater (see Chapter 13) since it has not previously been used in washing machines, baths or showers, but could be used to water gardens and hanging baskets, or, like greywater, to wash cars, flush toilets and, depending on its quality, to wash clothes. In the United Kingdom, due to the strict guidelines for potable water, rainwater cannot yet be used as a drinking water supply, but some countries already do this (e.g. Australia, see Section 12.8) and systems can be designed to include water treatment (see Section 12.5). Some developing countries, in particular, use harvested rainwater as a main potable supply and Fewkes (2006) has identified three main uses for harvested rainwater:

1. The main potable supply
2. To supplement potable supplies or
3. Non-potable uses.

Water can be collected in simple barrels or butts (see Figure 12.1) for individual dwellings, be designed into a tanked porous paving system (see Chapter 24) or can service larger commercial or agricultural premises via more complex infrastructures, which includes filtration and treatment of the stored water (EA, 2011).

Figure 12.1 An inexpensive domestic rainwater harvesting butt or barrel.

RwH need not necessarily be confined to urban environments, nor just be applied to domestic properties (EA, 2010); commercial buildings could also benefit, as could agricultural activities (EA, 2009). There are also studies proposing its use at sea, where rainfall is intercepted using large floating collectors feeding fresh water into tanks located on the seabed (Su, 2010). RwH is part of the Sustainable Drainage Systems (SUDS) approach (see Chapter 23), which gives equal value to water quality, water quantity, biodiversity and amenity (Charlesworth *et al.*, 2003). Whilst the benefits of RwH for the first two are argued in the following sections, biodiversity and amenity studies relating to SUDS in general are rare, and although it is possible to argue the biodiversity benefits of, for instance, a rain garden whose source of water is an RwH system, it is an indirect benefit of RwH, and is not easy to prove.

RwH reduces dependence on mains water that may benefit the individual householder if water use is metered, benefits the environment as less water is abstracted from lakes, rivers and groundwater, and stores excess stormwater, thus attenuating the storm peak in order to reduce flooding and, hence, reduces the pressure on the already overloaded storm sewer system (Charlesworth, 2010). By reducing surface water flow, RwH can potentially reduce the first foul flush effect (Charlesworth *et al.*, 2003), whereby polluted materials adhering to pavement surfaces during dry antecedent periods are removed during the first flush of a storm (Kim *et al.*, 2010) and are carried to a receiving watercourse where they can cause severe polluting effects. Where farm animals are housed and have access to outdoor hard standing, these yards can become covered in manure, which can be flushed into the drains or receiving watercourse during a storm event. RwH can reduce surface water runoff, enabling yard-scraping to be more efficient. In England and Wales uptake of SUDS generally, and RwH in particular, has been slow in comparison with other countries (such as the United States and Australia – see Section 12.8 for examples). As a result, it has taken recent policy and legislation to encourage its use. The following section examines the relevant recommendations and guidance that have seen increasing interest in SUDS and RwH in England, for example.

12.3 Policy

According to Ward *et al.* (2012), a changing global climate has highlighted the unreliability of infrastructure such as storm sewers, leading to the desirability of improved resilience and adaptability in the supply of water. An example of policy development can be seen in the United Kingdom, where, following the floods of 2007, the review of shortcomings in dealing with flooding and its effects (Pitt, 2008) recommended the use of SUDS, including rainwater harvesting. The subsequent Flood and Water Management Act 2010 (Department for Environment, Food and Rural Affairs (DEFRA), 2012) encouraged greater use of the SUDS approach to help address surface water flooding in particular. In terms of water quality, the European Union Water Framework Directive (2000/60/EC) requires all watercourses to be of good quality by 2015, and the pollutant remediation capability of SUDS to improve water quality has been demonstrated (see Charlesworth *et al.*, 2003). The EA, Construction Industry Research and Information Association (CIRIA) and the British Standards Institute (Code of Practice BS 8515:2009) have variously issued standards, best practice guidance and regulations relating to RwH, which includes input to the Code for Sustainable Homes (Department for Communities and Local Government (DCLG), 2006). In order to achieve the highest level (code 6), it is recommended that about a third of a dwelling's water requirements are sourced from rainwater harvesting or greywater recycling systems. One of the reasons why the uptake of SUDS and RwH needs to be encouraged by means of regulation is the negative history these approaches have inherited, frequently due to inadequate design of systems and lack of maintenance once installed. The following sections explore RwH design, cost and performance in order to emphasise the importance of these factors in encouraging the use of sustainable water resource management.

12.4 Rainwater Harvesting Design

Debord (2011) lists the four requirements of harvesting rainwater from collection to delivery where needed (Table 12.1). The nine possible components of an RwH system design are also listed in Table 12.1, which identifies the requirements from water falling on the roof of a building to the infrastructure needed to deliver it to its point of use. The EA (2010) states that the infrastructure required depends on the amount of water collected and the use to which it will be put. For instance, collected water can be pumped either directly to the point of use or to an elevated cistern and gravity-fed from there or directly to the point of use.

The design storm concept for drainage infrastructure is outlined in Chapter 24 (porous paving) and so will not be discussed further here. However, calculations specific to RwH need to be performed to ascertain the ratio of roof size to potential volume of harvested water, and these are then used to compute the size of the storage tank needed. Table 12.2 indicates the harvesting potential of a variety of roof types expressed as a drainage coefficient: a means of comparing incident rainfall with runoff, whereby a value of zero would indicate no runoff would occur and one represents a situation where all rain falling on the roof could be harvested.

Table 12.1 The requirements of an RwH system and the components that make up the system (adapted from Debord, 2011).

Requirements of RwH	Components of RwH system
Collection of water	Building roof
	Roof drains
	Gutters
Safe storage of water	Tanks
	Ponds
Disinfection (if necessary)	Filtration
Transport to where required	Pipes
	Pumps
	Labelling

Table 12.2 Drainage coefficients (K_D) from various roof types (adapted from EA, 2009, 2010).

Roof type	Drainage coefficient
Pitched	0.9
Pitched + tiles	0.8
Flat with tiles	0.8
Flat, smooth	0.6
Flat + gravel or thin turf	0.5

Table 12.3 Examples of harvestable volumes of rainwater for an example range of roof sizes and precipitation amounts (adapted from Bregulla *et al.*, 2010).

	Roof area (m^2)				
	50	75	100	125	150
Rainfall (mm/year)	Approximate annual yield of rainwater (m^3/year)				
500	15	22.5	30	37.5	45
1000	30	45	60	75	90
1500	45	67.5	90	112.5	135
2000	60	90	120	150	180

In order to calculate the volume of water that could potentially be collected, the area of the roof is required (A_R m^2). Based on some of the information given above, an approximate volume of water can be calculated for various roof sizes and rainfall amounts, and these values are given in Table 12.3 (Bregulla *et al.*, 2010).

Further factors to be taken into consideration include whether disinfection is to be carried out using a filter. The efficiency of the filter must be defined since usually the first flush is rejected and sent to the storm sewer or a SUDS treatment facility; for domestic properties a filter coefficient (K_F) of 0.9 is recommended (EA 2010), whereas the advised value for agricultural RwH is 0.8 (EA, 2009). Companies manufacturing RwH components should be able to identify an efficiency specific to their equipment. Whilst annual total rainfall (ATR in mm) does not take into account seasonal variations, nonetheless it is an acceptable approximation of the amount of rainwater that is likely to be available. Five per cent of the annual yield (~18 days) represents demand in terms of tank size. Tank size (litres) is thus calculated from:

$$A_R \times K_D \times K_F \times \mathrm{ATR} \times 0.05 \qquad (12.1)$$

where A_R = roof area (m^2), K_D = drainage coefficient (Table 12.2), K_F = filter coefficient and ATR = annual total rainfall (mm).

Whilst suitable in a domestic situation, for larger systems it is probably better to consult defined standards such as the British Code of Practice for RwH systems, BS 8515:2009. Formula (12.1) will give a reasonable approximation of tank size, but does not take account of specific rainfall amounts or the occupancy and use of the building, which will impact on water demand. Jones and Hunt (2010) emphasise that prior knowledge of the anticipated water requirement is the main driver in deciding sufficient tank size for RwH systems.

Tanks are available in various sizes (e.g. 1600, 2600, 3750, 4800 and 6500 litres). Determining the appropriate storage tank capacity for an RwH system can be ascertained from various methods (www.ciria.org): (i) a simplified approach where no calculations have to be carried out, except knowledge of the roof plan area and the annual rainfall, plus the number of occupants, which are then used to translate information onto available graphs to reveal an appropriate storage tank size; (ii) an intermediate approach that uses a simple formula (see above); and (iii) a detailed approach where storage capacity is estimated from continuous time series rainfall data (collected for 3–5 years).

An example of how to calculate the tank size. A building offering a roof area of 130 m^2, with a 0.85 pitch tile roof collection coefficient and a 0.65 rainwater pre-filter efficiency, situated in a region where the annual rainfall is 950 mm, would need a tank size of 3412 litres:

$$130 \times 0.85 \times 0.65 \times 950 \times 0.05 = 3412 \text{ litres}$$

However, it is important to also allow ~10% wet volume at the base of the tank and ~10% dead space at the top tank. This is because the tank needs to occasionally overflow and self-clean. In this case, an appropriate tank size would be 4100 litres.

12.5 Water Quality

Harvested rainwater in the United Kingdom cannot be used for drinking, but there are countries in which it does supplement potable supplies (Meera and Mansoor Ahammed, 2006). Thus, there have been concerns regarding the safety of harvested water where RwH systems are used. Pollution of the collected water can occur at any point in the system or associated processes listed in Table 12.1. Whilst the rain itself is usually considered to be reasonably unpolluted, its quality depends on the area on which it falls. Thus, it may be acidic, or may scavenge polluted particulates from the atmosphere as it falls, or may dissolve components, for example of pesticides and herbicides. Simmons (2001) reported that the surface of the roof itself may be the main source of contamination from bird and small mammal droppings and debris from overhanging vegetation. Treatments, such as that by filtration and UV disinfection, have proven effective in clearing total coliforms, *Esterichia coli* and *Enteroccoci* (Jordan *et al.*, 2008), but keeping the roof clear of rodents and overhanging branches (Mosley, 2005) is an obvious strategy for reducing the source of pathogenic bacteria. The concerns of harvested water quality thus centre on toxic elements dissolved or carried in the harvested water, such as metals and nutrients, but there are also microbiological safety concerns. The following subsections review these in turn.

12.5.1 Microbiological Safety of Harvested Rainwater

Whilst in the United Kingdom harvested rainwater cannot be used for potable purposes, in some developing countries the quality of drinking water sources such as shallow groundwater is so poor that stored rainwater is preferred and is in practice more accessible. However, concerns have been expressed regarding the microbial contamination that is sometimes associated with RwH in these countries. Meera and Mansoor Ahammed (2006) list four sources of microbial contamination including:

1. The quality of the roof materials. Runoff from metal roofs was less contaminated than other roof types reviewed since solar heating, particularly in the tropics, tended to have a disinfecting effect.
2. Rainfall characteristics also tended to influence the presence of microbes since the longer the period between rainfall events, the longer the time for bacterial populations to increase.

Table 12.4 Microbial guideline values (MGV) for domestic RwH systems.

Determinand (number/100 ml unless otherwise stated)	MGV by water use		RwH system (single site and communal domestic unless otherwise stated)
	Pressure washers and garden sprinklers	Garden watering and toilet flushing	
Escherichia coli	1	250	
Intestinal *Enterococci*	1	100	
Legionella (number/litre)	100	–	Monitored if indicated by risk assessment
Total coliforms	10	1000	

3. Studies of storage time on microbial populations are inconsistent, with some suggesting that the longer the storage time, the larger the bacterial population, while others show the opposite effect. For example, Vasudevan *et al.* (2001) found that faecal coliforms, total coliforms and faecal streptococci declined rapidly with time, possibly due to reduced nutrient availability.
4. Tank capacity, with smaller volume storage tanks containing higher levels of bacteria. Plazinska (2001) tested >100 storage tanks of various sizes utilised by indigenous populations in the Australian outback and found that, in comparison with larger tanks, smaller tanks were able to support relatively larger populations of microorganisms per unit volume.

The main indicators generally used to assess the microbial quality of water include total coliforms and thermotolerant (or faecal) coliforms (EA, 2010). The Market Transformation Programme (2007) recommends that harvested water use in the United Kingdom should refer to the EU Bathing Water Directives (1976/160/EEC and 2006/7/EC) when monitoring microbial quality. Table 12.4 gives the guidelines, which have been further adapted by BS 8515.

12.5.2 Physicochemical Water Quality

Harvested rainwater is generally collected from the roofs of buildings; they can be made of metal sheets, ceramic tiles, rock slate or ferrocement (Mosley, 2005). The runoff water quality of urban roofs has been examined by many studies; for example, Van Metre and Mahler (2003) found that the particulates adhering to them were polluted with metals such as Zn, Cd and Pb, and organic compounds such as chrysene and pyrene. Potentially, these pollutants could be washed off the roof surface into the RwH tank. Meera and Mansoor Ahammed (2006) list several factors associated with the roof itself that could potentially influence the quality of runoff, including the material the roof is constructed from, its aspect and location. Eletta and Oyeyipo (2008) suggest that the age of the roof is important, finding that samples collected from an old iron-sheeted roof contained higher concentrations of Fe, Zn and Cu than those from a new roof, or from directly falling rain. In contrast, Chang *et al.* (2004) found Zn was higher in runoff from new roofs in comparison with older ones. Since rain is generally acidic, varying between pH 4.5 and 6.5 with an average of 5.5 depending on location (Mendez *et al.*, 2010), the make-up of the roof and tank materials is important. For example, it is possible that cement will be dissolved by the acidic nature of rain and that, as a result, the harvested water will be alkaline (Meera and Mansoor Ahammed, 2006). Acid conditions also encourage the dissolution of metals so that Zn, Cu and Pb can appear in relatively high concentrations in harvested water from metal roofs, particularly those that are galvanised (Mosley, 2005).

Whilst it is recommended that gutters and pipes are made of PVC by some organisations (e.g. SOPAC (Secretariat of the Pacific Community Applied Geoscience and Technology Division); see Mosley, 2005), some studies (e.g. Morrow *et al.*, 2010) have found that materials making up the piping connections contributed 'significantly', for example, to Pb levels, where the pipes are extruded in Pb during the manufacturing process. Taps can also contribute to Cu levels when fixtures are based on Cu. Much of the focus in the literature would appear to be on roof conditions and subsequent water quality, with only a few studies (e.g. Morrow *et al.*, 2010) reporting on the contaminants found throughout the RwH system. Recommendations, such as EA (2010), suggest that larger tanks should be made of glass-reinforced plastic, polyethylene or concrete, and that they are best sited underground to keep contamination (e.g. biofilm, bacterial and algal growth) to a minimum.

Studies are contradictory in providing evidence of whether harvested rainwater is contaminated or not, and results probably reflect site-specific conditions that impact water quality. There are, however, various treatments that can be applied to the water, such as chlorination, filtration or UV, which may prove useful in improving water quality prior to use (Jordan *et al.*, 2008; Lye, 2009; EA, 2010).

12.6 Water Quantity

The EA (2011) state that RwH systems can be used as part of an overall SUDS approach and can reduce stormwater runoff in the short term, particularly in the early stages of a storm. In doing so, they can reduce the amount of water the storm sewers have to cope with, thereby increasing the overall capacity of the sewerage infrastructure, enabling the system to cope better with the additional foul flows emanating from new-build development. Kellagher and Maneiro Franco (2007) modelled domestic RwH and found that, by increasing the size of the tank to between 1.5 and 2.5 times that normally used to harvest water for standard uses in the home, considerable benefits would accrue from the point of view of reducing stormwater volumes and attenuating peak flows. Specifically they found that increasing the tank size resulted in a 50% reduction in peak flow for 80% of extreme events, although for the remaining 20% there was no discernable improvement. It was concluded that by retrofitting RwH systems to areas that already suffer from frequent flooding, significant reductions in both flood frequency and flood volume could be made.

At the individual domestic building scale, flood risk reduction by RwH alone is based on tank size and antecedent conditions (Gerolin, 2009). At the larger scale, studies of community-wide schemes, for example in Australia, Mitchell *et al.* (2007) concluded that stormwater harvesting, when designed into a treatment train, offered the potential to control flooding due to excess surface water. In New York, Basinger *et al.* (2010) found that by installing RwH on the most commonly occurring rooftop sizes (51–75 m^2), inputs of surface water to the storm sewers could be reduced on average by 28% in a typical rainfall year and potable water demand reduced by 53%.

12.7 Cost–Benefit Analysis and Whole Life Costs

Different valuation methods have been used for financial assessment of RwH. Cost–benefit analysis (CBA) is a means of assessing whether the benefits of a scheme outweigh the costs of implementation and hence whether the public, local authorities or governments would be prepared to invest in them as a strategy (Jianbing *et al.*, 2010). Whole life costing (WLC), on the other hand, estimates the total cost over the lifetime of the system and takes a longer term view by identifying future outgoings such as operational costs and maintenance (HR Wallingford, 2004).

Roebuck (2007) concluded that the financial performance of RwH systems had been carried out in an ad hoc, simplistic way, by attempting to predict the cost of the harvested supply and then comparing it against an equivalent volume from the mains water supply. In an evaluation of a variety of financial assessment approaches, the author determined that WLC appeared to be the most relevant approach for RwH. Roebuck's (2007) study of 3840 simulated permutations of a domestic, new-build scenario revealed that high initial capital costs of RwH systems lead to poor overall financial performance; hence, the smaller the tank, the more viable, in monetary terms, the system would be. The study also suggested that long payback periods reported in the literature of up to 267 years (Brewer *et al.*, 2001) were probably unlikely. Communal RwH systems, on the other hand, performed far more effectively in economic terms, and Roebuck (2007) states that '*substantial*' (p. 403) financial savings are possible for such installations.

Most other studies either specified CBA or attempted to construct different financial approaches, which makes comparison between studies difficult. In a CBA of the retrofit of SUDS surface water management strategies to UK urban areas, Gordon-Walker *et al.* (2007) state that ~75% of industrial and commercial buildings could retrofit RwH systems and ~50% of public buildings could do so. The main barriers to implementation, adoption and maintenance have cost implications, which at the local level may be balanced by benefits such as: reduction in energy costs at the sewerage treatment works due to reduced pumping; deferral of upsizing the present 309 000 km of storm sewage networks that OFWAT (Worsfold, 2010) cost at ~ £174 billion at 2007–2008 prices and that would take centuries to deliver; and more efficient use of potable water resources that would have cost savings. However, Gordon-Walker *et al.* (2007) admit that estimation of some of the benefits is not possible. On the small scale in the United Kingdom, however, they do estimate that if those householders with a metered water supply were to install water butts or barrels (see Figure 12.1) and make regular use of these during the summer months, a national outlay of £325 million would realise savings approaching £1 billion. A study in Beijing, China, by Jianbing *et al.* (2010) calculated the benefit index, α, a ratio of present value of the asset, in this case RwH, throughout its life to its benefit present value, whereby the higher the value of α the more benefit accrues, with a value of one equating to no benefit. Of the 267 RwH installations in the 18 districts examined, the mean α was 2.0 with two-thirds of the calculated α values >1.0 and two of the districts having benefit indices >6.0.

The experience of RwH users in Barcelona, Spain, was that all their toilet flushing needs were met by using a relatively small tank (Domènech and Saurí, 2011). The long payback period of RwH installations was seen as a drawback. Even if discount rates were applied to the water used, payback was ~60 years; nevertheless, residents of multifamily buildings appeared to be unconcerned with such costs. Farreny *et al.* (2011) examined the cost effectiveness of RwH in the city of Granollers, eastern Spain. The result of their study was similar to that of Domènech and Saurí (2011) in that, at current prices for water, RwH was not cost-effective since existing water supplies were subsidised. However, differences were associated with the scale of installation: at the individual building scale RwH was 'unrewarding' (Farreny *et al.*, 2011, p. 692), but at the neighbourhood scale, applied to new-build housing, RwH had a more favourable payback period of 27 years. They also found costing what was termed 'less tangible' benefits difficult, such as the deferral of costly flood resilience projects due to the storm attenuation abilities of some RwH systems (see Section 12.6); lessening the need for desalinisation, dam construction and other water supply infrastructure; and reduction in the need for water conveyance and treatment systems. Farreny *et al.* (2011) concluded that cost compared to savings was not the only means of assessing the performance of RwH systems, but that environmental, social, as well as financial benefits should be taken into account as part of a wider sustainability assessment. In spite of the lack of information, particularly in the United Kingdom (Roebuck, 2007), of the costs and potential benefits of RwH, many countries and individual cities have encouraged its use. Some have already been discussed in this chapter (e.g. Beijing and Barcelona), but the following section presents an overview of some of the ways RwH has been either integrated into formal policy or the ways people are simply encouraged to adopt the technology.

12.8 Case Studies

Germany leads modern RwH technology and use, with 35% of new builds incorporating such systems; many city councils offer incentives and subsidies to encourage their use and the industry is worth €340 million. Studies have shown that RwH can reduce consumption of potable water considerably, as well as attenuating peak flow (Herrmann and Schmida, 1999). France, on the other hand, has few installed systems, which are often for garden irrigation alone with few extending beyond that to toilet flushing or washing clothes. The primary reason for this is the structure of the French water supply industry, where there is no incentive to use RwH techniques, because this would reduce water company profits (Koenig, 2001).

Australia is the driest inhabited continent on Earth, yet it has high use of water, but low and variable rainfall (Khouri, 2006). To address concerns associated with population growth and the worst drought in living memory, the approach of Total Water Cycle Management (Water by Design, 2010) integrates water quality and stormwater management (Khouri, 2006) to promote more sustainable water resource usage (van der Sterren *et al.*, 2012). The use of RwH by tank storage is encouraged via state legislation, development control plans and various subsidies. As a result, harvesting of rainwater is popular even though it was found to be financially non-viable at the single dwelling scale, but at the catchment scale the many benefits of RwH impart financial savings, increasing their viability. The Australian Rainwater Industry Development Group (ARID) (2008) promotes water harvesting, recycling and reuse, and calculates that use of potable water could be reduced by 30%, saving up to 60 000 litres of water per year per household. Upscaling to 7.4 million buildings over a 10-year period would lead to annual savings of up to 444 000 000 000 litres of water per annum.

There are many case studies of the implementation of RwH in Africa as a whole (see, for instance, World Agroforestry Centre, 2012) and UNEP (2006) has stated that Africa has 'massive potential' to make use of RwH to 'dramatically' increase their water resources. RwH can therefore aid in addressing the fact that, in Africa, approximately 300 million people (one-third of the continent's population) are living under 'water scarcity' and there is the possibility of 12 more African countries joining them by 2025. The African Water Vision, 2025 (UN Water Africa, no date) demands 'improving water wisdom' in which RwH would be a major focus. Organisations such as UNEP have commissioned reports (Regional Land Management Unit (RELMA) in International Centre for Research in Agroforestry/World Agroforestry Centre (ICRAF) and United Nations Environment Programme (UNEP), 2005) that have used geographical information system (GIS) techniques to map the harvestable areas present in 10 African countries, finding in Botswana, for example, that RwH introduction was a 'necessity' (p. 17). Rooftop collection could be supplemented by other approaches, such as runoff and storage into pans and ponds, the use of sand or subsurface dams and *in situ* harvesting employing microcatchments and terracing.

Publications such as that by WaterAid (no date) give further details of the efficiency and efficacy of these different systems, comparing their costs, yield, water quality and the most appropriate siting. RwH was classified as a low-cost technology, with low operating costs and medium volume water yield, obviously dependent on the area of the collecting surface with good bacterial water quality. In terms of the most appropriate locations, two wet seasons a year were seen as optimal, which would include many tropical countries. For instance, to avoid seasonal shortages, a recently published study in Ibadan, Nigeria (Lade and Oloke, 2013) after Nigeria has harvested roof-derived rainfall from a domestic dwelling to recharge the groundwater in a household well. To minimise groundwater contamination, the collection tank was layered with sand and gravel to filter and treat the harvested water before piping it into the well. Compared to a well in the neighbourhood, which served as a control for this experiment, results indicate no sizeable difference in water quality. However, that said, both wells fail to meet the potable water standards outlined by the World Health Organisation (WHO). This issue is a reflection of the circumstances facing local communities but, drawing a positive outcome, the recharged well yielded an available water supply for most of the year, unlike the control well, which was quickly exhausted during the dry season.

12.9 Conclusions

RwH systems alone can answer some of the problems associated with excess surface water in that they can offer volume reduction, storm peak attenuation and a limited amount of flood resilience to a property. However, performance depends on the size of the system installed and climatic conditions. Many of the studies discussed in this chapter have indicated that increasing the size of the volume capability increases flood resilience. However, in financial terms WLC and CBA both suggest that the smaller the installation, the increased likelihood that RwH will have a reasonable payback period, and in areas where water is metered, offer monetary savings on mains water used. Like all SUDS, RwH has to be properly designed and maintained in order to function effectively and to address concerns about water quality. A far better approach, however, has to be the integration of RwH into a SUDS management train (Charlesworth *et al.*, 2003; Charlesworth, 2010), addressing all aspects of the SUDS triangle and taking full advantage of the multiple benefits of the flexible SUDS approach.

References

African Development Bank (no date) *Rainwater Harvesting Handbook*, 72 pp.

Australian Rainwater Industry Development Group (ARID) (2008) *Implementation of a Sustainable National Urban Water and Energy Savings Program*. Available at: http://www.arid.asn.au/images/stories/documents/ARID%20-%20Discussion%20Paper%20-%20Nov%2008.pdf.

Basinger, M., Montalto, F. and Lall, U. (2010) A rainwater harvesting system reliability model based on nonparametric stochastic rainfall generator. *Journal of Hydrology*, **392**, 105–118.

Bregulla J., Powell, J. and Yu, C. (2010) *A Simple Guide to Sustainable Drainage Systems for Housing*. IHS BRE Press for the NHBC Foundation.

Brewer, D., Brown, R. and Stanfield, G. (2001) *Rainwater and Greywater in Buildings: Project Report and Case Studies*. Technical Note TN 7/2001, BSRIA, Berkshire.

Chang, M., McBroom, M. and Beasley, R. (2004) Roofing as a source of nonpoint water pollution. *Journal of Environmental Management*, **73**, 307–315.

Charlesworth, S.M. (2010) A review of the adaptation and mitigation of Global Climate Change using Sustainable Drainage in cities. *Journal of Water and Climate Change*, **1**, 165–180.

Charlesworth, S.M, Harker, E. and Rickard, S. (2003) Sustainable Drainage Systems (SUDS): A soft option for hard drainage questions? *Geography*, **88**, 99–107.

Debord, D.E. (2011) *Rainwater Harvesting Design*. Consulting-Specifying Engineer. November **2011**, pp. 38–42.

Department for Communities and Local Government (DCLG) (2006) *Code for Sustainable Homes: A Step-Change in Sustainable Home Building Practice*.

Department for Environment, Food and Rural Affairs (DEFRA) (2012) *Flood and Water Management Act 2010*. Available at: http://www.defra.gov.uk/environment/flooding/legislation/.

Domènech, L. and Saurí, D. (2011) A comparative appraisal of the use of rainwater harvesting in single and multi-family buildings of the Metropolitan Area of Barcelona (Spain): social experience, drinking water savings and economic costs. *Journal of Cleaner Production*, **19**, 598–608.

Eletta, O.A.A. and Oyeyipo, J.O. (2008) Rain water harvesting: effect of age of roof on water quality. *International Journal of Applied Chemistry*, **4**, 157–162.

Environment Agency (EA) (2009) *Rainwater Harvesting: An On-Farm Guide*, 42 pp.

Environment Agency (2010) *Harvesting Rainwater for Domestic Uses: An Information Guide*, 31 pp.

Environment Agency (2011) *The Use of Rainwater Harvesting Systems: A Position Statement*. Available at: http://publications.environment-agency.gov.uk/PDF/GEHO0611BTYB-E-E.pdf.

Farreny, R., Gabarrell, X. and Rieradevall, J. (2011) Cost-efficiency of rainwater harvesting strategies in dense Mediterranean neighbourhoods. *Resources Conservation and Recycling*, **5**, 686–694.

Fewkes, A. (2006) The technology, design and utility of rainwater catchment systems. In: Butler, D. and Memon, F.A. (eds), *Water Demand Management*. IWA Publishing, London, pp. 27–61.

Gerolin, A. (2009) *Rainwater Harvesting Systems for Stormwater Management – An investigation of the benefits provided by domestic rainwater tanks during extreme rainfall events: factorisation into design guidelines*. MSc dissertation, Centre d'Etudes Techniques de l'Equipment de l'Est, University of Grenoble, France.

Gordon-Walker, S., Harle, T. and Naismith, I. (2007) *Cost–Benefit of SUDS Retrofit in Urban Areas*. Environment Agency Science Report SC060024.

Herrmann, T. and Schmida, U. (1999) Rainwater utilisation in Germany: efficiency, dimensioning, hydraulic and environmental aspects. *Urban Water*, **1**, 307–316.

HR Wallingford (2004) *Whole Life Costing for Sustainable Drainage*. Available at: http://www.ciria.org.uk/suds/pdf/whole_life_cost_summary.pdf.

Jianbing, Z., Changming, L. and Hongxing, Z. (2010) Cost–benefit analysis for urban rainwater harvesting in Beijing. *Water International*, **35**, 195–209.

Jones, M.P. and Hunt, W.F. (2010) Performance of rainwater harvesting systems in the southeastern United States. *Resources, Conservation and Recycling*, **54**, 623–629.

Jordan F.L., Seaman, R., Riley, J.J. and Yorke, M.R. (2008) Effective removal of microbial contamination from harvested rainwater using a simple point of use filtration and UV-disinfection device. *Urban Water Journal*, **5**, 209–218.

Kellagher, R.B.B. and Maneiro Franco, E. (2007) *Water Cycle Management for New Developments (WaND), Rainfall Collection and Use in Developments; Benefits for Yield and Stormwater Control.* WaND Briefing Note 19, WP2 Briefing Note 2.15, Report SR 677, Release 3.0.

Kim, W.J., Managaki, S., Furumai, H. and Nakajima, F. (2010) Washoff behaviour of in-sewer deposit in combined sewer system through artificial flushing experiment. *Water Science and Technology*, **61**, 2835–2842.

Khouri, G. (2006) Integrated water cycle management, an Australian perspective. *European Water*, **15/16**, 23–32.

Koenig, K.W. (2001) *The Rainwater Technology Handbook: Rainharvesting in Building*. Wilo-Brain, Dortmund, Germany.

Lade, O. and Oloke, D. (2013) Assessment of rainwater harvesting potential in Ibadan, Nigeria. *Environmental Engineering Research*, **18**, 91–94.

Lye, D.J. (2009) Rooftop runoff as a source of contamination: a review. *Science of the Total Environment*, **407**, 5429–5434.

Market Transformation Programme (2007) *Alternative Sources of Water – Greywater and Rainwater Reuse*. Innovation Briefing Note BNWAT 19.

Mbilinyi, B.P., Tumbo, S.D., Mahoo, H.F., Senkondo, E.M. and Hatibu, N. (2005) Indigenous knowledge as decision support tool in rainwater harvesting. *Physics and Chemistry of the Earth*, **30**, 792–798.

Meera, V. and Mansoor Ahammed, M. (2006) Water quality of rooftop rainwater harvesting systems: a review. *Journal of Water Supply: Research and Technology – AQUA*, 55.4, 257–268.

Mendez, C.B., Afshar, B.R., Kinney, K., Barrett, M.E. and Kirisits, M.J. (2010) *Effect of Roof Material on Water Quality for Rainwater Harvesting Systems*. Texas Water Development Board, Austin, Texas.

Mitchell, V.G., Deletic, A., Fletcher, T.D., Hatt, B.E. and McCarthy, D.T. (2007) Achieving multiple benefits from stormwater harvesting. *Water Science and Technology*, **55**, 135–144.

Morrow, A.C., Dunstan, R.H. and Coombes, P.J. (2010) Elemental composition at different points of the rainwater harvesting system. *Science of the Total Environment*, **408**, 4542–4548.

Mosley, L. (2005) *Water Quality of Rainwater Harvesting Systems*. SOPAC Miscellaneous Report **579**.

Nanzala, E. (2008) *Maasai Women Adapt to a Changing Climate*. Chartered Institution of Water and Environmental Management. Available at: http://www.ciwem.org/hot-topics/climate-change/2008/sep/maasai-women-adapt-to-a-changing-climate.aspx.

Pitt, M. (2008) *Learning Lessons from the 2007 Floods*. The Cabinet Office, London.

Plazinska, A. (2001) Microbiological quality of rainwater in selected indigenous communities in Central Australia. In: *10th International Rainwater Catchment Systems Conference: 'Rainwater International, 2001'*, Germany.

Regional Land Management Unit (RELMA) in International Centre for Research in Agroforestry/World Agroforestry Centre (ICRAF) and United Nations Environment Programme (UNEP) (2005) *Potential for Rainwater Harvesting in Africa: A GIS Overview*. Available at: http://www.unep.org/pdf/RWH_in_Africa-final.pdf.

Roebuck, R. (2007) *A Whole Life Costing Approach for Rainwater Harvesting Systems: An investigation into the whole life cost implications of using rainwater harvesting systems for non-potable applications in new-build developments in the UK*. Unpublished PhD thesis, Bradford University.

Simmons, G. (2001) Contamination of potable roof collected rainwater in Auckland, New Zealand. *Water Research*, **35**, 1518–1524.

Su, C.Q. (2010) Rainwater harvesting on the sea: a new sustainable water resource. *Water International*, **35**, 779–785.

United Nations Environment Programme (UNEP) (2006) *Harvesting Rainfall a Key Climate Adaptation Opportunity for Africa*. Available at: http://www.unep.org/Documents.Multilingual/Default.asp?ArticleID=5420&DocumentID=485&l=en.

United Nations Environment Programme (2009) *Rainwater Harvesting: A Lifeline for Human Well-being*, J. Barron (ed.). UNEP/SEI.

United Nations Water Africa (no date) *The Africa Water Vision for 2025: Equitable and Sustainable Use of Water for Socioeconomic Development*. Available at: http://www.uneca.org/awich/african%20water%20vision%202025.pdf.

van der Sterren, M., Rahman, A. and Dennis, G.R. (2012) Rainwater harvesting systems in Australia. In: K. Voudouris (ed.), *Ecological Water Quality – Water Treatment and Reuse*. InTech. Available at: http://www.intechopen.com/books/ecological-water-quality-water-treatment-and-reuse/rainwater-harvesting-systems-in-australia.

Van Metre, P.C. and Mahler, B.J. (2003) The contribution of particles washed from rooftops to contaminant loading to urban streams. *Chemosphere*, 52, 1727–1741.

Vasudevan, P., Tandon, M., Krishnan, C. and Thomas, T. (2001) Bacteriological quality of water in DRWH. In: *RWH Conference, IITD*, New Delhi. Available at: http://besharp.archidev.org/IMG/pdf/bacterial_quality.pdf.

Ward, S., Barr, S., Butler, D. and Memon, F.A. (2012) Rainwater harvesting in the UK: socio-technical theory and practice. In: *Technological Forecasting and Social Change*. DOI: 10.1016/j.techfore.2012.04.001.

WaterAid (no date) *Rainwater Harvesting: Technical Brief*. Available at: https://www.wateraid.org/documents/plugin_documents/rainwater_harvesting.pdf.

Water by Design (2010) *Total Water Cycle Management Planning Guideline for South East Queensland*. Department of Environment and Resource Management, Brisbane, Queensland. Available at: http://www.derm.qld.gov.au/environmental_management/water/environmental_values_environmental_protection_water_policy/pdf/twcmp-guideline.pdf.

WHO/UNICEF (2010) *Progress on Sanitation and Drinking-Water: 2010 Update*. Available at: http://whqlibdoc.who.int/publications/2010/9789241563956_eng_full_text.pdf.

World Agroforestry Centre (2012) *World Agroforestry Centre*. Available at: http://www.worldagroforestry.org/.

Worsfold, M. (2010) *Sustainable Drainage*. Available at: http://www.ofwat.gov.uk/mediacentre/speeches/prs_spe_100303drainage.pdf.

13

Greywater Harvesting – Reusing, Recycling and Saving Household Water

Emma Hatfield, Colin A. Booth and Susanne M. Charlesworth

13.1 Introduction

Dramatic increases in societal water demand have been reported in recent decades (Brookshire *et al.*, 2002), which are chiefly attributed to an increase in the number of households and a decrease in occupancy numbers, together with changes in lifestyles, resulting in a dramatic increase in personal water use (Chang *et al.*, 2010). However, since water meters have become a mandatory requirement in many countries (e.g. compulsorily installed in all new UK homes since 1997), it is widely acknowledged that metering can reduce household water usage through raised awareness of the financial incentive of using less water (Hoffmann *et al.*, 2006). Similarly, improved awareness of waste management also highlights homeowners to the opportunities of reusing and recycling 'grey' wastewater already having been used inside their homes.

Greywater harvesting (GwH) has traditionally been used in countries with experience of water scarcity. In Madhya Pradesh, India, treated greywater is used in residential schools to flush toilets and irrigate crops. Savings are made from the reduction in water supplied by delivery tankers, on water infrastructure, from environmental benefits and reduction in health costs, demonstrating that benefits outweigh the costs of the system (Godfrey *et al.*, 2009). Wastewater from kitchen and bathroom sinks, showers and laundering are used in Jordan for landscaping and irrigation (Al-Jayyousi, 2003). Treated greywater was used more extensively in Australia in the 1990s due to limits imposed on sewage discharge by Environment Protection Authorities. After interest grew in greywater recycling due to the 2001–2003 droughts, it has been included in strategies to reduce per capita water demand (Radcliffe, 2006). A large number of urban areas in the world are water-scarce, leading to studies of the economic feasibility of GwH systems, for example in Israel (Friedler and Hadari, 2006) and Germany (Nolde, 2000).

Water Resources in the Built Environment: Management Issues and Solutions, First Edition.
Edited by Colin A. Booth and Susanne M. Charlesworth.

This chapter provides insights and observations about GwH, from sources of greywater to types of systems, including issues due to uncertainty and reliability of its use. It also details case studies where GwH systems have been installed and outlines operational user feedback, before critically assessing the future of GwH.

13.2 Insights into Greywater Harvesting

GwH is a relatively new concept in some Western countries and, as such, available technology is widely considered to be at a youthful stage of development. In fact, many people are even unaware of it – in terms of what the initiative means: where to collect greywater, how to store and use it and what advantages it can offer individuals, society and the environment.

13.2.1 Greywater Sources

In terms of sources of available waste water in domestic buildings it is important to know which streams offer the largest volumes of least contaminated water. This is because the chemical and biological composition of wastewater varies depending on where it is harvested in the home. For instance, wastewater from toilets is plentiful but is far too contaminated with bacteria and pathogens to be reused or recycled, although advances in technology may make this possible in the future; wastewater from washing machines contains detergents and soaps (such as sodium hydroxide), which can be detrimental to soils and plants; and wastewater from kitchen sinks contains high amounts of fats and foods from washing-up. Therefore the most plentiful and least contaminated wastewater can be derived from baths, showers and hand basins. However, even this will often contain human intestinal bacteria and viruses, together with organic matter (such as skin particles and hair), which means the water quality can rapidly deteriorate in storage, so it should always be reused within a few hours or disposed of, unless it is treated and later recycled. Domestic GwH in the United Kingdom was a virtually untested technology in the late 1990s; any information available at that time proved to be unreliable. Absence of water quality standards and limited understanding of the constituents of greywater has resulted in a range of different types of treatment systems (Jefferson *et al.*, 2000). Examples of these are described in the following sections.

13.2.2 Greywater Reuse Systems

Simply moving greywater from one place to another without it undergoing any processesing is termed 'reuse'. For instance, the Watergreen Syphon can be used to drain bathwater to use on plants in the garden or wash the car; it does not require permanent installation or regular maintenance (Droughtbuster UK Ltd, 2007). Similarly, the Water Two valve can be used for watering the garden, but this is a permanent fixture that is fitted by the customer between the bathroom greywater waste water pipe and an outside water butt. Regular maintenance is not required (Water Two UK Ltd, 2007). However, the greywater harvested by these two systems must be used within a few hours because the water has not undergone treatment to remove organic materials (Griggs and Burns, 2009).

13.2.3 Greywater Recycling Systems

Processing greywater by changing its composition and improving its quality before it is used again is termed 'recycling' (Leggett *et al.*, 2001a). For instance, Ecoplay is a simple storage system and its treatment processes include skimming and settlement of particles from the water. To safeguard users from contamination arising from stagnant water, stored water is purged from the system if the toilet is not flushed within 24 hours. The system can be located in the same room as the water source, cutting plumbing costs, and does not require regular maintenance (CME Sanitary Systems Limited, no date). Filtration systems include the Aquastore and the Ezy-Filter. The Aquastore is appropriate for irrigation only; the water is diverted into a filter placed over a water butt and does not require electricity as it is gravity fed (Combined Harvesters Ltd, 2007). The Ezy-Filter is also used for irrigation; the filter unit is attached to the wall on the outside of a house and is gravity fed via the bathroom greywater waste pipe (Kingspan Environmental, 2008). The Aquawiser chemical treatment system uses two tanks plumbed into the pipework. The water is collected and treated with bromine tablets in the first tank sunk into the ground and then transported to the second tank inside the home for storage, from where the water is drawn for toilet flushing (Aquawiser, no date). Similarly, BRAC Systems Inc. supplies a filtration and chemical system, which is a tank installed inside the house that collects greywater from the bathroom and supplies the toilet (BRAC Systems Inc., 2009).

Biomechanical systems include the Green Roof Water Recycling System (GROW and GROW 2), Aquacycle 900, and ARC4 and WME–4 systems. The GROW and GROW 2 systems work in a similar way to a reed bed (see Chapter 25) by introducing oxygen into the wastewater so that microorganisms around the roots can break down any waste products in the water. It uses a tiered garden of indigenous plants; water from the bathroom is pumped to treatment beds and the treated water is used for watering the garden and/or flushing the toilet. GROW 2 is installed in the gardens of individual housing, whereas GROW is a communal system installed on the roof of an apartment block (Water Works UK Ltd, 2006a, 2006b). The Aquacycle 900 allows domestic householders to recycle bathroom greywater to flush toilets, wash clothes and water plants. The system treats the water by microfiltration, settlement of solids, use of biocultures and UV sterilisation (Freewater UK Ltd, no date). Aquality's ARC 4 and WME–4 systems rely on bacteria to break down biodegradable particles in the water, followed by filtration to remove any particles remaining (Aquality Trading and Consulting Ltd, 2008a, 2008b).

13.3 The Potential for Using Greywater Harvesting Systems

High-population densities in urban areas, growing water demand and increased water metering make GwH a potential alternative to using high-quality drinking water to flush the toilet or irrigate the garden. Reuse of water would prevent overloading of the sewers, reduce the volume of wastewater treated, decrease processing costs and reduce energy consumption (Eriksson *et al.*, 2002; Hunt and Rogers, 2005). Systems that recycle water have mainly been installed in domestic housing as an individual system, although some water utility companies have been assessing the potential of using communal systems (Jefferson *et al.*, 2000), which would be more cost-effective but more difficult to install in Britain because of dwelling sizes (Burkhard *et al.*, 2000). However, a study by Dixon *et al.* (1999a) concluded that small-capacity domestic systems have the potential to be used in urban areas to achieve a more sustainable water supply. They assessed the potential of combining greywater and rainwater harvesting systems and found that household occupancy, roof area, appliance type and storage capacity affect the efficiency of individual household systems.

A range of factors exist that prevent social acceptability of GwHS. Firstly, it would be a challenge to change the mindset of people living in the United Kingdom who are not responsible for sourcing their own water and do not want to start doing so. The population are used to, and expect, a constant supply of drinking quality water for all purposes in the home. Water scarcity in the United Kingdom is not perceived as a problem due to frequent rainfall; it is viewed as an issue applicable to Third World countries (Hunt and Rogers, 2005). Other factors that determine acceptance of GwH include cost–benefit uncertainty, system reliability, lack of existing water quality standards and consequential risk to public health.

13.3.1 Cost–Benefit Uncertainty

The cost–benefit of GwHSs has not been successfully demonstrated because payback periods have a range of variants, including number of system users, volume of greywater produced, initial and operational running costs of the system, as well as current and future metered water charges. System lifespan is also a factor in the effectiveness of payback periods; it is generally perceived that money saved from water charges on metered properties would not be enough to recoup the costs of installing a GwHS. The cost of retrofitting a property with a GwHS is higher than installation in new build because dual pipework is required, as well as tanks and pumps to transfer the water around the system.

In terms of whole life cost (WLC) (see Chapter 12, which differentiates between WLC and cost–benefit analysis – CBA) GwHS is affected by its lifespan, water prices, system efficiency, size and energy consumed to treat the water (Memon *et al.*, 2005). The UK Environment Agency has calculated the payback period of GwHS in areas of high and low water charges, finding that the initial cost of installation of the treatment system was estimated to be £3000. Payback periods for areas of high water charges ranged from 30 to 40 years for older toilets and dual-flush toilets, whereas payback periods for areas of low water charges ranged from 83 to 113 years for older and dual-flush toilets (Environment Agency (EA), 2008). Most GwHS are not, therefore, economically viable in the United Kingdom because of low water charges (Mustow *et al.*, 1997). However, installation of communal GwH systems for a group of houses could reduce the cost per house and make it more financially viable (Marshallsay *et al.*, 2007). Research into people's decision as to whether to use greywater and rainwater harvesting systems (RwHS) concluded that most GwHS were not cost-effective, due to long payback periods (Leggett and Shaffer, 2002). The latter required more maintenance, were more difficult to install and less reliable than RwHS. In terms of effectiveness, financial benefit and reliability, RwHS would probably be more cost-effective than greywater, although GwHS are becoming less expensive and more reliable, which will decrease payback periods (Leggett and Shaffer, 2002). Whilst research has shown that house prices can benefit from being close to green infrastructure (Environment Agency, 2006), installation of GwH could have a positive or negative impact on house sales, due to the purchaser's perception of the technology (Leggett *et al.*, 2001b). For instance, the need to hire a plumber to avoid cross-contamination between potable and nonpotable water supplies and backflow from the sewer may deter people from using GwHS. Continued maintenance of the system after installation can be difficult for the homeowner to manage, as well as expensive (Marshallsay *et al.*, 2007) and reliability has yet to be proven for most greywater supplies (Hunt and Rogers, 2005), as discussed in the following section.

13.3.2 System Reliability and Water Quality

Some GwHS failed to achieve satisfactory reliability during test projects such as the Buildings that Save Water Project, which has meant that some systems have been abandoned due to reliability issues mainly associated with poor installation (Brewer *et al.*, 2001). GwHS cannot be installed and forgotten; they require frequent checks and maintenance to ensure that they are running efficiently. However, no suitable accreditation schemes exist to provide interested parties a reliable indication of system performance (Leggett *et al.*, 2001a).

In terms of water quality, standards for greywater treatment and storage need to be established to provide clear criteria for nonpotable water quality. For example, it is known that nutrients in greywater pose contamination risks due to the encouragement of microorganism growth (Butler and Fayaz, 2006). A study by Bixler and Floyd (1997) highlighted concerns about the presence of pathogenic microbes in greywater; their impact was found to be not only dependent on the concentration of microorganisms found but also on the degree of exposure, health and age of individuals in contact with the greywater. A risk assessment, together with recommendations from the Building Services Research and Information Association (BSRIA) to limit certain determinands, were used to formulate proposed changes to the guidelines for greywater use (Dixon and Fewkes, 1999). However, for drinking, cooking, bathing and irrigation of crops to be eaten raw, the guidelines recommend that the UK Water Supply (Water Quality) Regulations 1989 (amended 1991) are followed. For all other uses it states solely that faecal coliforms must not be detectable (Mustow and Grey, 1997). Regardless of the tightening of guidelines for greywater use, lack of knowledge of GwHS can impede its acceptance as a source of water, and in particular disgust sensitivity (Po *et al.*, 2003) can have a negative effect on the idea of using water that has already been used for other purposes.

Due to the settlement of organic particles in the water, the quality of greywater can be improved by storing for up to 24 hours, but it must not be stored for longer than 48 hours as dissolved oxygen levels reduce and aesthetic problems occur (Dixon *et al.*, 1999b), the stored water becoming 'blackwater'. The source of the greywater can impact on its composition since kitchen and bathroom greywater are different; bathroom greywater contains bacteria, hair, soaps and nutrients and kitchen greywater can also contain fats, oils, salt, food and detergents (Australian Capital Territory, 2007). Solids found in greywater can be reasonably efficiently removed by installing filter membranes, but these cannot fully remove organics; however, biological or complex systems can efficiently remove organics (Pidou *et al.*, 2007). When disinfection of greywater is required, perhaps when it is stored for longer than the recommended 24 hours, and to prevent transmission of diseases from microorganisms, chlorine is the most commonly used agent (Winward *et al.*, 2008). Chemical treatment processes can also remove suspended solids and organic materials, but a combination of microbiological processes, filtration and disinfection is considered to be the most economical and effective way of processing the water (Li *et al.*, 2009).

In terms of efficiency of mechanical treatments, some authors believe that processes utilising membranes are key to greywater treatment (Melin *et al.*, 2006). Membrane bioreactors (MBRs), membrane chemical reactors and constructed wetlands (see Chapter 25 on constructed wetlands) have all been used in treating greywater (Stephenson *et al.*, 2006). It is felt, however, that MBRs were the most efficient, producing the highest quality of treated water and robustly dealing with greywater containing different concentrations of contaminants (Melin *et al.*, 2006). However, a vertical flow reed bed was the most reliable constructed wetland for pathogen removal (Winward *et al.*, 2008). Research by Ottosson and Strenström (2003) to ascertain the effect of bacterial competition and temperature on pathogens in sediments from a GwHS found that competition from the indigenous microbiota in the nutrient-rich sediment can reduce the pathogenic bacteria count considerably in the first day. After two weeks nutrients in the sediments had decreased and sediment kept at 20 °C experienced a larger depletion of pathogens than that maintained at 4 °C. Ottosson and Strenström (2003) concluded that microbiota and temperature could be used effectively to reduce pathogen levels in greywater.

13.3.3 Guidance on Greywater Harvesting Systems (GwHS)

The Construction Industry Research and Information Association (CIRIA) have developed Guidance Note C539: *Rainwater and Greywater Use in Buildings: Best Practice Guidance*, which aims to help users and build confidence in using these systems. Common issues are discussed and specific guidance for rainwater, greywater and combined systems are set out (Leggett *et al.*, 2001a). The Building Services Research and Information Association (BSRIA) produced Technical Note TN6/2002: *Water Reclamation Guidance: Design and Construction of Systems using Greywater*, which specifies safeguards required in packaged GwHS (Brown and Palmer, 2002a). The former states that sewer backflow must be prevented by this technology and stagnant water must be flushed out. A reliable back-up system is required to ensure that loss of power does not occur and to ensure that disinfection of the water is carried out. Packaged GwHS must include a means of monitoring its own status and should incorporate a warning system to indicate problems (Brown and Palmer, 2002a). The BSRIA Technical Note TN7/2002: *Water Reclamation Guidance: Laboratory Testing of Systems Using Greywater* refers to and compliments the TN6/2002 guidance, providing information on testing procedures for GwHS (Brown and Palmer, 2002b).

The Water Regulations Advisory Scheme (WRAS) Information and Guidance Note 9-02-04) discusses installation or modification of a reclaimed water system. It details contamination risks and hazard assessment protocols; the latter classifies the water quality required at end use (WRAS, 1999a). WRAS Guidance Note 9-02-05 details requirements for marking pipes conveying greywater or reclaimed water to ensure that the potable water mains supply is not contaminated. Pipe markings inside and outside of buildings, as well as below ground, are managed differently, with diagrams to show the types of labelling to use (WRAS, 1999b). Approved codes of practice and guidances have been published by the Health and Safety Commission to control *Legionella* bacteria in water storage systems. The presence of organic matter such as sludge and temperatures of 20 °C or more can create suitable conditions for multiplication of bacteria (Health and Safety Commission, 2000). Guidance Note TM13:2002: *Minimising the Risk of Legionnaires Disease* is written for designers, installers, operators and those maintaining building services. It details risks in hot and cold water systems, as well as in storage cisterns, and how they should be avoided (Chartered Institution of Building Services Engineers, 2002).

13.3.4 Greywater Harvesting Legislation

Section 14 of the Water Supply (Water Fittings) Regulations 1999 stipulates that nonpotable water is not to be cross-connected to potable water supplies and that nonpotable pipework must be clearly identifiable from drinking water pipes. Requirements for adequate devices to prevent backflow from a water fitting are detailed in Section 15 of the same regulations (Her Majesty's Government, 1999).

The Green Alliance (2005) recommended that water conservation should become a fundamental part of the buildings regulations and could only be an exception in places where the water supply was plentiful. Proposals for regulation changes to implement water efficiency in new homes were made by the Department for Communities and Local Government (DCLG); however, these focused on water metering and water-saving fittings, as opposed to installation of greywater harvesting. The former involved amendments to the Buildings Regulations and Water Supply (Water Fittings) Regulations 1999 (DCLG and Department for the Environment, Food and Rural Affairs (DEFRA), 2006).

As stated earlier, GwH has had a slow take-up in the United Kingdom, but there are demonstration projects that show the utility of this source for water other than for potable uses. The following sections describe two case studies in which GwH has been successfully used, the discussion of which leads into considerations of the future of this approach to meeting increasing water demand.

13.4 Case Studies in the United Kingdom

Whichelo Place, in Brighton (OSGR 532185E, 104956N), is a privately owned bungalow (Figure 13.1), which was built in 2007. Originally, planning permission was granted subject to the achievement of the Eco-homes 'good' rating. However, the owners of the bungalow were keen to build a home that exceeded expectations and to demonstrate their commitment

Figure 13.1 Outside the front door of the bungalow at *Whichelo Place*, Brighton.

Figure 13.2 Plan view of the exterior greywater tank at the bungalow at *Whichelo Place*, Brighton.

they built their home to an Eco-homes 'excellent' standard. In recognition of the build, they became Green Apple National Silver Award winners for environmental best practice (Eco Open Houses, no date). The GwHS installed in the bungalow was one of the technologies that contributed to this rating.

The GwHS installed at the bungalow is supplied from the bath and shower; a short pipe directs water from the bathroom into a tank sunk into the ground outside (Figure 13.2), where bromine tablets are used to treat the stored water. Each time the toilet is flushed inside the property, the system draws on water from the exterior tank, sending it to an interior tank located in a bathroom cupboard, which then feeds the toilet cistern with the treated greywater. The aesthetic quality of the water supplied to the toilet appears clear and smells faintly like swimming pool water.

An interview with the owners revealed that they have experienced some minor difficulties with the system since installation as the system does not use a filter and sometimes (about once a year) the small pipes in the dual-flush toilet cistern block. However, simple maintenance by removing the cistern lid and flushing out the pipes cures the problem. The owners believe their dual-flush toilet is ill-equipped for greywater flushing and in the longer term they plan on creating their own do-it-yourself filter to minimise blockages. In terms of water savings, six months water usage at the bungalow totalled 22.10 m^3 of potable water, which equates to ~60 litres of water use per person per day, and recycling of greywater has reduced wastewater sewage to ~55 litres per person per day. Furthermore, it is estimated that total water efficiency savings equate to about £100 per year (Eco Open Houses, no date).

Earthship, in Brighton (OSGR 533025E, 109790N), is a low-carbon bungalow built (2003–2006) by the Low Carbon Network (now known as the Low Carbon Trust). After

Figure 13.3 *Earthship*, Brighton.

gaining draft planning permission, finalising the business plan and gaining funding, construction of *Earthship* started in spring 2003. The hut module was built first, and then tyre ramming and the wooden frame of the building were completed in summer 2003. The building was completed in autumn 2006 (Eco Open Houses, no date). Finally, *Earthship* was opened to the public as an 'eco show home' in early 2007 (Figure 13.3).

In terms of water, *Earthship* is not connected to the water mains and uses rainwater and GwH and even processes its blackwater (i.e. water from toilets, sewage). Rainwater harvesting produces water supplies and the GwHS is supplied by showers, the kitchen and bathroom sinks. The greywater is cleaned by both filtration and biological treatment by means of a filter, which removes grease and particles, after which the water percolates between rocks (~75 mm) into eco-system planters lined with a rubber membrane, which aids treatment via transpiration, evaporation and oxygenation from the soil and by bacteria living around the roots of the plants. These processes remove suspended solids and reduce bacteria levels in the water. The recycled greywater is stored in a well at the end of the GwHS until it is needed and is pumped for toilet flushing.

There are two planters at *Earthship* Brighton located in the conservatory (Figure 13.4) and meeting room (Figure 13.5). The total area of the planters is 12.75 m^2; they are made up in layers starting with ~20 mm of pea shingle, ~75 mm of sand and ~150 mm of topsoil. The greywater entering the system reaches the pea shingle level only and the sand is used as a separating layer to stop the topsoil from clogging up the pea shingle. There is no risk of backwash of greywater back into the system and, since the pea shingle is buried, no human contact is possible with the greywater. However, it is noteworthy that care is taken

Figure 13.4 Greywater harvesting system planters in *Earthship*'s conservatory.

Figure 13.5 Greywater harvesting system planters at *Earthship*'s meeting room.

to limit the use of detergents and cleaners because they can upset the natural balance of the biological treatment system. After construction the system took a considerable time to establish itself; water was discoloured during the first year of use (Hewitt and Telfer, 2007; Hewitt, 2010).

Both greywater case studies consume additional energy compared with dwellings without GwH installed. This is because they both require pumps to move water around the systems. However, the extra energy consumption in these cases is offset by solar thermal water heating at *Whichelo Place* (Strube and Strube, 2010) and wind turbines and photovoltaic panels at *Earthship* (Hewitt, 2010).

13.5 The Future of Greywater Harvesting

Greywater harvesting has a high potential in reducing water demand (Memon *et al.*, 2005). Most literary sources confirm that between 29 and 35% of tap water can be conserved by employing GwH for toilet flushing alone and approximately half of daily water consumption can be reduced by using greywater for irrigation, WC flushing, car washing and laundry (Environment Agency, 2007; Woking Borough Council, 2008; Market Transformation Programme, 2008a; Waterwise, 2009). The most popular greywater end use in England is toilet flushing, which reduces water demand by 28–38%.

People will only consider installation of such systems if they can see an achievable payback period, but because the majority of UK households are unmetered, defining the payback period of GwHS is difficult (Environment Agency, 2009). Payback periods for GwHS are longer than their lifespan because tap water is inexpensive (Leggett *et al.*, 2001a), but this is dependent on the number of users; with a minimum of three people in a household, it can be reduced. The bungalow at *Whichelo Place* is an exception, achieving a payback period of 18 years with two inhabitants, which is within the system lifespan; although this is still lengthy, it is evidence that they can be reduced. The Environment Agency (2008) has quoted minimum and maximum payback periods of 30–40 years and 83–111 years respectively, in which case it is unlikely that GwHS will become the norm in the near future unless this can be reduced considerably.

The Market Transformation Programme (2008b) highlights two main concerns affecting the future uptake of GwHS. These are focused on health and the perceived costs of maintaining the system. However, an interview with the owners of *Whichelo Place* revealed that their chemical system was inexpensive to maintain (£25 per annum) and, apart from an annual flush-through, the system has been reliable (Strube and Strube, 2010). Similarly, *Earthship* Brighton has also proven to be inexpensive to maintain. Both studies demonstrate that chemical greywater systems and biomechanical systems using plants are reliable but, undoubtedly, these systems will need to be proven for much longer to ensure their reliability and gain peoples' confidence. The Market Transformation Programme (2008b) suggests that various financial and regulatory incentives could be used to encourage uptake, including the subsidisation of installation in the form of a rebate, a lower VAT rate, a reduced council tax on the property, reduced water charges and increasing the threshold liable for stamp duty on homes with GwHS installed. Regulatory incentives could be sought through changes to the Building Regulations Approved Document G, Strategic Regional Plans, site-specific regulatory changes, for example the London Assembly Water Plan, and through the Code for Sustainable Homes.

Planning permission can require water conservation methods to be installed (Leggett *et al.*, 2001b), as demonstrated by the development of the bungalow at *Whichelo Place.*

The Code for Sustainable Homes could encourage GwH because household water consumption must be reduced to 80 litres per person per day to achieve excellence ratings of 5 or 6 stars (Department for Communities and Local Government, 2006). The Environment Agency is investigating the operation of smart meters to include energy use, as well as water consumption (Environment Agency, 2009). The influence of the Code for Sustainable Homes on developments could be reinforced by the proposed integration of energy and water measurements in smart meter technology.

13.6 Conclusions

Water demand is reduced significantly by GwH used for toilet flushing; household water use is reduced by approximately one-third. Reliability of water availability for toilet flushing is constant because its source is bathing and showering. It does not depend on the weather as is the case with RwH. However, payback periods are often longer than the lifespan of the GwHS due to low water charges in comparison to the initial outlay and maintenance costs of these systems. Water charges would need to increase and system costs decrease to render this technology financially viable.

The reliability of GwH is unproven as it is a newly used technology. Extensive research into GwH is required by the homeowner to ensure that suitable and reliable systems are installed, although there is now a variety of GwHS available that include simple or complex treatment processes and householders can select a system to suits their needs, including those that require very little maintenance, or use little space in the home.

Communal systems may decrease the overall impact of the installation and maintenance costs, although more systems need to become available, installed and used before this can be determined. Payback periods may not be available to property owners using communal systems because costs recouped from reduction in water and sewerage bills do not exist; tenants normally pay their own water and sewerage bills. However, the communal supply of greywater could be workable, for example, in student accommodation, because water and sewerage bills are usually paid by the owner of the building(s). Installation of more complex systems and regular maintenance regimes in agreement with the system supplier would ensure reliability of communal systems.

Water quality standards for greywater do not yet exist; consumers and suppliers of greywater systems have to rely on the Bathing Waters Regulations 2008 to provide the most relevant guidance. These regulations are designed to protect bathers in cases of accidental ingestion of water. It is less probable that the greywater would be ingested than bathing waters; therefore, the Bathing Waters Regulations 2008 are sufficient for the interim.

Awareness of the environmental benefits is increasing and the Code for Sustainable Homes has prompted implementation of GwHS. It is more straightforward and economically feasible to put GwHS in new builds, as in the case of the bungalow at *Whichelo Place* and *Earthship*. Retrofitting for a GwHS causes disruption due to the necessity of a second pipe for nonpotable water. It would be less disruptive if a GwHS was installed during full refurbishment of a dwelling. Greywater systems must be fitted by a competent person who can follow WRAS guidance for pipe labelling of potable and nonpotable supplies. Adherence to the Water Supply (Water Fittings) Regulations 1999 is required to ensure that the drinking water supply is not contaminated by cross-connection with GwHS and GwHS are not contaminated by backflow from sewers. Consultation with the consumer is required to ensure that they are made aware of how the system functions and its safeguards to avoid contamination.

Most GwHS use more of the household's energy because of mechanical processes including aerating the water to enable microbial cleansing, tank flushing, self-cleaning and use of pumps to transfer the greywater from one place to another within the system. Use of greywater for irrigation does not consume extra energy and reduces energy spent in processing the water at sewerage treatment plants. However, it is not advisable to store the water for long because of bacterial growth, due to the presence of organic material in the water and the ambient temperature at which it is stored.

References

Al-Jayyousi (2003) Greywater reuse – towards sustainable water management. *Desalination*, **156** (1–3), 181–192.

Aquality Trading and Consulting Ltd (2008a) *Greywater Recycling Unit WME–4: Item G17001*. Aquality Trading and Consulting Ltd, London.

Aquality Trading and Consulting Ltd (2008b) *Aquality: Intelligent Water Management: Aqua Recycling Control System with ISB Filtration System*. Aquality Trading and Consulting Ltd, London.

Aquawiser (no date) http://www.aquaco.co.uk/images/Aquawiser-Direct-Grey-Water-Datasheet.pdf [October 2012].

Australian Capital Territory (2007) *Greywater Use: Guidelines for Residential Properties in Canberra*. Publication 07/1380, 24 pp. Available at: www.health.act.gov.au/c/health [October 2012].

Bixler, R.D. and Floyd, M.F. (1997) Nature is scary, disgusting and uncomfortable. *Environment and Behaviour*, **29** (4), 443–467.

BRAC Systems Inc. (2009) *Model RGW–350* [online]. Quebec, Canada http://www.bracsystems.com/ [October 2012].

Brewer, D., Brown, R. and Stanfield, G. (2001) *Technical Note TN7/2001: Rainwater and Greywater Use in Buildings: Project Report and Case Studies*. BSRIA, Bracknell, Berkshire.

Brookshire, D.S., Burness, H.S., Chermak, J.M. and Krause, K. (2002) Western water demand. *Natural Resources Journal*, **42**, 873–877.

Brown, R. and Palmer, A. (2002a) *Technical Note TN6/2002: Water Reclamation Guidance Design and Construction of Systems Using Grey Water*. BSRIA, Bracknell, Berkshire.

Brown, R. and Palmer, A. (2002b) *Technical Note TN7/2002: Water Reclamation Standard: Laboratory Testing of Systems Using Greywater*. BSRIA, Bracknell, Berkshire.

Burkhard, R., Deletic, A. and Craig, A. (2000) Techniques for water and wastewater management: a review of techniques and their integration in planning. *Urban Water*, **2** (3), 197–221.

Butler, D. and Fayaz, A. (2006) *Water Demand Management*. IWA Publishing, London.

Chang, H., Parandvash, G.H. and Shandas, V. (2010) Spatial variations of single-family residential water consumption in Portland, Oregon. *Urban Geography*, **31** (7), 953–972.

Chartered Institution of Building Services Engineers (2002) *Minimising the Risk of Legionnaires Disease*. Chartered Institute of Building Surveyors, London.

CME Sanitary Systems Limited (no date) *The Ecoplay System*. CME Sanitary Systems Limited. http://www.macdee-systems.com/flushing-systems/pdf/ecoplaybrochure.pdf [October 2012].

Combinedharvesters Ltd (2007) *Aquastore – Grey Water Recycling System*. Combined Harvesters, Bromborough, Wirral. http://www.combinedharvesters.co.uk/productcart/pc/viewPrd.asp?idproduct=999&idcategory=0 [October 2012].

Department for Communities and Local Government (2006) *Code for Sustainable Homes: A Step Change in Sustainable Home Building Practice*. Department for Communities and Local Government, London.

Department for Communities and Local Government and Department for Food, Environment and Rural Affairs (2006) *Water Efficiency in New Buildings: A Consultation Document*. Department for Communities and Local Government, London.

Dixon, A.M. and Fewkes, A. (1999) Guidelines for greywater re-use: health issues. *Water and Environment*, **13** (5), 322–326.

Dixon, A., Butler, D. and Fewkes, A. (1999a) Water saving potential of domestic water reuse systems using greywater and rainwater in combination. *Water, Science and Technology*, **39** (5), 25–32.
Dixon A., Butler, D., Fewkes A. and Robinson, M. (1999b) Measurement and modelling of quality changes in stored untreated grey water. *Urban Water*, **1** (4), 293–306.
Droughtbuster UK Ltd (2007) *Watergreen: Move Water with a Watergreen from Droughtbuster UK Ltd.* http://www.reuk.co.uk/Watergreen-Drought-Buster-Product-Review.htm [October 2012].
Eco Open Houses (no date) [homepage] [online]. Eco Open Houses. http://www.ecoopenhouses.org/houses2008/house10.html.
Environment Agency (2006) *SUDS – A Practical Guide. Development Control*, Thames Region. 12 pp. http://www.brent.gov.uk/tps.nsf/Files/LBBA-377/$FILE/SUDS%20A%20Practical%20Guide%20Thames.pdf [October 2012].
Environment Agency (2007) *Conserving Water in Buildings*. Environment Agency, Bristol.
Environment Agency (2008) *Greywater: An Information Guide*. Environment Agency, Bristol.
Environment Agency (2009) *Water for People and the Environment: Water Resources Strategy for England and Wales*. Environment Agency, Bristol.
Eriksson, E., Auffarth K., Henze, M. and Ledin, A. (2002) Characteristics of grey wastewater. *Urban Water*, **4** (1), 85–104.
Freewater UK Ltd (no date) *Use Your Water Twice – It's the Smart Way!* Freewater UK Ltd, Lincoln, 10 pp. http://www.freewateruk.co.uk/pdf/aquacycle.pdf [October 2012].
Friedler, E. and Hadari, M. (2006) Economic feasibility of on-site greywater reuse in multistory buildings. *Desalination*, **190** (1–3), 221–234.
Godfrey, S., Labhasetwar, P. and Wate, S. (2009) Greywater reuse in residential schools in Madyha Pradesh, India – a case study of cost benefit analysis. *Resources, Conservation and Recycling*, **53** (5), 287–293.
Green Alliance (2005) *Better Buildings: Designing for Water Efficiency*. Green Alliance, London.
Griggs, J. and Burns, J. (2009) *Water Efficiency in New Homes: An Introductory Guide for Housebuilders*. National House Building Council Foundation, Amersham, Buckinghamshire.
Health and Safety Commission (2000) *Legionnaires Disease: The Control of Legionella Bacteria in Water Systems*. Her Majesty's Stationery Office, Norwich.
Her Majesty's Government (1999) *Statutory Instrument 1999 No. 1148: The Water Supply (Water Fittings) Regulations 1999*. Her Majesty's Stationery Office, London.
Hewitt, M. (2010) *Personal communication*, Project Manager, Carbon Trust.
Hewitt, M. and Telfer, K. (2007) *Earthships: Building a Zero Carbon Future for Homes*. Building Research Establishment, Bracknell.
Hoffmann, M., Worthington, A.C. and Higgs, H. (2006) Urban water demand with fixed volumetric charging in a large municipality: the case of Brisbane, Australia. *Australian Journal of Agricultural and Resource Economics*, **50** (3), 347–359.
Hunt, D.V.L. and Rogers, C.D.F. (2005) Barriers to sustainable infrastructure in urban regeneration. *Engineering Sustainability*, **158** (2), 67–81.
Jefferson, B., Laine, A., Parsons, S., Stephenson, T. and Judd, S. (2000) Technologies for domestic water recycling. *Urban Water*, **1** (4), 285–292.
Kingspan Environmental (2008) *Ezy-Filter*. Kingspan Environmental. http://www.kingspan.com/kingspangroup/products_solutions/by_sector/self_build_housing/environmental/ [October 2012].
Leggett, D., Brown, R., Stanfield, G., Brewer, D. and Holliday, E. (2001a) *Rainwater and Greywater Use in Buildings: Best Practice Guidance*. Construction Industry Research and Information Association, London.
Leggett, D., Brown, R., Stanfield, G., Brewer, D. and Holliday, E. (2001b) *Rainwater and Greywater Use in Buildings: Decision-Making for Water Conservation*. Construction Industry Research and Information Association, London.
Leggett, D.J. and Shaffer, P. (2002) Buildings that save water – rainwater and greywater use. *Municipal Engineer*, **151** (3), 189–196.
Li, F., Wichmann, K. and Otterpohl, R. (2009) Review of the technological approaches for grey water treatment and reuses. *Science of the Total Environment*, **407** (11), 3439–3449.

Market Transformation Programme (2008a) *BNWAT28: Water Consumption in New and Existing Homes: Version 1.0*. Market Transformation Programme, Didcot.
Market Transformation Programme (2008b) *BNWAT 19: Alternative Sources of Water – Greywater and Rainwater Reuse: Innovation Briefing Note: Version 1.3*. Market Transformation Programme, Didcot.
Marshallsay, D., Trew, J. and Waylen, C. (2007) *Assessing the Cost of Compliance with the Code for Sustainable Homes*. Environment Agency, Bristol.
Melin, T., Jefferson, B., Bixio, D., Thoeye, C., De Wilde, W., De Koning, J., van der Graaf, J. and Wintgens, T. (2006) Membrane bioreactor technology for wastewater treatment and reuse. *Salinisation*, **187**, 271–282.
Memon, F.A., Butler, D., Han, W., Liu, S., Makropoulos, C., Avery, L.M. and Pidou, M. (2005) Economic assessment tool for greywater recycling systems. *Engineering Sustainability*, **158** (3), 155–161.
Mustow, S. and Grey, R. (1997) *Greywater and Rainwater Systems: Recommended UK Requirements: Final Report 14034/2: March 1997*. Building Services Research and Information Association, Bracknell.
Mustow, S., Grey, R., Smerdon, T., Pinney, C. and Waggett, R. (1997) *Water Conservation: Implications of Using Recycled Greywater and Stored Rainwater in the UK: Final Report 13034/1*. Building Services Research and Information Association, Bracknell, Berkshire.
Nolde, E. (2000) Greywater reuse systems for toilet flushing in multi-storey buildings – over ten years experience in Berlin. *Urban Water*, **1** (4), 275–284.
Ottosson, J. and Strenström, T.A. (2003) Growth and reduction of microorganisms in sediments collected from a greywater treatments system. *Letters in Applied Microbiology*, **36** (3), 168–172.
Pidou, M., Memon, F.A., Stephenson, T., Jefferson, B. and Jeffrey, P. (2007) Greywater recycling: treatment options and applications. *Engineering Sustainability*, **160** (3), 119–131.
Po, M., Kaercher, J.D. and Nancarrow, B.E. (2003) *Literature Review of Factors Influencing Public Perceptions of Water Reuse*. CSIRO Land and Water Technical Report 54/03, 44 pp.
Radcliffe, J.C. (2006) Future directions for greywater recycling in Australia. *Desalination*, **187** (1–3), 77–87.
Stephenson, T., Judd, S. Jefferson, B. and Brindle, K. (2006) *Membrane Bioreactors for Wastewater Treatment*. IWA Publishing, 175 pp.
Strube, J. and Strube, A. (2010) Personal communication. Homeowner, *Whichelo Place*, Brighton.
Water Regulations Advisory Scheme (1999a) *Information and Guidance Note No. 9-02-04: Information about Installing, Modifying or Maintaining Reclaimed Water Systems*. Water Regulations Advisory Scheme.
Water Regulations Advisory Scheme (1999b) *Information and Guidance Note No. 9-02-05: Marking and Identification of Pipework for Reclaimed (Greywater) Systems*. Water Regulations Advisory Scheme.
Water Two UK Ltd (2007) *Grey Water Recycling with a Water Two Valve: Water Two Grey Water Recycling System, let us show you how to legally use your hosepipe, during any hosepipe bans or drought orders*. Water Two, Kirkby Stephen, Cumbria. http://www.reuk.co.uk/Water-Two-Greywater-Diverter-Review.htm [October 2012].
Waterwise (2009) *Evaluation of the Water Saving Potential of Social Housing Stock in the Greater London Area*. Waterwise, London.
Water Works UK Ltd (2006a) *GROW*. Water Works UK Ltd. http://www.wwuk.co.uk/grow.htm [October 2012].
Water Works UK Ltd (2006b) *GROW 2*. Water Works UK Ltd. http://www.wwuk.co.uk/grow2.htm [October 2012].
Winward, G.P., Avery, L., Fraser-Williams, R., Pidou, M., Jeffrey, P. Stephenson, T. and Jefferson, B. (2008) A study of microbial quality of grey water and an evaluation of treatment technologies for use. *Ecological Engineering*, **32** (2), 187–197.
Woking Borough Council (2008) *Water Conservation and Recycling: Good Practice*. Woking Borough Council, Woking, Surrey.

14

Inland Waterway Systems – A Solution to Drought and Flooding Issues

Carly B. Rose and Luke Walker

14.1 Introduction

Bending and shaping the landscape to suit the needs of humankind has been a feature of our history for millennia. Just as the growth of agriculture once led to a need for irrigation and drainage channels in the ancient world, the more recent past saw the Industrial Revolution in the United Kingdom drive the creation of reliable transport arteries, initially in the form of canals. The growth and development of this network gave rise to many of the technological advances that would later be used in railway and road construction, as well as being the birthplace of a new profession: Civil (as opposed to military) Engineering.

The transport function of the UK canal system was largely superseded by the newer and more efficient methods of moving goods and people; a new leisure-boating industry based, in part, on the network's heritage value then gave the remaining canals a new lease of life and continues to prompt the restoration of some derelict waterways. As we look towards a future that includes the challenges of climate change, the canals promise to offer innovative solutions to some of the problems we face, in terms of droughts, floods and heatwaves. The story of rise, fall and resurgence of the inland waterways looks set to continue for many years to come.

14.2 The Past

Humankind has been creating artificial channels for the conveyance of water for millennia: these range from the drainage systems of the Bronze Age Indus valley civilisation to those used for irrigation purposes, such as the Qanat system in ancient Persia, the similar Turpan system in ancient China and irrigation canals in Peru (Burke, 2009). The first known 'artificial river', the Grand Canal, intended for navigation purposes, was constructed in China

Water Resources in the Built Environment: Management Issues and Solutions, First Edition.
Edited by Colin A. Booth and Susanne M. Charlesworth.

around the 5th century BCE as a military supply route; this waterway also includes the first recorded use of the pound lock in the 10th century, to safely overcome height differences in the terrain it crossed (Gascoigne, 2001). The summit section of this canal, however, suffered from water shortage problems for many centuries, highlighting the fact that artificial waterways have always needed methods of obtaining, and retaining, water for optimal functioning.

The Romans, although better known for their water supply infrastructure, also created some canals for navigation in the 1st century CE, including one connecting the Rhine and the IJssel in the Netherlands; they are also believed to have been responsible for the first artificial waterway in Britain, the Foss Dyke, which joins the Rivers Trent and Witham (British Waterways, no date a). In the United Kingdom, navigable rivers were the predominant trading routes for many centuries, as the roads, usually unpaved, could be impassable in winter due to deep mud and equally problematic in summer when the deep wheel ruts would be baked hard. The long-distance turnpike roads did not become practical alternatives until major improvements were begun in the 18th century and, even then, water transport was more efficient for heavy goods (Rosevear, 2008). Rivers can, however, be subject to low water levels, particularly in the summer months; to avoid interruption to trade, this meant laden craft had to be manually towed. For example, it could take around 50 men (or, at a later date, 10–12 horses) to haul a fully laden barge upstream on the River Thames. Artificial waterways could, therefore, offer a more reliable alternative, provided adequate water supply systems were in place.

14.2.1 Early Canals

Some authorities suggest that the earliest post-Roman canal in Europe was the Fossa Carolina, created to link the Rhine and Danube catchments in the early Middle Ages; recent archeological work has suggested, however, that in design terms this may have been an interim stage, consisting of a chain of linked ponds, rather than a continuous canal (Leitholdt *et al.*, 2012). A short section of the Stecknitz Canal in present-day Germany, built in the late 14th century to serve the salt trade by connecting the Trave and Elbe catchments, is the first documented canal as the term is now generally understood. Others followed, for example building of the Brussels–Scheldt Maritime Canal in present-day Belgium was commenced in 1550 and the Canal de Briaré (France), specifically for grain transport, opened in 1642, linking the rivers Loire and Seine (United Nations Educational, Scientific and Cultural Organization, no date). The European country perhaps most strongly linked with canals, the Netherlands, also embarked upon its canal-construction programme in the 17th century (the earlier works being drainage-related, rather than for navigation purposes).

All the aforementioned waterways obtained their water supplies from the rivers that they connected, but this was not the case with the Canal du Midi (France), constructed between 1667 and 1681; this canal provided a shortcut between the Bay of Biscay and the Mediterranean by connecting the navigable River Garonne at Toulouse with Sète on the southern coast (United Nations Educational, Scientific and Cultural Organization, no date). It was the first canal to use a series (or 'flight') of locks to allow craft to climb up and down hills in order to cross a watershed (Creme de Languedoc, 2007); this method, however, meant that water would be lost from the summit level in both directions each time a vessel made the passage. Not only did this canal not have a ready-made water supply but it also passed through an area prone to low rainfall and, therefore, novel solutions to the water supply problem were required. The engineer responsible for the project designed a feeder canal, to convey water from the mountains near Carcassonne, as well as creating the first artificial reservoir to supply the summit level of a canal (United Nations Educational, Scientific and Cultural Organization, no date). Within a century, however, an increase in traffic meant additional supplies of water were required and a second reservoir had to be built; thus, even a well-designed canal can sometimes face water shortages, as will be discussed later in this chapter.

The Newry Canal (Ireland) was constructed by architect Richard Cassels' who had studied navigation in both England and the Netherlands; it was the first summit level canal in the British Isles, being completed in 1742 (Gladwin, 1988). The idea was not taken up in the rest of the United Kingdom for almost two decades, but it was then enthusiastically adopted, with engineers developing a high level of expertise driven by the needs of the industrial revolution.

14.2.2 Canal Growth in the Industrial Revolution

During the second half of the 18th century, good transport links became increasingly important, both for moving raw materials to the factories and for delivering finished products to the consumer. In England, the navigable rivers did not serve the needs of the new industrial economy, as they failed to connect the industrial areas of the north and the Midlands with either the south or the ports through which goods could be exported; railway development was not yet widespread and viable road networks were still in their infancy. Despite Cassels earlier success in Ireland, canals did not become widespread in England until the Duke of Bridgewater, who owned coal mines in the north-west of England, recognised the value of building a short canal to link his mines with a navigable river, which led directly to the primary market for his coal: the major industrial city of Manchester. The Bridgewater Canal opened in 1761 and provided a catalyst for a century of canal building, as other industrialists adopted the idea; during the period of 'canal mania' between 1840 and 1850 the system rapidly expanded, nearing 4000 miles in length (Burton and Pratt, 2002). The technology and expertise developed by the leading canal engineers were also in demand elsewhere in the world: Thomas Telford drew up the original plans for the Göta Canal in Sweden, which opened in 1832 (Rolt, 2007) and James Brindley's eponymous nephew was responsible for many early canals in the United States (Kapsch and Long, 2011).

The UK inland waterway system was, for two centuries, the equivalent of the modern motorway network for the widespread conveyance of goods, estimated to exceed 30 million tonnes per year at its peak (Hadfield, 1964). It was unable to compete, however, with other transport methods that overtook it in the 20th century, due to a failure to modernise; the European system had a very different experience, as will now be discussed.

14.3 The Present – Canals in the 20th Century

14.3.1 The UK Experience

The commercial demise of UK canals was brought about initially by the expansion of the railways, which offered superior capacity and speed for passengers as well as for freight. From the early 20th century, when the canals were still in private ownership, some waterways began to be abandoned, due to falling traffic and the consequent lack of revenue for maintenance and repair purposes; the main canal network, however, continued to carry a substantial amount of freight until the 1950s, when competition from road transport began to have adverse impacts upon both the waterway and (somewhat ironically) the railway systems alike (Hadfield, 1964).

Although modernisation of the waterways could have mitigated these problems, the majority of UK canals had been built to a narrow gauge specification: the infrastructure was designed for vessels less than two metres wide and with a draught of just over one metre. During the 1930s, with the aid of a government grant, the Grand Union Canal Company embarked upon a scheme to deepen and widen the northern sections of the major Midlands–London link in order to accommodate vessels of a size that would have been more cost-effective. Government funding was not, however, forthcoming for the initiative to be applied to the entirety of that canal, severely limiting the effectiveness of the initial grant; nor were other canal companies successful in obtaining such funds (Paget-Tomlinson, 2006, p. 128). In some areas of the country, water supply problems imposed an additional constraint, militating against the creation of larger lock chambers by the private companies involved. Nationalisation of both the waterways and railways in 1948 saw attempts to increase commercial traffic, but this met with a lack of success and in 1963 the British Waterways Board formally ceased their commercial canal carrying operations, although some independent operators continued to trade under license (Paget-Tomlinson, 2006).

14.3.2 The European Experience

The European approach, where much of the canal system was subjected to modernisation and development via public funding, contrasts markedly with the UK experience: in France, for example, upgrading to the standard Freycinet gauge of 300 tonnes had commenced in 1879 (Inland Waterways International, 2010). The interconnected nature of the European waterways aided the success of such initiatives: in 1954, the European Council of Transport Ministers set minimum standards for waterways of international importance, in order to create a coherent network (Jordan, 1988). This led, in the following decade, to the canalisation of 275 km of the Moselle River linking Germany, Luxembourg and France and, in more recent times, to the construction of the Strépy–Thieu boat lift in 2002 on the Canal du Centre (Belgium) to permit usage by craft of 1350 tonnes (Directorate General for Waterways –Ministry of Equipment and Transport (Wallone Belgium), no date).

14.3.3 Resurgence, Restoration and Regeneration

In the United Kingdom in the 1960s, whilst the waterborne freight industry was declining, there was a surge of interest in pleasure boating on the canals with leisure users keen to enjoy the experience of peace and remoteness. This eventually gave rise to new industries, offering canal holidays and building craft designed specifically for pleasure boating purposes; a similar shift also occurred on some of the lesser used European canals, the Canal du Midi (France) and the Göta Canal (Sweden) being examples (Jordan, 1988). The income from tourism can be an important source of revenue, as well as providing employment in rural areas in need of economic regeneration: in recent years, therefore, European Union funding has been made available for the restoration of previously abandoned waterways, with the Droitwich Canal (UK) and some French canals having benefited from such support (EU INTERREG VNE Canal Link, 2007; Blue Links, 2009). An added benefit, from an environmental perspective, is the creation of green 'wildlife corridors' reaching into the hearts of towns and cities, thus enhancing biodiversity.

The change from freight to leisure use does, however, have implications for water level management, including the maintenance regimes to be applied to the waterways.

14.3.4 Water Level Management

Water levels need to be maintained as appropriate for the draught of vessels using a given waterway: laden freight craft require a greater depth of water than leisure craft, which necessitates regular (costly) dredging to remove silt and other debris in order to retain the design profile. It is, therefore, possible to maintain waterways used solely by leisure craft at less expense than commercial navigations: for example, the Canal and River Trust (hereafter C&RT), the body responsible for the majority of waterways in England and Wales, adopts a risk-based approach by defining 'minimum open channel' dimensions for each of its navigations and prioritises dredging works accordingly (Holland, 2011). The C&RT monitor and control the water in the canals within their remit via a sophisticated telemetry system (SCADA); this governs flow (lock bypasses and reservoir outflows) and levels in reservoirs and pounds (British Waterways, 2005). Control of pumps and sluices are also automated; excess water is drained off via sluice gates, overflow weirs and spillways (typically set to allow around 20 cm of freeboard to the top of the banks). Where there are planned changes of usage to waterways, both the water level management and maintenance regimes may, therefore, need to be reviewed and amended accordingly: this can pose a challenge to initiatives designed to reintroduce waterborne freight. An example is provided by the plan to transfer waste and construction materials by water during the creation of the Olympic Park for the London 2012 Games (Olympic Delivery Authority, 2012).

Canals rarely pose a flood risk in themselves, provided structures and equipment function correctly; the chief exceptions are where breaches occur (due to structural failure, for example embankments weakened by burrowing animals) or in the event of damage or vandalism, typically where mechanisms are triggered accidentally or maliciously (for example, BBC News, 2010). Where such escapes do occur, or where access is required for planned maintenance, the water is usually controlled by means of 'stop planks' (effectively a temporary dam) inserted into pre-formed vertical channels in the canal structures. Some canals may, however, be overwhelmed by flooding from adjacent rivers or streams, particularly during extreme weather events (Pennine Waterways, 2012).

Waterborne transport can be environmentally beneficial, efficient and cost-effective for appropriate cargoes; in the light of climate change predictions, governments are increasingly keen to examine methods that can offer reduced environmental impacts (e.g. Blaauw, 2008; European Transport Forum, 2010) and these issues will now be discussed in more detail.

14.3.5 Carbon Emissions and Climate Change Issues

An analysis undertaken for the UK Government's Commission for Integrated Transport of CO_2 emissions (McKinnon, 2010) demonstrated that the movement of freight on inland waterways is relatively energy efficient: it generates around 35 gm of CO_2 per tonne-km, compared to the weighted averages of 14.5 gm for rail freight, 200 gm for road freight and in excess of 1500 gm for air freight. A report by the Inland Waterways Advisory Council (Inland Waterways Advisory Council (IWAC), 2008, p. 3) stated that: '... waterborne freight transport could make a useful contribution towards meeting the UK Government's commitment to reducing carbon emissions by 60% by 2050.'

The UK government has supported such initiatives via financial incentives such as the Waterborne Freight Grant scheme (Department for Transport, 2010). This provided assistance to companies with the operating costs associated with running water freight transport, where the latter was more expensive than the road-based alternative. As an example, Lille is one of the most important inland ports in France, with over a million tonnes of waterborne freight passing through the port annually; given adequate infrastructure, transport of containers by barge over relatively short distances can be successful and could be adopted in the United Kingdom (IWAC, 2008). The European Union funded 'Creating' project looked at optimal technical solutions and innovative ship designs to strengthen the position of inland shipping within the logistics chain (Blaauw, 2008; European Commission Research and Innovation (Transport Projects), no date).

Water transport is suitable for a variety of nonperishable cargoes: steel and other metal products, forest products and bulk cargoes such as grain, aggregates, coal, petroleum products, chemicals, waste and cement (e.g. Commercial Boat Operators Association, no date; Inland Waterways Association, 2010) as well as indivisible abnormal loads on some of the larger canals and river navigations (British Waterways, 2003). Whilst unable to rival the European model, within the limitations of the existing UK network there are niche opportunities to be exploited, where the origin and destination of goods are both situated on waterways. An example is the supermarket chain Tesco, which began to use barges on the Manchester Ship Canal to transport wine from Liverpool in 2007; it is estimated that this has taken 50 lorries off the roads each week and has reduced the associated carbon emissions by 80% (Tesco, no date).

Some unusual nontransport uses for the existing canal system and its infrastructure have also been developed, such as the installation of a 400 mile fibre-optic network under canal towpaths (Sim, 2003); not only did this offer the advantage of a reduced risk of the cables being damaged by other utilities, compared to highway installation, but also conveyance of construction materials and subsequent maintenance activities could be conducted using waterborne transport, again reducing emissions (Wood Hall and Heward Engineering, no date). Another example is the use of canal water to cool waterside buildings, as an alternative to traditional air conditioning: it is reported that a multinational company with headquarters in London saved over £100k and 276 tons of carbon per year by switching to the new method (Linden Environmental, 2012).

Whether providing waterborne transport, leisure activities or these more novel uses, canal systems, being artificial waterways, require adequate water supplies on a continual basis. A wide variety of methods has been developed to achieve this, and these will be examined in detail in the next section.

14.4 Sourcing and Conserving Water Supplies

14.4.1 Sources of Supply

Canals have been designed with, or undergone post-construction, adaptations to make use of, many different sources of water:

1. Natural sources, at or above the summit level:
    - From natural lakes, such as Lough Shark on the Newry Canal (Northern Ireland) (Paget-Tomlinson, 2006).
    - From natural rivers, such as the River Dee (UK), which supplies both the Llangollen Canal (5 km/3 miles distant) and, fed by gravity alone, the Shropshire Union Canal at Hurleston Junction (66 km/41 miles distant) (Fisher, 2009).

- Feeders, such as that from the mountains near Carcassonne (20 km/12.5 miles distant) supplying the Canal du Midi (France) (United Nations Educational, Scientific and Cultural Organization, no date).

2. Dedicated reservoirs, at or near the summit level:
   - The Bassin de Saint-Ferréol on the Canal du Midi (France), created by damming the Laudot River (United Nations Educational, Scientific and Cultural Organization, no date); Bosley reservoir supplying the Macclesfield Canal (UK), which is fed by a total of 13 streams that drain from the surrounding hills (Macclesfield Canal Society, no date).
3. Water pumped from natural sources below the summit level:
   - Crofton pumping station on the Kennet and Avon Canal (UK) lifts water from a lake, Wilton Water (Burton and Pratt, 2002).
4. Input from mine pumping:
   - The Bridgewater Canal (UK) originally used water drained from the Worsley mines (Paget-Tomlinson, 2006); the Birmingham Canal Navigations (UK) benefitted from numerous mine water supplies, including Bradley, Deepfields and Stow Heath (Shill, 2006).
5. Input from treated effluent:
   - Barnhurst Sewage Treatment Works, Wolverhampton (UK) supplies both the Shropshire Union and the Staffordshire and Worcestershire Canals (Severn–Trent Water, 2005) (although such sources need to comply, in Europe, with the Water Framework Directive, 2000). A similar arrangement applies to the restored Canal de Roubaix, which uses water from the Wattrelos waste water treatment plant after it has passed through a reed-bed filtration system (The Barge Association (DBA), 2012).
6. Input from land drainage and storm drainage:
   - The Gloucester and Sharpness Canal (UK) makes use of both natural feeder streams (River Cam and River Frome) but other inputs to the canal come from smaller watercourses, land drains and stormwater drains in Gloucester (British Waterways, no date b).
7. Groundwater abstraction:
   - Groundwater from a new borehole supplies the restored Ashby Canal (UK) at Moira (Sutton, 2005).

14.4.2 Water Conservation

The major water losses from a canal are from seepage and evaporation, depending upon the construction methods and surface area/ambient temperature respectively; in England and Wales, for example, the top 20 cm of the canal lining poses the greatest risk of leaks, as it is constantly wetted then dried, which compromises its structural integrity. The activities of burrowing animals, together with the action of the wash from passing boats, add to the stresses on the lining (Waterscape, 2012). An additional challenge to water conservation arises from lock use: conventional lock design means that each descending boat takes a lock full of water from the summit level (upper pound) down to the sump level (trough pound) where any excess water will have to be discharged to avoid flooding. Typically, a narrow lock in the United Kingdom uses around 140 cubic metres of water each time a boat descends; volumes are, of course, greater for the wider lock chambers on the 'broad'

Figure 14.1 Water storage area (to the right) situated next to each lock chamber (black and white gated structures on left) at the Caen Hill flight, Devizes (UK). Copyright Chris Talbot and licensed for reuse under the Creative Commons Licence.

canal network, these being designed to accommodate a pair of standard narrowboats or wide-beam vessels (Waterscape, 2012). Another type of water loss occurs even when no boats are using a lock, as there is usually a residual flow from the upper level; this is either routed around the chamber by means of a bypass channel (termed a bywash) or flows through the chamber itself, via leakage through the lock gates themselves. The latter problem is exacerbated where gates have sustained damage from inexpertly steered boats.

Where canals lack reliable supplies from major rivers, various methods have been developed to conserve water, ranging from structural and mechanical solutions to changes in custom and practice for those using the system:

1. Structural remedies for use with locks
   - Storage within modified canal pounds, such as those constructed on the lock flight at Devizes (UK). This rises over 70 m in just over 3 km, with the intervals between adjacent locks being, of necessity, very short. Each of the 16 locks was built with, in effect, its own reservoir (the intervening pounds being extended sideways) to store the water needed for its operation (Figure 14.1).

Figure 14.2 Twinned lock chambers, Camden, London (UK). Copyright Alan Murray-Rust and licensed for reuse under the Creative Commons Attribution-ShareAlike 2.0 license.

- Side ponds and related techniques, which employ connecting culverts to direct the water emptied from a lock into a separate storage area, or 'water saving basin' (e.g. Canal World, 2012). This area is then drained to partly fill the chamber for use by a boat climbing the flight; this method is also useful where the pounds between lock chambers are exceptionally short. A variation uses paired, or twinned, locks with a connection between the two, again to allow water being drained from one chamber to partly fill the other (Figure 14.2).

2. Mechanical remedies:
   - Back-pumping schemes, whereby water is recycled back to the top of a flight of locks using a pump or a sequence of pumps at each lock, an example being that on the Canal de Roubaix (The Barge Association (DBA), 2012). This is often necessary on restored canals where the original water sources are no longer available, or are inadequate, but can prove costly. The scheme installed at Foxhangers Wharf to serve the 29-lock Caen Hill flight as part of the restoration of the Kennet and Avon Canal cost £1 million in 1996 (Lindley-Jones, 2002).
   - Boat lifts, as an alternative to locks, whereby boats can be raised and lowered. These have the advantage of saving water by using a water-filled tank, or caisson, to transfer boats between levels. An example is the iconic Falkirk Wheel (Figure 14.3), which replaced a flight of 11 locks to overcome a height difference of 35 m; each rotation takes around 20 minutes, compared to the two hours that would have been required to negotiate the lock flight. As well as being a functional structure on a newly restored canal, this was also designed to be a major tourist attraction, providing economic regeneration in the local area (Falkirk Wheel, 2012).

Figure 14.3 The Falkirk Wheel, Scotland (UK), with the visitor centre on the left.

- Inclined planes may also use water-filled tanks, often in pairs acting as counterweights; others lift boats out of the water entirely, transporting them on wheeled cradles (sometimes termed marine railways). These methods also offer much shorter transit times than are possible with a sequence of locks; an example is the Strepy–Thieu lift on the Canal du Centre (Belgium), which was constructed in the 1960s as part of the work necessary to bring the Belgian network up to the European 1350-ton barge standard.

3. Custom and practice. The standard practices that applied to the UK canal system when commercial freight was the norm and waterway staffing levels were far higher have now been modified, to reflect the predominant usage by the leisure boating community who are perceived as being less adept at managing complex mechanisms:
   - Side pond use has largely been discontinued, or retained only where experienced staff members are available to operate the mechanisms, as incorrect operation by inexperienced users can cause localised flooding.
   - Sharing broad locks. Some lock chambers were built to take two standard narrowboats side by side.
   - Anti-vandal devices. In some urban areas there have been repeated instances where locks have been deliberately misused in order to drain the water from canals. To combat such unauthorised use of lock mechanisms, anti-vandal devices may be fitted, requiring the use of a special key, which can only be purchased from the relevant authorities.

The canal systems of both the United Kingdom and Europe have seen major changes since their inception, and this shows no sign of abating. As we move forward with the challenges of climate change in mind, canals may be able to offer yet more additional functions and services over and above those envisaged by their original architects. We will conclude by considering some of the projected developments in this area.

14.5 A Climate Resilient Future

To develop climate resilient infrastructure in the most cost-effective manner, designs can incorporate two or more functions: road and rail embankments can act as flood defences and reservoirs can be used for flood control as well as water storage (Secretary of State for Environment, Food and Rural Affairs, 2011). Canals can also have additional uses over and above, or in some cases as an alternative, to transport usage: for example, the Tavistock Canal (UK) is used to convey water for a hydropower scheme at Morwellham Quay (British Hydropower Association, no date). Other climate resilience uses will now be considered.

14.5.1 Flood Alleviation

Canals can, in some instances, provide flood flow routes or storage (Dun, 2005). An example is the Gloucester and Sharpness Canal (UK), which provides some flood storage to attenuate peak flows on the River Severn at Gloucester (Gloucester Docks and Sharpness Canal, 2012), and also the Flood Risk Assessment for the proposed restoration of the Droitwich Canal (UK) predicted that alleviation of flood risk for over 30 properties (at a 1/100 return period) would result (Halcrow Group Limited, 2006). Similarly, the Dahme Flood Relief Canal (Germany) diverts water from the upper reaches of the River Spree, but is also a navigable channel.

One of the key challenges for the United Kingdom in the coming decades will be the management of both water supplies and flood risk. Water shortages are expected to be exacerbated by climate change (UK Climate Impacts Programme, 2009; Department for Environment, Food and Rural Affairs (DEFRA), 2011; Hall *et al.*, 2012), but it is anticipated that these effects will impact more upon the south, particularly the south-east, of the country in the summer months. In contrast, the north-west is likely to experience additional rainfall in winter, with concomitant increases in flood risk. This combination of circumstances has led to renewed interest in the concept of water transfer between river basins.

14.5.2 Water Transfers

Proposals for a piped national water grid in the United Kingdom have been put forward in the past, but were ruled out on the grounds of cost and practicality. There are precedents for the use of regional water grids, however: pipeline transfers already exist between Wales and England, with the Elan Valley dams supplying Birmingham and the Vyrnwy Reservoir supplying Liverpool. Pipelines are not the only means of transferring water over distance, however, and making use of existing infrastructure such as the canal network may provide a more cost-effective solution in some instances. For example, the Bridgwater and Taunton Canal has conveyed water to Taunton via the Durleigh Reservoir since 1962, by transferring water from the River Tone to the River Parrett (Canals and Waterways Roots and Routes, no date); likewise, the Llangollen Canal has been used to transfer supplies from the River Dee to the Nantwich area, via the Hurleston Reservoir, since 1955.

Transfers of water from the River Severn to the River Thames have been examined in the past, though none has been pursued (e.g. Cascade Consulting, 1992; Sheriff *et al.*, 1996; Ringham *et al.*, 1996). More recent analyses of water shortages in the south-east of England have differed in their conclusions; for example, Rodda (2008, p. 125) asserted that: 'A transfer of water from the lower Severn into the upper Thames must be the first step in augmenting resources.'

The Environment Agency, however, took the opposite view, with a report stating (Environment Agency, 2006, p. 20): 'The cost and environmental impact mean that large scale transfers of water from the north of England or Wales to the south east are unnecessary and inappropriate.'

The latter report was, however, predicated on the creation of a number of new reservoirs, in particular a very large (150 billion litres) one at Abingdon, Oxfordshire; proposals for the latter were submitted by Thames Water as part of its Water Resources Management Plan (WRMP) but this was subsequently rejected by the relevant Government Inspector (Burden, 2011; Spelman, 2011). Part of the criticism of the WRMP was that all possible alternatives had not been adequately considered: one option had been a proposal for water transfers from the River Severn via a combination of pipes and the Cotswold Canals (which are in the process of being restored). The estimated cost of the latter was reported as around £330 million compared with £1 billion for the reservoir. The Inspector's report includes (Burden, 2011, p. 275) the recommendation that: '... (the) scheme using the Cotswold Canals be added to the feasible options list.'

Since this development, it seems the policy-making sector may be showing signs of warming to the idea of water transfers: for example, at a conference in October 2011, it was stated that the Environment Agency was 'open-minded' to the future movement of water over long distances, including schemes from the River Severn to the Thames, and wished to see all the potential options assessed (Bishop, 2011). During a period of severe drought in early 2012, this was further underlined in a statement from the head of water resources at the Environment Agency (Gray, 2012):

> Water companies are starting to plan to see if there are any prudent actions that can be taken, which include further transfers of water between companies and even from river basin to river basin. Water takes a lot of energy to transfer, so it won't be something that happens every day, but it can be very useful for meeting drought demands. There are risks, however, such as transferring invasive species and changing river chemistry, so we have to weigh this up.

It remains to be seen how this situation will develop in the future, with the needs of consumers and the requirement to protect the natural environment potentially in opposition. The canal systems across Europe have already undergone many changes since their original construction as transport arteries. Their future may, however, lie in their ability to provide assistance in attaining climate resilience.

14.6 Conclusions

The creation of artificial waterways to convey water itself, people and cargoes has been a key element in the development of humankind and our current environment. From our early agriculturally driven needs to the construction of extensive waterborne transport networks, the means whereby the natural world could be modified and its geographic challenges overcome have developed inexorably.

Today's Civil Engineers can look back to the Industrial Revolution for the inception of their profession's expertise in large-scale construction projects motivated by aims other than the needs of the military and the clergy. The variety of methods developed for water supply and conservation purposes is also illustrative of the inventiveness of the human mind in the face of problems.

Canals, as a transport system, may have major advantages compared to natural watercourses but have not always been able to compete with more modern methods. Their success varies from country to country, however, as the continued use of the European waterways for freight demonstrates. The leisure boating industry is, currently, the principal alternative use that has been applied within the United Kingom, but other opportunities have been developed in recent years, notably the use of towpaths for accommodating the fibreway network and, where viable, providing a means of flood alleviation. Other innovative usages continue to arise, many driven by the need to adapt to climate change. The changing face of canal usage may provide an incentive for further restoration as well as new construction projects in the future.

Canals have provided the solutions to some of our past and present problems. There is the potential for them to continue to serve us for many years to come.

References

BBC News (2010) *Water Floods A38 as Vandals Target Birmingham Canal* [online]. Available at: http://www.bbc.co.uk/news/10486116.

Bishop, T. (2011) Water transfers: the Environment Agency's view. In: *Engineering the Future – Water Security Challenges – Is Water Transfer the Answer?* Institution of Mechanical Engineers, One Birdcage Walk, Westminster, 25 October 2011.

Blaauw, H.G. (2008) *CREATING Project No. FP6–506542 (Concepts to Reduce Environmental Impact and Attain Optimal Transport Performance by Inland NaviGation)*, 19 May 2008.

Blue Links (2009) *Blue Links, A Leading EU-Funded Project for Sustainable Canal Restoration* [online]. Available at: http://www.bluelinks2008.org/default.html.

British Hydropower Association (no date) *Hydro in the UK – Installations – Morwellham Power Station* [online]. Available at: http://www.british-hydro.org/installations/m/morwellham_power_station.html.

British Waterways (2003) *Freight on our Waterways*. British Waterways, Watford.

British Waterways (2005) *Managing Water*. Waterway User and Special Interest Group (WUSIG) – Central Shires, Tamworth.

British Waterways (no date a) *History of the Fossdyke Navigation* [online]. [Accessed 01/11/2009]. Available at: http://www.waterscape.com/canals-and-rivers/fossdyke-navigation/history.

British Waterways (no date b) *Gloucester and Sharpness Canal* [online]. Available at: http://www.britishwaterways.co.uk/south-west/environment-and-heritage-projects/gloucester-and-sharpness-canal.

Burden, W.J. (2011) *Water Resources Management Plan Regulations 2007 – Inquiry into the Thames Water Revised Draft Water Resources Management Plan 2010–2035, September 2009*. The Planning Inspectorate, London.

Burke, E.I. (2009) Islam at the center: technological complexes and the roots of modernity. *Journal of World History*, **20** (2), 165–186.

Burton, A. and Pratt, D. (2002) *The Anatomy of Canals – The Mania Years*. Tempus Publishing Ltd, Stroud.

Canals and Waterways Roots and Routes (no date) *Bridgwater and Taunton Canal* [online], [Accessed 27/05/12]. Available at: http://www.canals.btinternet.co.uk/canals/bridgwatertaunton.htm.

Canal World (2012) *Droitwich Canals Restoration: Hanbury Locks, Side Pound Operation* [online]. Available at: http://www.canalworld.net/forums/index.php?showtopic=35354.

Cascade Consulting (1992) *Assessment of Alternative Schemes for Draft Final Water Resources Management Plan – Severn–Thames Transfer: Longdon Marsh A*. Thames Water Utilities Ltd, Reading.

Commercial Boat Operators Association (no date) *Small Canals* [online], [Accessed 17/07/12]. Available at: http://www.cboa.org.uk/small-canals.html.

Creme de Languedoc (2007) *The Canal du Midi – Origin and History of the Canal du Midi* [online], [Accessed 14.06.12]. Available at: http://www.creme-de-languedoc.com/Languedoc/sightseeing/canal-du-midi.php.

Department for Environment, Food and Rural Affairs (DEFRA) (2011) *Water for Life*. DEFRA, London.

Department for Transport (2010) *Grant Funding to Transport Freight by Rail and Water* [online]. Available at: http://www.dft.gov.uk/topics/freight/grants.

Directorate General for Waterways – Ministry of Equipment and Transport (Wallone Belgium) (no date) *Canal du Centre – Strepy–Thieu Lift* [online], [Accessed 28 April 2012]. Available at: http://www.canalducentre.be/Guide/Sites/En/strepythieu.html.

Dun, R.W. (2005) *The Hydraulic Response of Canals to Water Transfers – Final Report – March 2005*. British Waterways, Watford.

Environment Agency (2006) *Do We Need Large-Scale Water Transfers for South East England?* Environment Agency.

EU INTERREG VNE Canal Link (2007) *Final Report – France* [online]. Available at: http://www.vne-waterways.eu/fileadmin/media/publications/071213VN_Final_Report_Classification_Chapter_France.pdf.

European Commission Research and Innovation (Transport Projects) (no date) *'CREATING' Shifting Cargo from Road to Water* [online], [Accessed 28 April 2012]. Available at: http://ec.europa.eu/research/transport/projects/items/_creating_shifting_cargo_from_road_to_water_en.htm.

European Transport Forum (2010) *Why the EU Wants More Water Freight* [online], [Accessed 28 April 2012]. Available at: http://europeantransportforum.eu/transport-efficiency/6-inland-waterways/88-antwerpfreight.html.

Falkirk Wheel (2012) [online]. Available at: http://www.falkirk-wheel.com/faq/32-how-long-is-the-trip-on-the-falkirk-wheel.

Fisher, S. (2009) *Canals of Britain – A Comprehensive Guide*. Adlard Coles Nautical, London.

Gascoigne, B. (2001) *History of Canals* [online], [Accessed 6 May 2012]. Available at: http://www.historyworld.net/wrldhis/PlainTextHistories.asp?groupid=141&HistoryID=aa19&gtrack=pthc#ixzz1u6tc3Nwc.

Gladwin, D. (1988) *Building Britain's Canals*. Brewin Books Ltd, Studley.

Gloucester Docks and Sharpness Canal (2012) *History* [online]. Available at: http://www.gloucesterdocks.me.uk/studies/twofloods.htm.

Gray, R. (2012) Rivers to be linked to ease drought under water company plans. *The Daily Telegraph* [online], 1 April 2012, [Accessed 4 April 2012]. Available at: http://www.telegraph.co.uk/earth/drought/9178003/Rivers-to-be-linked-to-ease-drought-under-water-company-plans.html.

Hadfield, C. (1964) *Canals of the World*. Basil Blackwell, Oxford.

Halcrow Group Limited (2006) *Droitwich Junction and Barge Canals Restoration Flood Risk Assessment*. British Waterways, Watford.

Hall, J.W., Watts, G., Keil, M., de Vial, L., Street, R., Conlan, K., O'Connell, P.E., Beven, K.J. and Kilsby, C.G. (2012) Towards risk-based water resources planning in England and Wales under a changing climate. *Journal of the Chartered Institution of Water and Environmental Management*, **26** (1).

Holland, G. (2011) *Dredging Prioritisation Paper (WUSIG Meeting 13 April 2011)*. British Waterways, Watford.

Inland Waterways Advisory Council (IWAC) (2008) *Decreasing Our Carbon Footprint – Moving More Freight onto the Inland Waterways of England and Wales*. IWAC, London.

Inland Waterways Association (2010) *Advantages of Waterways Freight* [online]. Available at: https://www.waterways.org.uk/activities/freight/advantages_of_freight.

Inland Waterways International (2010) *World Canals* [online]. Available at: http://www.worldcanals.com/.

Jordan, T.G. (1988) *The European Cultural Area – A Systematic Geography*, 2nd Edition. Harper & Row Publishers Inc., New York.

Kapsch, R. and Long, Y. (2011) James Brindley, American canal engineer. *International Journal for the History of Engineering and Technology*, **81** (1), 22–59.

Leitholdt, E., Zielhofer, C., Berg–Hobohm, S., Schnabl, K., Kopecky–Hermanns, B., Bussmann, J.W., Hartling, J., Reicherter, K. and Unger, K. (2012) Fossa Carolina: the first attempt to bridge the central European watershed – a review, new findings, and geoarchaeological challenges. *Geoarchaeology: An International Journal*, **27**, 88–104.

Linden Environmental (2012) *The Use of Canal Water to Cool Waterside Buildings* [online], [Accessed 27/05/12]. Available at: http://www.lindenenvironmental.co.uk/News/1/Linden-to-be-reseller-for-British-Waterways/.

Lindley-Jones, P. (2002) *Restoring the Kennet and Avon Canal*. Tempus, Stroud.

Macclesfield Canal Society (no date) *The Macclesfield Canal – Reservoirs and Feeders* [online], [Accessed 7 May 2012]. Available at: http://www.macclesfieldcanal.org.uk/reservoirs.htm.

McKinnon, A. (2010) *CO_2 Emissions from Freight Transport: An Analysis of UK Data* [online]. Available at: http://www.greenlogistics.org/SiteResources/d82cc048-4b92-4c2a-a014-af1eea7d76d0_CO2%20Emissions%20from%20Freight%20Transport%20-%20An%20Analysis%20of%20UK%20Data.pdf.

Olympic Delivery Authority (2012) *Transport of construction materials by sustainable means*. London: Olympic Delivery Authority.

Paget-Tomlinson, E. (2006) *The Illustrated History of Canal and River Navigations*. Landmark Publishing Ltd, Ashbourne.

Pennine Waterways (2012) *Waterways News* [online], [Accessed 25/06/2012]. Available at: http://waterwaynews.blogspot.co.uk/2012/06/rochdale-canal-remains-blocked-after.html.

Ringham, J.E., Walker, S. and Wyatt, T. (1996) Evaluation of Lake Vyrnwy within the integrated water supply systems of North West Water. *Water and Environment Journal*, **10** (2), 105–110.

Rodda, J.C. (2008) How can water resources in South East England be sustained? In: *British Hydrological Society – 10th National Hydrology Symposium: Sustainable Hydrology for the 21st Century*, University of Exeter, 15–17 September 2008. British Hydrological Society.

Rolt, L.T.C. (2007) *Thomas Telford*. Sutton Publishing Ltd, Stroud.

Rosevear, A. (2008) *Turnpike Roads in England* [online], [Accessed 7 May 2012]. Available at: http://www.turnpikes.org.uk/.

Secretary of State for Environment, Food and Rural Affairs (2011) *Climate Resilient Infrastructure: Preparing for a Changing Climate* [online]. The Stationery Office Ltd, London. Available at: http://www.defra.gov.uk/publications/files/climate-resilient-infrastructure-full.pdf>.

Severn–Trent Water (2005) *Barnhurst Sewage Treatment Works*.

Sheriff, J.D., Lawson, J.D. and Askew, T.E.A. (1996) Strategic resource development options in England and Wales. *Water and Environment Journal*, **10** (3), 160–169.

Shill, R. (2006) *Water Sources on the Birmingham Canal Navigations* [online]. Available at: http://allensregister.com/bp188_chasewater.php.

Sim, S. (2003) *Regenerating the Waterways*. British Waterways, Watford.

Spelman, C. (2011) *Thames Water – Water Resources Management Plan ('WRMP')* [online], [Accessed 04/03/12]. Available at: http://archive.defra.gov.uk/environment/quality/water/resources/documents/thames-inquiry-decision-letter.pdf.

Sutton, R. (2005) Water management on inland waterways, Chapter 22. In: IWA (ed.) *IWA Technical Restoration Handbook*. Inland Waterways Association, London.

Tesco (no date) *Greener Living – Using Greener Transportation* [online], [Accessed 11 July 2012]. Available at: http://www.tesco.com/greenerliving/greener_tesco/what_tesco_is_doing/greener_transportation.page.

The Barge Association (DBA) (2012) *Canal de Roubaix Gets a New Water Supply System (7–Jan–07)* [online]. Available at: http://www.barges.org/main.php?section=2411.

UK Climate Impacts Programme (2009) *Adapting to Climate Change – UK Climate Projections*, June 2009. Department for Environment, Food and Rural Affairs, London.

United Nations Educational, Scientific and Cultural Organization (no date) *Canal du Midi* [online], [Accessed 7 May 2012]. Available at: http://whc.unesco.org/en/list/770/.

Waterscape (2012) *Water Management FAQs* [online]. Available at: http://www.waterscape.com/features-and-articles/features/water-management-faqs.

Wood Hall and Heward Engineering (no date) *Infrastructure Maintenance – Cables* [online]. Available at: http://www.whhbarges.co.uk/infrastructure_maintenance.htm.

Section 5 Flooding Responses and Reinstatement

15

Urban Precipitation: Measurements, Monitoring and Processes

Omolara O. Lade, Michael A. Fullen, David Oloke, Madhu Subedi and Colin A. Booth

'You have spread out the heavens like a tent and built your home on the waters above. You use the clouds as your chariot and ride on the wings of the wind. You use the winds as your messengers and flashes of lightning as your servants'

(Psalm 104, verses 2–4).

15.1 Introduction

Precipitation is a major component of the water cycle and is responsible for depositing fresh water on the planet. Some 505 000 km^3 of water falls as precipitation each year; ~398 000 km^3 falls over the oceans, while 107 000 km^3 falls over land (Chowdhury, 2005). Given the Earth's surface area, that means the global mean annual precipitation is 990 mm, but over land it is only 715 mm. However, global warming is changing global precipitation patterns. These changing patterns can cause more extreme weather conditions, leading to the apparently contradictory conditions of both increased incidence and severity of droughts and heavy rainfall. Intense and voluminous rainfall falling on the largely sealed and, thus, impermeable urban surfaces can induce floods. These entrain attendant problems of flooding of residential, commercial and industrial premises and drainage and sewer systems. Thus, in our 'warming world' urban managers must tackle these problems (Houghton, 2009).

This chapter reviews the main forms and mechanisms of precipitation and considers how they interact with the urban fabric. This leads to an exploration of the causes and consequences of precipitation falling on urban areas and approaches to precipitation measurement and the assessment of local precipitation inputs. The chapter concludes with a consideration of appropriate response strategies in our 'warming world'.

Water Resources in the Built Environment: Management Issues and Solutions, First Edition.
Edited by Colin A. Booth and Susanne M. Charlesworth.

15.2 Types of Precipitation

The mechanisms producing precipitation include convective, stratiform and orographic rainfall (Anagnostou, 2004). Convective processes involve strong vertical motions that can cause the rapid (within an hour or less) overturning of the atmosphere in that location and, thus, induce heavy, often highly localized, precipitation (Jones, 2002). Stratiform processes involve weaker upward motions and less intense precipitation. Precipitation can be divided into three categories, based on whether it falls as liquid water (rain, dew), liquid water that freezes on contact with cold surfaces (frost) or ice (snow, sleet and hail).

15.3 Urban Climate and Potential Impacts

Cities are becoming increasingly vulnerable to flooding because of rapid urbanisation, installation of complex infrastructures and changes in precipitation patterns, probably caused by climate change. Large-scale sewer systems have been constructed across cities worldwide to reduce the vulnerability of 'hot' cities to flood damage. This could make these cities more vulnerable to rainfall extremes, due to lack of consideration of the occurrence and frequency of flooding exceeding the design criteria. Due to global warming, the probabilities and risks of sewer surcharge and flooding are increasing. The Intergovernmental Panel on Climate Change (IPCC) (2007) reported a global increase in the frequency of extreme rainstorms, believed to be largely attributed to global warming. IPCC (2007) further concluded that it is very likely (>90% likelihood) that, based on climate model simulations with different future greenhouse gas emission scenarios, this trend will continue. Hence, water managers must develop intelligent and appropriate adaptation strategies to respond to these effects.

Studies of climate change, many of which focus on risks of floods and droughts on the river catchment scale, have markedly increased in recent years. However, the number of

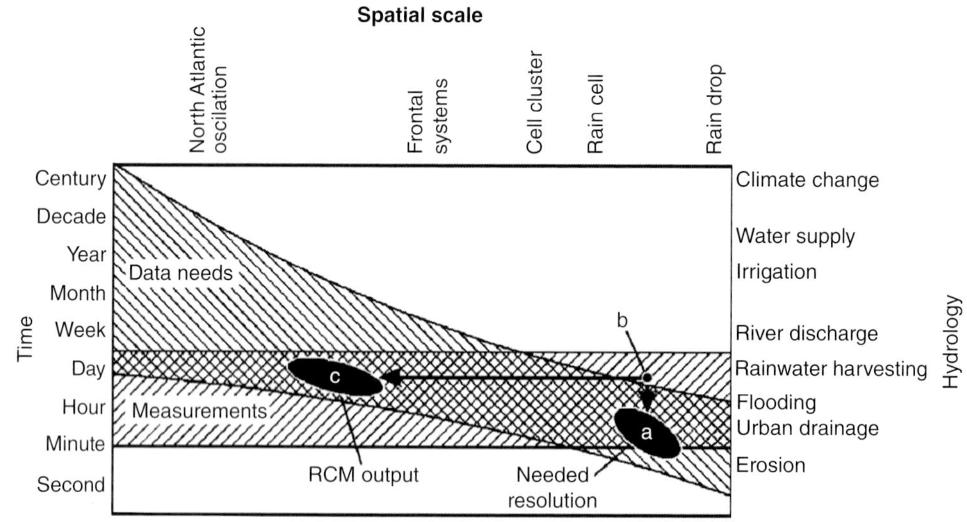

Figure 15.1 Scale mismatch between the global circulation model/regional circulation model (GCM/RCM) outputs and the needs for urban hydrological impact studies (adapted from Arnbjerg-Nielsen, 2008).

climate change studies dealing with urban drainage impacts remains limited. In part, this is because they require a specific focus on small urban catchment scales (<500 km^2) and short-duration precipitation extremes (<1 day). To evaluate the regional impacts of climate change on urban drainage, extreme and short duration rainfall statistics for the period and the geographical region of interest must be estimated. For historic conditions, climate change effects can be investigated by analysing trends in long-term historical records of rainfall. For future conditions, projected changes in rainfall statistics are based on future scenarios in greenhouse gas emissions simulated in climate models (generally described as 'global circulation models', GCMs) or statistical extrapolation based on historical observations. Despite the significant increase in computational power in recent years, climate models still remain relatively coarse in space and time resolution and are unable to resolve significant features at the fine scales of urban drainage systems (Figure 15.1). However, other drivers will also have large impacts on the performance of sewer systems, particularly urbanisation, changes in sewer systems and changes in performance criteria (Arnbjerg-Nielsen, 2010).

15.4 Urban Irrigation

Urban irrigation is an important component of the hydrological cycle in many arid and semi-arid areas. Urban irrigation can:

1. Influence the groundwater hydrological cycle, by increasing the amount of water available to infiltration through return flow (Grimmond *et al.*, 1986).
2. Influence the surface water hydrologic cycle by affecting runoff rates from rainfall events (Agnew and Anderson, 1992).

An understanding of how urban irrigation affects the hydrological cycle will enable water managers to:

- Be better prepared for alterations in the water supply, whether due to climatic, engineered or catastrophic reasons (Ward, 1989).
- Model and track changes in water chemistry (Lee and Longhurst, 1992).
- Test future consumptive use scenarios (Hurd *et al.*, 1999; Vörösmarty *et al.*, 2000; Jacobs *et al.*, 2001; Alcamo *et al.*, 2008).

Studies have shown that >50% of the water used in a typical household per year is used outside the home (Grimmond and Oke, 1991; Mayer *et al.*, 1999). For example, in residential areas within the City of Los Angeles *circa* 225 million cubic metres of water are used for irrigation per year, which inevitably influences the urban hydrological cycle (Southern California Area Governments, 2005). Hence, cost-effective and repeatable techniques for estimating urban irrigation at local and regional scales are needed. Several methods for estimating urban irrigation include:

1. The use of dataloggers, minimum month methods and energy balance formulae (Mayer *et al.*, 1999; Gleick *et al.*, 2003). However, many of these methods can be costly and time-consuming. Hence, they can be difficult to implement on a regional scale and may be inaccurate at the local scale.

2. An alternative method is to use satellite remote sensing. Remote sensing is an effective method for quantifying vegetation cover and density, and for estimating evapotranspiration, both of which influence residential and commercial irrigation (Mayer *et al.*, 1999; Keith *et al.*, 2002).

15.5 Urban Effects on Rainfall Variability

The US National Science Foundation published *Water: Challenges at the Intersection of Human and Natural Systems* (NSF, 2005). Two overarching questions were posed:

1. Can human intervention in the water cycle set processes and trajectories in motion that we cannot readily see, because they are masked by complexity and variation?
2. Can we begin to assemble the pieces in a way in which we can begin to forecast environmental changes related to water and, therefore, intervene before irreparable environmental or social damage is done?

Humans have traditionally settled near water. For instance, 42 out of 44 US metropolitan areas are adjacent to a major water resource. Human activities, such as landscaping and pumping of water, affect the water cycle and ecosystems. The IPCC (2007) noted a growing interest in understanding what role urban land cover, land use and pollution has on climate change (Trenberth *et al.*, 2007). IPPC Report Chapter 3 (Section 3.3.2.4) reviewed the linkages between urban-related processes and regional precipitation changes. Recent studies were reviewed on urban rainfall anomalies around Tokyo, Beijing and other Chinese cities, Taipei, Jerusalem, Sydney, Kolkata, Houston, St Louis, Atlanta, Paris and other European cities (Shepherd *et al.*, 2008).

There is no consistency in the literature of enhanced precipitation by urban environments. An econometric analysis of rain gauge data in the Pearl River Delta of China concluded that urbanisation reduced local precipitation (Kaufmann *et al.*, 2007). Although the possible effects of pollution were not considered in this study, smaller cloud droplet distributions and suppressed rainfall have occurred due to increased aerosol concentrations from anthropogenic sources above and downwind of urban areas. However, some studies suggest that urban areas can induce increased precipitation. Figure 15.2 presents an idealised diagram of the typical geographical pattern of the urban anomaly. Locations 25–75 km downwind of the city centre will, typically, experience greatest increases in rainfall (Shepherd *et al.*, 2002; Diem and Mote, 2005; Shepherd, 2006; Mote *et al.*, 2007; Hand and Shepherd, 2009). Shepherd (2005) reviewed the possible mechanisms for urbanisation to enhance or initiate precipitation or convection, which included one or a combination of the following:

1. Enhanced convergence due to increased surface roughness in the urban environment (Bornstein and Lin, 2000).
2. Enhanced sensible heat fluxes (Thielen *et al.*, 2000).
3. Destabilisation due to the urban heat island (UHI), i.e. thermal perturbation of the planetary boundary layer (i.e. the lower ~500 m of the atmosphere) and resulting downwind translation of the UHI circulation or UHI-generated convective clouds (Shepherd *et al.*, 2002; Shepherd and Burian, 2003; Baik *et al.*, 2007).
4. Enhanced aerosols in the urban environment for cloud condensation nuclei (CCN) sources (Rosenfeld *et al.*, 2007; Muller *et al.*, 2010).
5. Bifurcation or diversion of precipitation systems by the urban canopy or related processes (Bornstein and LeRoy, 1990).

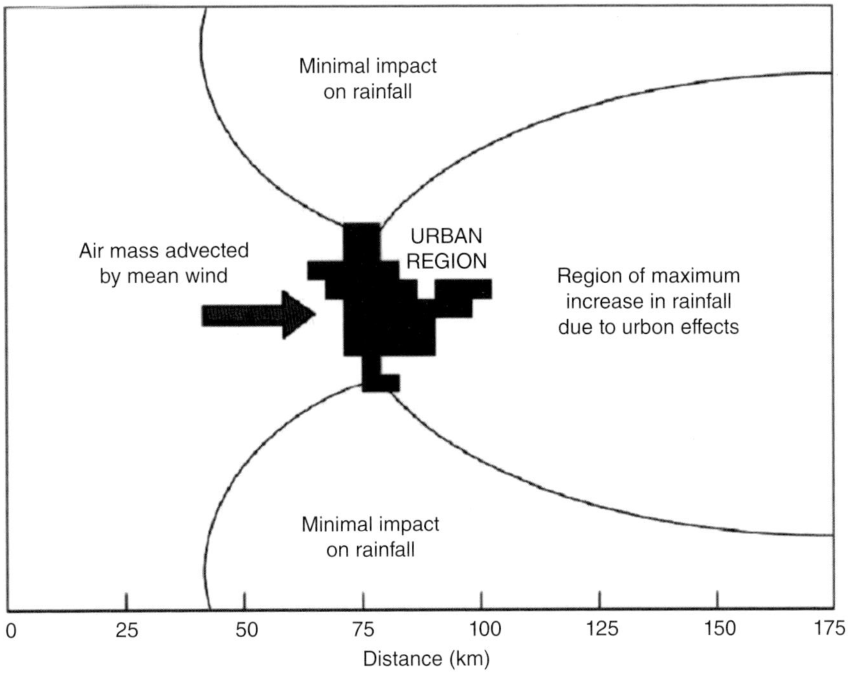

Figure 15.2 Conceptualisation of the spatial extent of the urban rainfall effect (adapted from Shepherd *et al.*, 2002, in Simpson, 2006).

Emissions of water vapour and dust particles (which can become CCN) can cause localised increases in cloud cover and precipitation around the cooling towers of power stations. This is known as 'the La Porte effect' and is named after the town of La Porte, east of the steelmills of Gary, Michigan, where this effect was first observed (Changnon, 1979).

15.6 Precipitation Variability in Thunderstorms

Society no longer believes that thunderstorms are produced by the Norse god Thor hammering the 'anvil' of cloud (the glaciated upper part of mature cumulonimbus (Cb) clouds). However, we have named 'thunder' after Thor (McIlveen, 1986). Extreme weather is a regular occurrence. At any one time, *circa* 2000 thunderstorms are active across the globe. The violent nature of these storms can lead to serious damage, on both local and larger scales. Intense thunderstorm-related precipitation often triggers local flooding, landslides and damage to infrastructure, such as roads and bridges (Looper and Vieux, 2012). They may also accompany synoptic flood situations over large areas, such as the case of the 1997 flood in Poland. Thunderstorm precipitation is the third most serious cause of natural disaster damage, after floods and tropical cyclones (hurricanes or typhoons) (Changnon, 2001a). However, extremely intense rainfall, hailstorms and wind are the main direct causes of damage. Tornadoes are one of the most destructive atmospheric phenomena and are always linked to Cb clouds, which are both the source of thunderstorms and the 'parent' of tornadoes. In specific conditions associated with massive atmospheric instability, thunderstorms can spawn tornadoes.

While it is relatively easy to identify the occurrence of thunderstorms and hailstorms, it is very difficult to determine the amount and intensity of accompanying precipitation. This is due to unavailability of detailed daily data on precipitation and meteorological phenomena, compounded by the often localised nature of convective storms (Looper and Vieux, 2012). Thus, most literature on precipitation in relation to thunderstorms employ statistics on numbers of thunderstorm days, days with a certain precipitation type and daily precipitation amounts (Winkler, 1987; Dupuy, 1995; Dai, 2001a, 2001b).

Few studies are available on the use of hourly precipitation records (usually expressed in mm of precipitation per hour, or mm/h) to determine thunderstorm precipitation parameters (Changnon, 2001b). Changnon compared thunderstorm precipitation with annual and seasonal precipitation totals, estimated its influence on the occurrence of dry and wet years and determined long-term trends in annual thunderstorm rainfall depths. Twardosz (2005) presented synoptic and probabilistic aspects of diurnal precipitation variation in Krakow and differentiated daily patterns of both frontal thunderstorms and mass thunderstorms.

In hailstorm studies, spatial and temporal hailstorm variability is discussed in relation to atmospheric circulation. Examples of hailstorm studies include investigations of hailstorm occurrence in the Czech Republic (Brázdil *et al.*, 1998; Chromá *et al.*, 2005) and a synoptic analysis of hailstorm occurrence in Bulgaria (Simeonov and Georgiev, 2003).

15.7 Measuring Precipitation

Precipitation is usually measured using rain gauges and has been measured for centuries. The first known mention of rainfall measurements were in India in 400 BC (Shaw, 1988). In the United Kingdom the standard rain gauge is the UK Meteorological Office rain gauge, which has a gauge orifice 12 inches (30.48 cm) above the ground surface and an orifice of 5 inches (127 mm) (Meteorological Office, 1982, 2010). Measurements are usually taken at 0900 GMT (Greenwich Mean Time (since 1972 also referred to as UTC, or Coordinated Universal Time). The gauge rim is above the surface to prevent splash-in of precipitation. However, the projection increases turbulence around the gauge orifice, which tends to decrease precipitation totals, typically by *circa* 5%. However, other countries often use taller gauges (Shaw, 1988). For instance, the US gauge is two feet above the surface, so that it probably projects above any snow pack. Similar tall gauges are used in Central European countries. Of course, different countries measuring precipitation using slightly different techniques poses problems of international comparability of data sets. This is especially important given the need for international comparisons to analyse climate change and is an issue given serious consideration by the World Meteorological Organization (WMO), based in Geneva.

Some gauges are automatic and record precipitation amount, duration and intensity (McIlveen, 1986; Shaw, 1988). The Hellman and Dines rain gauges are examples of such 'autographic gauges'. Tipping bucket gauges send a timed electronic signal when internal buckets tip after a specific amount of precipitation (usually 0.1, 0.2 or 0.5 mm).

A network of gauges is needed to produce an areal estimate of precipitation input. These point data are converted to areal estimates using various techniques. Rain gauge data are considered representative of much larger areas. For instance, a Mark II gauge (orifice area 150 cm^2) considered representative of a 15 km^2 area (a fairly typical value in the relatively dense rain gauge network of the United Kingdom) is only sampling 10^{-9} of the specific area.

In point-to-areal calculations, the simplest approach is to calculate the arithmetic mean precipitation of the network. Other approaches give weighted means, adjusting for contours (certain gauges being considered representative of specific altitudes – the hypsometric method) or represent adjacent 'nearest-neighbour' territory, which is divided between gauges (these are known as 'Thiessen polygons') (Shaw, 1988). Sometimes, rain gauges are considered representative of specific identified rainfall regimes (the isohyetal method). Another approach preferentially weighs the data from gauges nearer the centre of a catchment (the Bethlamy two-axis method). Radar technology is being increasingly used to estimate areal precipitation (Muller *et al.*, 2010).

The basic input for urban sewer system simulation models is precipitation, so proper measures should be taken to ensure data accuracy for simulation models. Rain gauges and radar are commonly used to measure precipitation; the former is a point measurement, while the latter measures the spatial distribution of rain. Rain radar measurements adjusted with ground measurements are found to be superior to gauge measurements, as it incorporates the advantages of both radar and gauge measurements (Looper and Vieux, 2012). Radar measures the reflectivity of rain droplets and this reflectivity is converted into precipitation data using specific algorithms (Anagnostou and Krajewski, 1999). There are limitations in the technology for measuring reflectivity. For instance, sleet gives particularly intense signals due to particularly strong radar reflection, a phenomenon known as the 'white band'. Thus, the accuracy of precipitation measured using radar depends on how much errors are corrected due to these factors (Gorgucci *et al.*, 1996). Satellites are becoming increasingly used as platforms for measuring precipitation using radar, especially the TRMM (Tropical Rain Measuring Mission) system (Shepherd *et al.*, 2002). Further information can be gained from the NASA (National Aeronautics and Space Administration) web site: http://earthobservatory.nasa.gov/Features/UrbanRain/urbanrain3.php.

15.8 Spatial and Temporal Precipitation Monitoring

Convective rains are responsible for most flooding in urban areas, which makes spatial precipitation patterns essential in rainfall runoff modelling of urban areas (Desa and Niemczynowicz, 1996). The temporal and spatial variation of precipitation is very important for surface and subsurface flows in catchments. These estimates are essential for hydrological models estimating soil moisture and for forecasting floods. Rain gauge stations provide better estimates of rain at a single geographical point. In cases of inadequate rain gauge networks, because of few available measurements points, a more accurate spatial distribution of precipitation can be provided with grid cell averaged radar, which leads to improved calibration of rainfall runoff models (Lobbrecht and Andel, 2005). Urban areas require very large spatial and temporal resolution of rain data. To achieve this high resolution using rain gauges, very dense gauge networks are needed, which are difficult to build and maintain, from both financial and logistical perspectives (Bernea *et al.*, 2004).

Precipitation inputs into the urban infrastructure follow complex pathways. Some precipitation will evaporate back into the atmosphere. Some will flow into topographic hollows and enter temporary storage (depression storage). Some precipitation will run off from impervious surfaces (buildings, roads and pavements) and enter into the complex interconnected nexus of the sewage system (see Chapter 26). Some precipitation will infiltrate into bare or vegetated soils. Given the complex and poorly understood nature and properties or urban soils (Webb *et al.*, 2012), infiltration behaviour is extremely difficult to predict. However, strategies that promote infiltration (i.e. leaving extensive and interconnected tracts of vegetated land) rather than runoff (i.e. impervious surfaces) will tend to diminish the 'flashiness' of urban floods.

In response, urban managers need to monitor precipitation patterns and the greater the quality and density of the monitoring network, the greater the accuracy and precision of the database. Then managers need to develop predictive computer-based models of urban hydrology. Each urban unit (town or city) is, to some extent, unique and so universal models are not applicable. Thus, models need to be calibrated and continually updated and improved. Given the dynamic nature of the weather, climate and urban development, there are complex challenges in realistic model simulation.

15.9 Case Study of Small-Scale Variability: Spatial Variability in Precipitation within the Hilton Experimental Site, Shropshire, UK

A 25-year study (1982–2006) revealed the highly variable nature of precipitation within a small area. The study was conducted on a 0.5 hectare undulating site in east Shropshire. Precipitation variability within the site was measured using a network of 11 standard UK Meteorological Office rain gauges. Three aspects of variability were assessed: (a) differences in precipitation inputs between two adjacent standard rain gauges; (b) differences between gauges with standard exposure in the United Kingdom compared with surface-level gauges; and (c) the variability of precipitation over a ~15° steep slope (relative relief 16.3 m).

Comparison of the two gauges, only 6.24 m apart, over 25 years (1982–2006 inclusive) showed that the northerly (N) gauge tended to receive slightly more precipitation (total 44.6 mm) than the southerly (S) gauge (Subedi and Fullen, 2009). Taking the southerly gauge as the control, the northerly gauge received 0.3% more precipitation.

The surface-level (SL) gauge at the bottom of the slope received 2.5% more precipitation than the rain gauge at the top of the slope. The difference between monthly precipitation totals was statistically significant (paired $t = 8.17$, $P < 0.001$, degrees of freedom (d.f.) = 178) and the trend was consistent throughout the study period (1992–2006). Mean annual precipitation received by SL-Top and SL-Bottom were 666.3 and 682.7 mm respectively. Similarly, the standard gauge at the bottom of the slope collected more precipitation during most of the study period than the standard gauge at the top. The mean annual precipitation collected by the standard gauges at the bottom and top of the slope were 681.9 and 630.5 mm, respectively. The difference in monthly precipitation collected by the two gauges was statistically significant (paired $t = 4.22$, $P < 0.001$, d.f. = 130) and ranged between 7.3 and 111.7 mm (1.2 and 17.5%).

Comparison of precipitation along the slope revealed that precipitation was more at the bottom of the slope than the top and generally increased downslope. The difference in the mean highest precipitation was received by the rain gauge at the bottom (681.9 mm/year) and lowest was received by one of the upper rain gauges (622.3 mm/year) during 1996–2006, with the maximum difference being 59.6 mm/year. Precipitation totals were negatively associated with altitude ($r = -0.839$, $P < 0.01$, $n = 8$). The upper section experienced the highest wind velocities and, thus precipitation could be carried over the gauge orifice. Higher wind velocity and turbulence were probably the critical factors contributing to lower precipitation totals at the top of the slope than at the bottom.

At the hilltop, surface-level gauges received more precipitation than the standard rain gauge. However, this trend was not apparent at the bottom of the slope. This contrasting behaviour of standard versus surface-level gauges between the top and bottom of the slope may be indirect evidence of the importance of air turbulence. Turbulent conditions at the top of the slope exacerbated the contrast. However, the sheltered conditions on the basal slope meant standard and surface-level gauges received similar amounts of precipitation. After intense convective rains (characterized by calm surface-wind conditions), it was noteworthy that the rain gauges received very similar precipitation amounts. The role of wind turbulence in decreasing rainfall catches was eloquently explained by Stevenson (1842) as '*while torrents pour down from the heavens, an eddy plays about the rim of the basin, deranging the regularity of the discharge*'.

Although Hilton lies outside city limits, it does illustrate the small-scale variability of precipitation, with differences in annual precipitation being ≤8%. It is likely that urban areas experience even greater precipitation variability, with added effects of differential heating of buildings contributing to convection and air turbulence around buildings.

15.10 Conclusions

Precipitation is an 'umbrella term' covering different forms, including rain, drizzle, frost, snow, hail and dew. There are different causal mechanisms, including frontal, convectional and orographic processes. Precipitation is usually measured at a geographical point and then extrapolated to a larger scale. There are several procedures to produce local to regional estimates of precipitation. Radar and satellite remote sensing is increasingly used to study precipitation patterns, and accuracy is being progressively improved by careful calibration of point and 'sensed' (radar and satellite remote sensing) measurements. Precipitation falling on urban areas is influenced by many factors, including localised heating, topography, turbulence and dust emissions. These can produce precipitation patterns, which are complex, both in time and space. Precipitation often occurs as brief, intense and localised convectional storms. In turn, this poses challenges in monitoring and modelling these storms. Thus, it is extremely difficult to predict the effects of such convectional storms on the urban fabric.

The interaction of precipitation and the presence of extensive impermeable sealed urban surfaces are conducive to floods, which places stress on the urban infrastructure. Precipitation routes through the urban fabric are complex (consisting of evaporation, depression storage, runoff and infiltration) and unique to each urban unit. Thus, urban managers need accurate and predictive computer models, which must be calibrated and continually updated and improved. Given the dynamic nature of the weather, climate and urban development, there are complex challenges in realistic model simulation. Therefore, urban managers need to develop appropriate and intelligent response and adaptation strategies to effectively handle the challenges posed by urban precipitation in our warming world. In this complex interface of science, technology and management the scientific axiom 'the only certainty is uncertainty' is especially applicable.

References

Agnew, C. and Anderson, E. (1992) *Water Resources in the Arid Realm*. Routledge, London, 329 pp. (ISBN 0 415 07969 1).

Alcamo, J.M., Vörösmarty, C.J., Naiman, R.J., Lettenmaier, D.P. and Pahl-Wostl, C. (2008) *A Grand Challenge for Freshwater Research: Understanding the Global Water System*. Environmental Resources Letter 3, 010202, 6 pp. (DOI: 10.1088/1748-9326/3/1/010202).

Anagnostou, E.N. (2004) A convective/stratiform precipitation classification algorithm for volume scanning weather radar observations. *Meteorological Applications, Cambridge University Press*, **11** (4), 291–300.

Anagnostou, E.N. and Krajewski, W.F. (1999) Real-time radar rainfall estimation. Part I: Algorithm formulation. *Journal of Atmospheric and Oceanic Technology*, **16** (2), 189–197.

Arnbjerg-Nielsen, K. (2008) Quantification of climate change impacts on extreme precipitation used for design of sewer systems. In: *Proceedings of the 11th International Conference on Urban Drainage*, 31 August–5 September, Edinburgh, UK.

Arnbjerg-Nielsen, K. (2010) Past, present and future design of urban drainage systems with focus on Danish experiences. *Water Science and Technology*, **63** (3), 527–535.

Baik, J.J., Kim, Y.H., Kim, J.J. and Han, J.Y. (2007) Effects of boundary layer stability on urban heat island induced circulation. *Theoretical and Applied Climatology*, **89**, 73–81.

Bernea, A., Delrieub, G., Creutinb, J.-D. and Obledb, C. (2004) Temporal and spatial resolution of rainfall measurements required for urban hydrology. *Journal of Hydrology*, **299**, 166–179.

Bornstein, R. and LeRoy, M. (1990) Urban barrier effects on convective and frontal thunderstorms. In: *American Meteorological Society Conference on Mesoscale Processes*, Boulder, Colorado, pp. 25–29.

Bornstein, R. and Lin, Q. (2000) Urban heat islands and summertime convective thunderstorms in Atlanta: three cases studies. *Atmospheric Environment*, **34**, 507–516.

Brázdil, R., Štěpánek, P. and Vais, T. (1998) Časová a prostorová analýza bouřek, krupobiti a extrémnich srážek v jižni části Moravy v obdobi 1994–1995. *Meteorologickié Zprávy*, **51** (2), 45–52.

Changnon, S.A. (1979) More on the La Porte anomaly, 171–185 In: Changnon, S.A., Gatz, D.F., Jameson, A.R., Dzurisin, G.L., Scott, R.W. and Grosh, R.C. (eds), *Studies of Urban and Lake Influences in Precipitation in the Chicago Area*, Volume **2**. Atmospheric Sciences Section, Illinois State Water Survey, Urbana, Illinois 61801, 190 pp.

Changnon, S.A. (2001a) Damaging thunderstorm activity in the United States. *Bulletin of the American Meteorological Society*, **82**, 597–608.

Changnon, S.A. (2001b) Thunderstorm rainfall in the conterminous United States. *Bulletin of the American Meteorological Society*, **82**, 1925–1940.

Chowdhury, A. (2005) *Guide to Planet Earth. The Water Cycle*, West Edition.

Chromá, K., Brázdil, R. and Tolasz, R. (2005) Spatio-temporal variability of hailstorms for Moravia and Silesia in the summer half-year of the period 1961–2000. *Meteorologický Časopis*, **8**, 65–74.

Dai, A. (2001a) Global precipitation and thunderstorm frequencies. Part I: seasonal and interannual variations. *Journal of Climate*, **14**, 1092–1111.

Dai, A. (2001b) Global precipitation and thunderstorm frequencies. Part II: diurnal variations. *Journal of Climate*, **14**, 1112–1128.

Desa, M.M.N. and Niemczynowicz, J. (1996) Spatial variability of rainfall in Kuala Lumpur, Malaysia: long and short term characteristics. *Journal of Hydrological Sciences*, **41** (3), 345–362.

Diem, J.E. and Mote, T.L. (2005) Inter-epochal changes in summer precipitation in the Southeastern United States: evidence of possible urban effects near Atlanta, Georgia. *Journal of Applied Meteorology*, **44**, 717–730.

Dupuy, J. (1995) Precipitations orageusses dans l'agglomeration Lilloise. *Cahiers de Geographie Physique, Université des Sciences et Technologies de Lille*, **10**, 17–32.

Gleick, P.H., Haasz, D., Henges-Jeck, C., Srinivasan, V., Cushing, K.K. and Mann, A. (2003) *Waste Not, Want Not: The Potential for Urban Water Conservation in California*. Pacific Institute for Studies in Development, Environment and Security, 165 pp.

Gorgucci, E., Scarchilli, G. and Chandrasekar, V. (1996) Error structure of radar rainfall measurement at C-band frequencies with dual polarization algorithm for attenuation correction. *Journal of Geophysical Research*, **101**, 461–471.

Grimmond, C.S.B. and Oke, T.R. (1986) Urban water balance. 2. Results from a suburb of Vancouver, British Columbia. *Water Resources Research*, **22** (10), 1404–1412.

Grimmond, C.S.B. and Oke, T.R. (1991) An evapotranspiration interception model for urban areas. *Water Resources Research*, **27** (7), 1739–1755.

Grimmond, C.S.B., Oke, T.R. and Steyn, D.G. (1986) Urban water balance. 1. A model for daily totals. *Water Resources Research*, **22** (10), 1397–1403.

Hand, L. and Shepherd, J.M. (2009) An investigation of warm season spatial rainfall variability in Oklahoma City: possible linkages to urbanization and prevailing wind. *Journal of Applied Meteorology and Climatology*, **48**, 251–269.

Houghton, Sir John (2009) *Global Warming. The Complete Briefing*, Fourth Edition. Cambridge University Press, Cambridge, 438 pp. (ISBN 978 0 521 70916 3).

Hurd, B., Leary, N., Jones, R. and Smith, J. (1999) Relative regional vulnerability of water resources to climate change. *Journal of the American Water Resources Association*, **35** (6), 1399–1409.

Intergovernmental Panel on Climate Change (IPCC) (2007) *Climate Change: The Physical Science Basis*. Summary for Policymakers, Contribution of Working Group I to the Fourth Assessment Report of the Intergovernmental Panel on Climate Change, IPCC Secretariat, Geneva, Switzerland.

Jacobs, K., Adams, D.B. and Gleick, P. (2001) Potential consequences of climate variability and change for the water resources of the United States. In: National Assessment Synthesis Team (Ed.), *Climate Change Impacts on the United States: The Potential Consequences of Climate Variability and Change*. Report for the US Global Change Research Program. Cambridge University Press, Cambridge, UK, 620 pp.

Jones, P.D. (2002) Changes in climate and variability over the last 100 years. In: Pearce, R.P. (Ed.), *Meteorology at the Millennium*, pp. 133–142. Academic Press, London.

Kaufmann, R.K., Seto, K.C., Schneider, A., Liu, Z., Zhou, L. and Wang, W. (2007) Climate response to rapid urban growth: evidence of a human-induced precipitation deficit. *Journal of Climate*, **20** (10), 2299.

Keith, D.J., Walker, H.A. and Paul, J.F. (2002) Terrestrial vegetation greenness of the lower Galveston Bay watershed from satellite remote sensing and its relation to water use and the salinity regime of the Galveston Bay Estuary (USA). *International Journal of Remote Sensing*, **23** (5), 905–916.

Lee, D.S. and Longhurst, J.W.S. (1992) A statistical intercomparison between 'urban' and 'rural' precipitation chemistry data from Greater Manchester and two nearby secondary national network sites in the United Kingdom. *Atmospheric Environment*, **26**, 2869–2883.

Lobbrecht, A.H. and Andel, S.J.V. (2005) Integrated urban and rural water management using modern meteorological data. In: *10th International Conference on Urban Drainage*, Copenhagen, Denmark.

Looper, J.P. and Vieux, B.E. (2012) An assessment of distributed flash flood forecasting using radar and rain gauge input for a physics-based distributed hydrologic model. *Journal of Hydrology*, **412–413**, 114–132.

Mayer, P.W., De Oreo, W.B., Opitz, E.M., Kiefer, J.C., Davis, W.Y., Dziegielewski, B. and Nelson, J.O. (1999) *Residential End Uses of Water*. American Water Works Association Research Foundation, Denver, Colorado, 109 pp.

McIlveen, R. (1986) *Basic Meteorology: A Physical Outline*. Van Nostrand Reinhold (UK), Wokingham, 457 pp. (ISBN 0 442 31769 7).

Meteorological Office (1982) *Observer's Handbook*, 4th Edition. HMSO, London, 220 pp. (ISBN 0 11 400329 7).

Meteorological Office (2010) *National Meteorological Library and Archive Fact Sheet 17 – Weather Observations over Land*, Version 01. Meteorological Office, Exeter, 24 pp. [Accessed 23/08/12]. Available at: http://www.metoffice.gov.uk/media/pdf/k/5/Fact_sheet_No._17.pdf.

Mote, T.L., Lacke, M.C. and Shepherd, J.M. (2007) Radar signatures of the urban effect on precipitation distribution: a case study for Atlanta, Georgia. *Geophysical Research Letters*, **34**, L20710.

Muller, C.L., Kidd, C., Fairfield, I.J. and Baker, A. (2010) Investigations into clouds and precipitation over an urban area using micro rain radars, satellite remote sensing and fluorescence spectrometry. *Atmospheric Research*, **96**, 241–255.

National Science Foundation (NSF) (2005) *Water: Challenges at the Intersection of Human and Natural Systems*. National Science Foundation (AC–ERE), Athens, Georgia, p. 50.

Rosenfeld, D., Dai, J., Yu, X., Yao Z., Xu, X., Yang, X. and Du, C. (2007) Inverse relations between amounts of air pollution and orographic precipitation. *Science*, **315**, 9.

Shaw, E.M. (1988) *Hydrology in Practice*, Second Edition. Van Nostrand Reinhold (International), London, 539 pp. (ISBN 0 278 00061 4).

Shepherd, J.M. (2005) A review of current investigations of urban–induced rainfall and recommendations for the future. *Earth Interactions*, **9** (12), 1–27.

Shepherd, J.M. (2006) Evidence of urban-induced precipitation variability in arid climate regimes. *Journal of Arid Environments*, **67**, 607–628.

Shepherd, J.M. and Burian, S.J. (2003) Detection of urban-induced rainfall anomalies in a major coastal city. *Earth Interactions*, **7**, 1–17.

Shepherd, J.M., Pierce, H. and Negri, A.J. (2002) Rainfall modification by major urban areas: observations from spaceborne rain radar on the TRMM satellite. *Journal of Applied Meteorology*, **41**, 689–701.

Shepherd, J.M., Shem, W., Hand, L. and Manyin, M. (2008) Modeling urban effects on the precipitation component of the water cycle. In: Yao, X. and Jiang, H. (eds), *Geospatial Analysis and Modeling of Urban Environments*. Springer Book Series and GIScience, Athens, Georgia.

Simeonov, P. and Georgiev, C.G. (2003) Severe wind/hail storms over Bulgaria in the 1999–2001 period: synoptic- and meso-scale factors for generation. *Atmospheric Research*, **67–68**, 629–643.

Simpson, M.D. (2006) *Role of Urban Land Use on Mesoscale Circulations and Precipitation*. PhD dissertation, North Carolina State University, 294 pp.

Southern California Area Governments (2005) *Los Angeles County Land Use*. Digital polygon dataset [Accessed 23/08/12]. Available at: http://www.aisgis.com/projects/SCAG.html.

Stevenson, T. (1842) Observation on the deficits of rain gauges, with description of one of an improved form. *Edinburgh New Philosophical Journal*, **33**, 12–19.

Subedi, M. and Fullen, M.A. (2009) Spatial variability in precipitation within the Hilton Experimental Site, Shropshire, UK (1982–2006). *Hydrological Processes*, **23**, 236–244.

Thielen, J., Wobrock, W., Gadian, A., Mestayer, P.G. and Creutin J.D. (2000) The possible influence of urban surfaces on rainfall development: a sensitivity study in 2D on the meso-gamma-scale. *Atmospheric Resolution*, **54**, 15–39.

Trenberth, K.E., Jones, P.D., Ambenje, P., Bojariu, R., Easterling, D., Klein Tank, A., Parker, D., Rahimzadeh, F., Renwick, J.A., Rusticucci, M., Soden, B. and Zhai, P. (2007) Observations: surface and atmospheric climate change. In: Solomon, S., Qin, D., Manning, M., Chen, Z., Marquis, M., Avery, K.B., Tignor, M. and Miller, H.L. (eds), *Climate Change 2007: The Physical Science Basis*. Contribution of Working Group I to the Fourth Assessment Report of the Intergovernmental Panel on Climate Change. Cambridge University Press, Cambridge, UK.

Twardosz, R. (2005) *The Synoptic and Probabilistic Aspects of Diurnal Precipitation Variation in Cracow (1886–2002)*. Institute of Geography and Spatial Management of Jagiellonian University, Krakow, 176 pp. [in Polish].

Vörösmarty, C.J., Green, P., Salisbury, J. and Lammers, R.B. (2000) Global water resources: vulnerability from climate change and population growth. *Science*, **289**, 284–288.

Ward, R.C. (1989) *Principles of Hydrology*. McGraw-Hill, London, 381 pp. (ISBN 13 978 0 0770 7204 9).

Webb, J., Fullen, M.A. and Blum, W.E.H. (2012) The pedological value of urban landscapes, Chapter **10**. In: Booth, C.A. Hammond, F.N. Lamond J.E. and Proverbs D.G. (eds), *Solutions to Climate Change in the Built Environment*. Blackwell Publishing, Chichester, pp. 113–126 (ISBN 978 1 4051 9507 2).

Winkler, J.A. (1987) Diurnal variations of summertime very heavy precipitation in the eastern and central United States. *Physical Geography*, **8** (3), 210–224.

16

Urbanisation and Stormwater

John W. Davies and Susanne M. Charlesworth

16.1 Introduction

It is arguable that the water cycle in cities has been so modified by urbanisation that it no longer relates to the 'natural' hydrological cycle and, in fact, operates separately to it. Addition of hard infrastructure, such as roads, paving and buildings, has effectively sealed the urbanised area from the natural soils and lithology beneath. Water is conveyed in a mostly pipe-based system, removing water from the city as soon as possible, as something that is unwanted rather than as a potential resource (Semadeni-Davies *et al.*, 2008).

The traditional process of urbanisation increases impermeable areas, provides piped drainage for stormwater, increases flood risk and causes pollution in rivers. There are many impacts associated with different approaches to urban drainage, not least of which is that the drainage system itself can become flooded. This chapter covers the effects of urbanisation on stormwater management and examines simple ways of determining the capacity required for various elements of drainage systems.

16.2 Urbanisation and Flood Risk

Urbanisation has a transforming effect on the water cycle. As greater proportions of the surface area of a catchment become impermeable, less rainwater infiltrates into the ground and more runs off the surface (Figure 16.1). The effect of this is greatest when the urban area is drained by a conventional system of pipes (sewers).

When water infiltrates through the surface of an undeveloped natural catchment, it percolates downwards until it joins a zone in which the ground is saturated. This 'groundwater' moves under the influence of the hydraulic gradient created by the slope of its surface. It may eventually come above ground again via a spring or it may join the flow in

Water Resources in the Built Environment: Management Issues and Solutions, First Edition.
Edited by Colin A. Booth and Susanne M. Charlesworth.

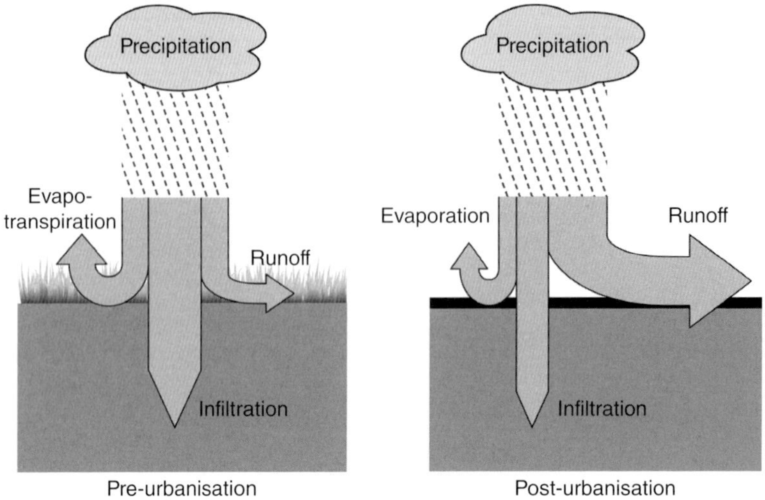

Figure 16.1 Effect of urbanisation on infiltration and runoff (adapted from Butler and Davies, 2011).

a natural watercourse (a river or stream) by some other route. Water moves relatively slowly when it follows this groundwater route; it can take rainwater weeks or months from the time it falls to find its way to a watercourse. Water travels much more quickly as surface runoff or as the flow in a sewer pipe; from falling as rain to reaching the river could take water a matter of minutes or hours. The faster the water travels to the river, the more chance there is that it will all arrive at about the same time. Therefore, in urban areas (especially with piped drainage) flow in a river builds up much more quickly as a result of rainfall and, consequently, achieves much higher peak flows (Figure 16.2) than in a natural catchment. For further details, the reader is referred to Zevenbergen *et al.* (2011).

The effect of urbanisation is, therefore, to increase peak flows in a river as a result of heavy rainstorms, and this has the effect of increasing the risk of flooding. However, the impact is not only on the quantity of the runoff but also on its quality. Surface runoff will cause pollutants and sediments to be washed off the urban surface or scoured by the river (De Miguel *et al.*, 2005). Studies have shown that the urbanised catchment is more polluted than in a rural or natural one; potential sources of pollution include overflows from gully pots, the urban aerosol and eroded catchment soils (Charlesworth *et al.*, 2011).

16.3 Urban Drainage Systems

The link between urbanisation and stormwater is affected significantly by the nature of the urban drainage system in place. As discussed earlier, a combination of conventional piped drainage and the increased proportion of impermeable surfaces associated with urbanisation cause an increase in flood risk. However, not all urban areas are drained in this way. Some urban areas in developing countries do not have an engineered drainage system for either stormwater or wastewater. More recently, an awareness of flood risk and other negative impacts of conventional piped drainage has led to the use in some urban areas of approaches to drainage that make use of semi-natural drainage features that increase

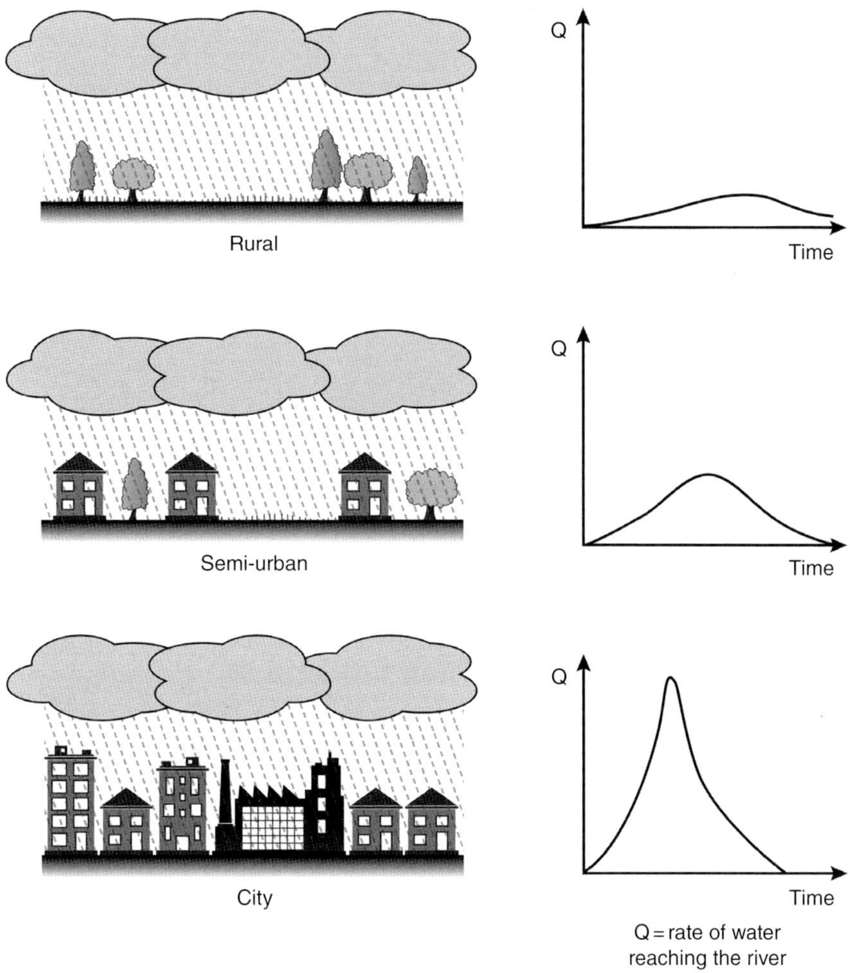

Figure 16.2 A series of stormwater hydrographs illustrating the effects of urbanisation on peak rate of runoff (adapted from Butler and Davies, 2011).

infiltration and storage and reduce the effect shown in Figure 16.2. These systems are known as SUDS (sustainable drainage systems) in the United Kingdom, and are the subject of Chapters 22 and 23 (and the reader is referred to the literature of Charlesworth *et al.*, 2003; Charlesworth, 2010).

In an urban area that does have conventional piped drainage, the type of drainage system has an effect on the environmental impact of stormwater. Broadly, two types of piped drainage system exist: combined and separate. There are some hybrid forms too, but it is the characteristics of combined and separate systems that are important in this context.

In a combined sewer system, stormwater is carried in the same pipe as the wastewater eventually to the water treatment plant (WTP) (see Chapter 26). About 70% of sewer systems in the United Kingdom – those constructed before 1945 – are combined. Figure 16.3 shows a simplified layout of a combined system. During heavy rain, the flow rate of

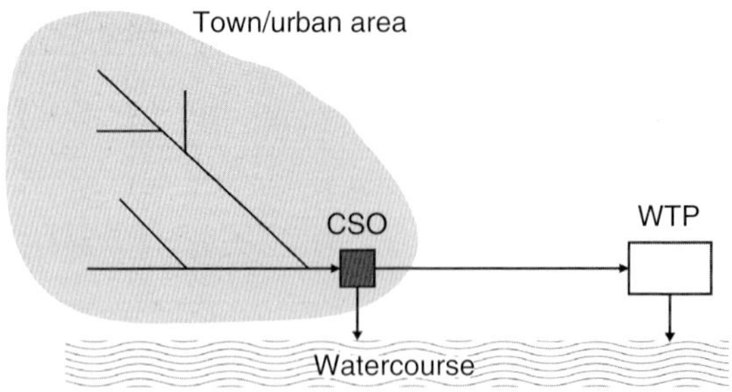

Figure 16.3 Combined sewer system (adapted from Butler and Davies, 2011).CSO = combined sewer overflow; WTP = water treatment plant.

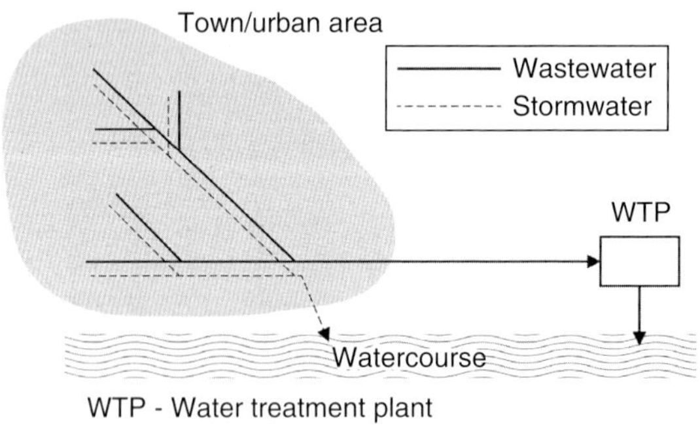

Figure 16.4 Separate sewer system (adapted from Butler and Davies, 2011).

stormwater greatly exceeds that of the wastewater, but it is not considered economic to build a sewer large enough to take the combined stormwater and wastewater flow to the WTP, even for moderate storms. So a structure called a combined sewer overflow (CSO) is provided to divert flows above the amount that will flow on to the WTP (typically 6 to 8 times higher than the average wastewater flow) to a natural watercourse. The flow entering the watercourse may be a dilute mixture of wastewater and stormwater, and the CSO is likely to be designed to prevent large solids from entering the watercourse, although some pollution from CSOs is inevitable.

In the separate system, stormwater is taken by a separate pipe, often constructed side by side with the wastewater sewer. This is shown in a simplified form in Figure 16.4. In this layout, stormwater is not taken to the WTP; rather it is discharged to the watercourse at a suitable point. Stormwater is not mixed with wastewater, so pollution from wastewater is not a problem, but stormwater may be polluted through contact with urban surfaces. Separate systems have been constructed in the United Kingdom since 1945 and now these make up 30% of all systems. However, legislation in various countries has encouraged the use of stormwater drainage based on SUDS rather than a piped system, where viable (with wastewater drained via a separate pipe system).

The combination of urbanisation and drainage cause an increase in flood risk, and the type of drainage system utilised influences the polluting potential of stormwater. The following sections consider these problems in more detail and discuss the design and proper function of storm sewer infrastructure, beginning with their overall problems.

16.4 Problems with Drainage Systems

Clearly a significant problem with the combined system is the pollution that is caused by CSOs (note that in normal UK usage, 'CSO' refers to the structure itself, whereas in US usage, 'CSO' refers to the phenomenon of the overflow discharge and its associated pollution; this chapter uses 'CSO' in the UK sense.) CSOs are considered in more detail in Section 16.9.

Whilst the separate system is effective in avoiding pollution due to the wastewater content of pollution caused by CSOs, it does have some problems of its own. Drainage systems are not sealed systems; they do not normally operate under pressure and are generally easily accessed via inspection chambers and manholes. Separate systems are hard to keep separate, and specifically it is hard to prevent stormwater entering the wastewater pipe due to either missing or wrong connections. This can be a significant problem when stormwater flows greatly exceed that of wastewater; it only takes a few wrong connections for the capacity of the wastewater pipe to be exceeded by unwanted stormwater. An example might involve directing the runoff from a newly paved patio to the wastewater connection at the back of a house (when it should be connected to the stormwater system, or better still to a SUDS device).

As well as this type of *direct inflow*, the capacity of a drainage pipe can be reduced by *infiltration*, where 'infiltration' refers to ingress from groundwater into a sewer at an imperfection or damaged pipe joint. Infiltration in some sewer systems, especially where the water table is high and the condition of the pipe is poor, can be high, certainly reaching as much as 50% of average wastewater flow (Stanley, 1975). This is rather like water leakage from a water main in reverse, and in some cases there can be a link: a leak from a water main can seep down into a sewer, causing no problem on the surface but creating unnecessary costs for water distribution and wastewater management (White *et al.*, 1997).

Some of these problems can lead to deterioration of the sewer systems themselves. For example, the leak from a water main that infiltrates into a sewer might slowly erode surrounding soils, causing a void around the sewer pipe, which might affect its stability. A collapse of the ground might damage or crush the pipe or, worse still, cause a collapse of the road above. For these reasons and others, including the fact that some sewers have been in the ground for a very long time, many sewer systems are in a poor condition. The engineering practice of remedying this situation, the rehabilitation of sewers, has developed into a sophisticated field, both in terms of the risk analysis involved in determining the extent of remedial work that may be needed and the construction techniques used to strengthen or replace sewers with minimum impact on the surface. The principles are presented as online guidance under the general title of *Sewerage Risk Management* (Water Research Centre, 2008). It is tempting to propose that the whole storm sewer infrastructure in the United Kingdom should be replaced. However, OFWAT (Worsfold, no date) have costed the upsizing of the 309 000 km of storm sewer network in the United Kingdom at a value of approximately £174 billion at 2007–2008 prices, which would take centuries to deliver, involve three months closure for every road and incur heavy carbon costs. Therefore, it is clearly not feasible. Water UK (2008) in any case have stated that 'bigger pipes are not the solution to bigger storms'.

16.5 Sewer Flooding

In extreme storm events the rate of runoff may exceed the capacity of the drainage system. Stormwater may flood on to the urban surface as 'exceedance flow'. Under these circumstances the drainage system may be seen as consisting of two components: the 'minor system' (consisting of the normal drainage system) and the 'major system' (on the surface). The latter may consist of 'default pathways' taken by the flood flow, such as roads, paths or incidental storage areas. Alternatively, 'design pathways' may have been created specifically to cope with exceedance flow. Design pathways include floodways, retention basins or designated areas of public open space for temporary storage. At the smaller scale, some adaptation to existing urban features like road profiles and curb heights can improve the effectiveness of pathways for extreme events. Significant components of a design pathway can be elements of urban infrastructure that have a dual role: for example as a road (at times when there is no flood) and as a flood channel (for extreme events). A particularly striking example of this is the Kuala Lumpur SMART (Stormwater Management and Road Tunnel) in Malaysia. This carries stormwater underneath a road carrying traffic. During extreme storm events the road is closed and the whole tunnel is used for stormwater (http://www.smarttunnel.com.my/).

Exceeding the capacity of the urban drainage system during extreme events is a significant cause of surface flooding. In the 2007 floods in the United Kingdom, it was estimated that the inundation of two-thirds of the 57 000 properties affected was a result of sewer flooding (Department for Environment, Food and Rural Affairs (DEFRA), 2008). However, in this case, the sewer system cannot be considered in isolation since its capacity is reduced by rising levels in the receiving waters. A good source of information on all aspects of urban flooding is Jha *et al.* (2012).

16.6 Drainage System Capacity

The simplest way of determining the capacity required for an urban drainage system is to consider a depth of rain d falling on an area A. This represents a total volume V of rainfall, where

$$V = d \times A \tag{16.1}$$

If the rain is falling at a constant intensity (depth/time) and the drainage system is required to accept this water without any build-up on the catchment surface, Equation (16.1) can be rewritten to take account of time:

$$\frac{V}{t} = \frac{d}{t} \times A$$

or

$$Q = i \times A \tag{16.2}$$

where

Q = flow rate (volume/time) of water that must be accepted by the drainage system
i = rainfall intensity (depth/time)

Equations (16.1) and (16.2) are relevant to the design of any drainage system: piped, SUDS-based, or other. For the specific case of a piped system draining an urban area, Equation (16.2) can be made more specific. The urban area is likely to have some surfaces that are connected to the drainage system, like roofs and roads, and some others, like grass verges or gardens that are not so connected. Only the impervious area needs to be considered when determining the required capacity of the drainage system. Not all rainwater falling on these impervious areas will find its way into the drainage system; some will form puddles to evaporate later or will seep through cracks. To reflect these issues, Equation (16.2) can be adjusted:

$$Q = i \times A_i \times C_v \tag{16.3}$$

where

A_i = the impervious area (not the total area but just the area of roofs and roads, etc., connected to the drainage system)
C_v = the proportion of rain falling on A_i that actually reaches the drainage system (the 'runoff coefficient')

Equation (16.3) can be adjusted to take account of the units normally used for each of these parameters. This introduces a constant of 2.78 to balance the units (when they are as specified below):

$$Q = 2.78 \times i \times A_i \times C_v \tag{16.4}$$

in which

Q is in litres per second (l/s)
i is in millimetres per hour (mm/h)
A_i is in hectares (where 1 ha = 10 000 m^2)

Although very simple, Equation (16.4) is the basis of much urban drainage design, especially for small catchments. By simply multiplying three numbers to determine the flow rate to be accepted by the drainage system, and therefore its capacity, it is effectively assumed that the values of rainfall intensity, impervious area and runoff coefficient can all be treated as if they were constant. Of course, in reality they are not. Rainfall intensity varies during a storm, typically building up to a maximum and then dying away. Area builds up with time; considering a pipe draining a small catchment, it will take time for rain from all parts of the catchment to reach the pipe, with longer times for more remote areas. The runoff coefficient, C_v, also increases with time; early rainfall will make the catchment surface wet or build up in puddles, with a greater proportion of rainfall finding its way to the drainage system later in the storm.

Since Equation (16.4) ignores these variations, a 'safety factor' can be used to allow for the fact that the maximum value of Q may be higher than that given by Equation (16.4). A value of 1.3 has become established for this safety factor. Multiplying 2.78 by 1.3 therefore gives

$$Q = 3.61 \times i \times A_i \times C_v \tag{16.5}$$

In designing a drainage system, the values of i, A_i and C_v therefore need to be determined. There are standard methods for determining C_v based on the catchment type, which takes account of soil type and antecedent rainfall conditions (Butler and Davies, 2011). A_i can be determined by measurements from a plan of the catchment and rainfall intensity, where i is related to the duration and frequency of a storm event. The relationship between intensity, duration and frequency is available in a number of different forms, from simple formulae to detailed methods (Butler and Davies, 2011), all based on empirical relationships derived from recorded rainfall data. Intensity can be determined if duration and frequency are known. The problem is in setting duration and frequency for a particular design.

Frequency is commonly expressed in the form 'once in N years', where N is the average period between storms of a particular severity, or alternatively it can also be referred to as the 'return period'. In urban drainage design, return periods of 2 years are common for residential areas, with 5 years used for areas at more risk of flooding. This means that the capacity of the drainage system is likely to be exceeded every 2 (or 5) years, which may cause the system to back up, but not necessarily cause flooding at the surface. However, it is more common to design systems such that surface flooding is prevented for return periods up to 30 years.

In terms of the duration of rainfall, a short duration would give a statistically higher rainfall intensity, but this may not be the worst case in design terms because the storm would be over quickly and there would not be enough time for runoff from the whole catchment to be contributing at the same time. The worst case is when the whole catchment begins to contribute together (at what is called the time of concentration). For durations longer than this, the whole catchment will be contributing together but the rainfall intensity will be statistically lower. In this scenario, duration is set to be equal to the time of concentration.

This approach, in conjunction with Equation (16.5), is referred to as the 'Modified Rational Method'. It is widely used for drainage design.

Example 16.1 Modified Rational Method

The stormwater drainage for a new residential development is being designed. The runoff from the whole catchment will be picked up by a stormwater pipe and discharged to a pond system. The impervious area being drained is 6000 m² and the value of the runoff coefficient C_v can be taken as 0.9. For a rainfall intensity of 50 mm/h, determine the capacity needed in the stormwater pipe (in litres per second) using the Modified Rational Method formula:

$$A_i = 6000\,\text{m}^2 \equiv 0.6\,\text{ha}$$

$$Q = 3.61 \times i \times A_i \times C_v$$

$$Q = 3.61 \times 50 \times 0.6 \times 0.9 = 97.5\,\text{l/s}$$

In practice the Modified Rational Method is embedded in design software, for example the package *Micro Drainge WinDes*, which is the most common tool for this purpose in current use in the United Kingdom. As well as designing new systems, software is also used to simulate behaviour in existing systems (to answer 'what if' questions about particular rainfall events or additional inflows, for example in an existing drainage system). A common package used in the United Kingdom for this is *InfoWorks*.

16.7 Increasing or Decreasing Impermeable Surfaces

The effect shown on Figure 16.2 is worsened if the coverage of an urban area by impermeable surface increases. Unfortunately, this is common, as increasing densification of towns and cities puts pressure on urban space. An example of an increase in an impermeable surface area is the paving of front gardens (or 'sealing') to create parking spaces at the front of properties, especially in areas where street parking is limited (London Assembly, 2005). This is a progressive trend; even in 2005, 68% of front gardens in the London Borough of Ealing were hard-surfaced and the figure was rising (Ealing's Local Agenda 21 Group, 2005). Cost-cutting measures to reduce maintenance of green spaces can also lead to the creation of additional impermeable surfaces. However, there are also strong moves in the opposite direction: to increase green space and permeable surfaces in cities, and therefore to move upwards on Figure 16.2 to a lower level of flood risk. *Future Water* (HM Government and DEFRA, 2008) sought to reduce front garden sealing by a change of permitted development rights, only allowing the retrofit of permeable surfaces without planning permission.

It is a common requirement in the United Kingdom now for the stormwater discharge from a new urban development to be no more than the runoff when the catchment was in its natural pre-developed state, and therefore to cause no increase in flood risk. There are established methods for determining this 'greenfield runoff' (Balmforth *et al.*, 2006) and an approach to drainage based on SUDS or including significant storage is called for in order to address this.

16.8 Storage

Sustainable drainage depends on storage, which can either be permanent (detention) or temporary (retention), infiltration and conveyance. Storage is, therefore, an important aspect of SUDS, and includes semi-natural devices, such as ponds, wetlands and basins (Chapters 24 and 25). Storage reduces the rate at which runoff reaches a natural watercourse, and can therefore reduce flood risk. Completely artificial storage in the form of a tank or an oversized pipe also has this effect, but does not provide the other benefits of SUDS, such as biodiversity or amenity.

To consider how this is caused, the governing relationship for storage can be expressed as:

$$I - O = \Delta S \tag{16.6}$$

where

I = inflow
O = outflow
ΔS = change in the volume of water stored

If inflow represents the runoff from an urban catchment, it is likely to exceed this amount during a storm. While inflow exceeds outflow, the volume of water stored increases (ΔS becomes positive). The total storage capacity required depends on the difference between inflow and outflow and the length of time over which this is maintained. In practice, inflow varies with time (and outflow may also). However, the rate of outflow can be limited by a flow control device (to ensure that it does not exceed the greenfield runoff, for example) such as a weir plate or hydrobrake. As inflow reduces after the storm, outflow will eventually exceed inflow and the storage will empty (ΔS becomes negative).

Infrastructure in which water can be stored can take a number of forms. Alternatives to conventional tanks include 'geocellular units' based on a three-dimensional plastic matrix with a high void ratio within which the water is stored, removing the need for the structural function of a tank. Storage can also be 'distributed' in the form of smaller tanks under the drives or garages of individual properties. 'Tanking' of porous paving using geotextile and the harvesting of rainfall is covered in Chapters 12 and 23.

16.9 Stormwater Quality

As set out in Section 16.2, apart from the risk of flooding, another effect of urbanisation on stormwater relates to its quality. Stormwater carries with it a complex mixture of natural and anthropogenic substances, in solution and suspension. These are typically derived from atmospheric pollution, from vehicles, from buildings and roads, animals, de-icing operations, litter, vegetation, and spills or leaks of oil or chemicals (Memon and Butler, 2005; Ellis and Mitchell, 2006).

When stormwater is mixed with wastewater in combined sewer systems the potential for pollution is even greater, as discussed in Section 16.3. The source of pollutants in this case is, of course, the wastewater. A combined sewer overflow (CSO) is required to achieve a particular flow split to ensure that an appropriate level of flow continues to the WTP, but in addition it must aim to control any potential polluting impact of the discharge on the receiving watercourse. The pollutants of concern could be dissolved or in suspension, yet most attention is paid to larger solids, especially those that are obviously derived from wastewater. The impact of these 'gross solids' (defined as solids larger than 6 mm in two dimensions) is often referred to as 'aesthetic pollution' (Butler *et al.*, 2003).

Aesthetic pollution is particularly offensive to the public and in order to control it modern CSOs are fitted with screens, commonly meshes or perforated plates, which retain gross solids within the sewer system (Figure 16.5). They may need to be mechanically cleaned to prevent blockage during storm flows, though a preferred arrangement is where the screen is self-cleansing as a result of the design of the screen itself and the flow patterns in the CSO.

Screens do not, however, control the discharge of dissolved or suspended pollutants. These pollutants can be retained in the sewer system by using a storage device. When the CSO diverts flow from the sewer, instead of discharging it to the watercourse, it can pass into a storage tank where particulate-associated pollutants can settle out. When the storm flows have subsided, the stored water can be returned to the sewer system. Any storm that exceeds the design criteria, either in terms of the inflow rate or the period over which it is sustained, will result in direct overflow to the watercourse. This will lead to negative environmental impacts, the details of which are beyond the scope of this chapter, but Butler and Davies (2011) have further details. In a SUDS management train, if ponds or wetlands are utilised that incorporate vegetation, physical settling will occur as in a standard tank, but the plants will also systemically take up many dissolved pollutants (Charlesworth, 2010).

A CSO with screens, and without significant storage, is considered the standard design, but other approaches are needed to cope with particular requirements (Butler and Davies, 2011).

Figure 16.5 Storm sewer screen at the outflow of a SUDS management train, North Hamilton, Leicester (UK).

16.10 Conclusions

This chapter has shown that urbanisation has a transforming effect on the water cycle. By increasing the proportion of the surface area that is impermeable, flood risk increases. This is made worse when a traditional piped drainage system is used. This also has the potential to increase the risk of pollution to the receiving watercourse. The impact of this is related to the type of drainage system in use. In a combined system, where stormwater is carried in the same pipe as wastewater, the pollution risks are particularly problematic as they result from contact with wastewater. A separate system, in which stormwater and wastewater are carried separately, avoids this, though stormwater itself can still be polluted. In practice, it is hard to achieve complete separation: stormwater is commonly found in wastewater pipes and can cause significant reduction in capacity.

Because of problems like this, and as a result of the age of many existing drainage systems, rehabilitation of sewer systems to solve hydraulic, environmental or structural problems is an important area of drainage engineering. When old storm sewers are in a poor condition, flooding problems can be made worse; however, their replacement can be prohibitively expensive.

Drainage systems of any type are designed to have sufficient capacity for a specified frequency of storm. It is common to design a drainage system for residential areas to have capacity for a storm that occurs on average once every two years. In extreme cases, stormwater may flood on to the urban surface, but it is common to design systems such that surface flooding is prevented for return periods of up to 30 years. Simple methods for designing drainage components, including storage, have been described and explained.

References

Balmforth, D., Digman, C., Kellagher, R. and Butler, D. (2006) *Designing for Exceedance in Urban Drainage – Good Practice*. CIRIA **C635**.

Butler, D. and Davies, J.W. (2011) *Urban Drainage*, 3rd Edition. Spon Press.

Butler, D., Davies, J.W., Jefferies, C. and Schütze, M. (2003) Gross solid transport in sewers. *Proceedings of the Institution of Civil Engineers, Water and Maritime Engineering*, **156** (WM2), 175–183.

Charlesworth, S.M. (2010) A review of the adaptation and mitigation of global climate change using sustainable drainage in cities. *Journal of Water and Climate Change*, **1**, 165–180.

Charlesworth, S.M., Harker, E. and Rickard, S. (2003) Sustainable Drainage Systems (SuDS): a soft option for hard drainage questions? *Geography*, **88**, 99–107.

Charlesworth, S., De Miguel, E. and Ordóñez, A. (2011) A review of the distribution of particulate trace elements in urban terrestrial environments and its application to considerations of risk. *Environmental Geochemistry and Health*, **33**, 103–123.

De Miguel, E., Charlesworth, S.M., Ordonez, A. and Seijas, E. (2005) Geochemical fingerprints and controls in the sediments of an urban river: River Manzanares, Madrid (Spain). *The Science of the Total Environment*, **340**, 137–148.

Department for Environment, Food and Rural Affairs (DEFRA) (2008) *Improving Surface Water Drainage*. Consultation to accompany proposals set out in the government's water strategy. DEFRA.

Ealing's Local Agenda 21 Group (2005) *The Extent of Hard Surfacing of Front Gardens in the London Borough of Ealing*. Report on a Community Survey Research Project, London Borough of Ealing.

Ellis, J.B. and Mitchell, G. (2006) Urban diffuse pollution: key data information approaches for the Water Framework Directive. *Water and Environment Journal*, **20**, 19–26.

HM Government and DEFRA (2008) *Future Water: The Government's Water Strategy for England*. 98 pp.

Jha, A.K., Bloch, R. and Lamond, J. (2012) *Cities and Flooding: A Guide to Integrated Urban Flood Risk Management for the 21st Century*. World Bank.

London Assembly (2005) *Crazy Paving – The Environmental Importance of London's Front Gardens*. Report published by the London Assembly Environment Committee.

Memon, F.A. and Butler, D. (2005) Characteristics of pollutants washed off from road surfaces during wet weather. *Urban Water Journal*, **2**, 171–182.

Semadeni-Davies, A., Hernebring, C., Svensson, G. and Gustafsson, L.-G. (2008) The impacts of climate change and urbanization on drainage in Helsingborg, Sweden: suburban stormwater. *Journal of Hydrology*, **350**, 114–125.

Stanley, G.D. (1975) *Design Flows in Foul Sewerage Systems*, Project Report No. **2**. Department of the Environment.

Water Research Centre (2008) *Sewerage Risk Management*. Available at: http://srm.wrcplc.co.uk.

Water UK (2008) *Lessons Learnt from Summer Floods 2007: Phase 2 Report, Long-Term Issues*. Water UK's Review Group on Flooding, London.

White, M., Johnson, H., Anderson, G. and Misstear, B. (1997) *Control of Infiltration to Sewers*, Report R175. CIRIA.

Worsfold, M. (no date) *Ofwat – Sustainable Drainage*. Available at: http://www.ofwat.gov.uk/media-centre/speeches/prs_spe_100303drainage.pdf.

Zevenbergen, C., Cashman, A., Evelpidou, N., Pasche, E., Garvin, S. and Ashley, R. (2011) *Urban Flood Management*. CRC Press.

17

River Flood Defences

Carly B. Rose

17.1 Introduction

Flooding is a natural process for many river systems and typically poses a problem only where the washlands and floodplains are put to use by humankind: as populations have grown, the area occupied, for agricultural, habitation and industrial purposes, has inevitably expanded. The oldest documented flood control works were constructed in China around 2500 years ago (Takeuchi, 2004); currently in England and Wales, over 5 million properties (both domestic and commercial) are believed to be at risk of flooding from rivers or the sea (Environment Agency, 2009) and almost 3 million properties, in addition to this, may be affected by surface water (pluvial) flooding. Both coastal and surface water floods are covered elsewhere in this volume. This chapter will consider the issues of floods from rivers only (fluvial flooding).

One of the reasons for building in a river's floodplain can be that the risk of flooding is unknown, or unacknowledged, particularly where long periods have elapsed between severe flood events. Even where the flood risk is known, the land may still appear attractive for development as it is typically flat and the existence of the hazard may also render it relatively cheap to acquire. In some countries, building regulations may have been brought into being in order to prevent further expansion, but these are not always enforced for reasons as varied as political expediency, financial corruption or the humanitarian pressures arising from existing informal settlements. The latter problem is particularly likely to be exacerbated by future population growth, as well as predicted climate change effects; the World Bank has recently published a handbook in which these issues, and some suggested solutions, are explored in detail (Jha *et al.*, 2012).

The methods deemed appropriate for managing larger-scale fluvial flood risks can be seen to change over time, not only in terms of the approaches adopted but also the financial and political frameworks that are in operation. The relevance of flood probability

Water Resources in the Built Environment: Management Issues and Solutions, First Edition.
Edited by Colin A. Booth and Susanne M. Charlesworth.

(return periods) in the context of design standards is also discussed, as the issue of 'how high is high enough?' gains prominence with reference to climate change predictions. In addition to community-level flood defence measures, there are now available mitigation techniques applicable to individual dwellings or small groups of properties. This chapter will, however, focus upon the larger alleviation schemes both past and present, including recent innovations such as temporary and demountable barriers. The alternative solutions to flood risk management, such as sustainable drainage systems and property level mitigation, are beyond the remit of this chapter, but possible options for dealing with flood risk in the future are also examined.

17.2 The Historical Context

> Although both Defra's and the Environment Agency's current strategies for flood risk management place significant emphasis on alternative means of reducing or mitigating the risks of flooding, floodwalls and embankments will continue to provide one of the most important means of protecting houses, businesses and infrastructure against flooding for the foreseeable future (Rickard, no date).

Flood management practices have, inevitably, changed over time: in the Netherlands, for example, prior to 1100 CE, individual villages were originally protected by dikes (*dorpspolders*); subsequently a widespread system of embankments to confine, or divert, the rivers themselves was constructed (Berendsen, 2007). In more recent times, these dike systems have been strengthened to withstand floods of a magnitude to be expected (on average) once in 1250 years (Berendsen, 2007).

Earthen embankment systems intended to restrict heightened flows to the desired channel can also be found in the United Kingdom, typically protecting agricultural land. An example is the system of 'argaes' in the area around the confluence of the rivers Severn and Vyrnwy on the border between England and Wales (the name is believed to stem from the Welsh phrase *ar gau* meaning closed or shut). In the winter months in particular, the confluence area is subject to frequent, but relatively minor, flood events for which these embankments provide adequate protection. The structures are, however, deliberately profiled to allow water to flow over the top of them in the more severe flood events: this allows the enclosed farmland to act as a flood storage area, thereby protecting downstream settlements. Strategically sited sluices are then operated to release the water after the rivers have returned to safe levels (Environment Agency, 2008a, p. xix). By contrast, the flood levees around New Orleans were not designed to be overtopped: one element of the catastrophic flood event arising from Hurricane Katrina in 2005 was that the storm surge flowing over the crest led to erosion and subsequent structural failure (van Heerden, 2005).

The use of other 'hard defence' methods, typically including walls of brick, masonry or concrete, together with sheet piling, weirs and pumping stations, can be found in many locations; for detailed information specific to the United Kingdom, in terms of design, operation and maintenance, reference should be made to the Environment Agency's online *Fluvial Design Guide* (Environment Agency, no date a). Other approaches have included deepening or widening channels, vegetation removal, regrading and dredging (Parker and Harding, 1978) but these methods are now deemed to be unsustainable and ecologically inappropriate (Hansard, 2007).

More innovative methods are now favoured, one of the most iconic being the Thames Barrier and its associated embankments; the Barrier itself consists of a series of movable gates, brought into operation only when the relevant weather conditions are forecast, as the river is a major shipping route. It was primarily intended to protect 125 km^2 of central London from extreme tidal flooding from the Thames estuary but the Barrier can also be used to alleviate fluvial flooding; for example:

> During the first week of 2010, the barrier was closed five times as high fluvial flows and the high spring tides coincided (Environment Agency, no date b).

In addition to the meteorological processes that give rise to flooding, it should be borne in mind that humankind's manipulation of the natural world can exacerbate flood risk, as can the political drivers underpinning them. As an example, both during and in the aftermath of, the Second World War the UK's farmers were exhorted to maximise land usage for food production, with government grants being made available to drain waterlogged areas (Hansard, 1940). In more recent times, however, it has become evident that such practices can contribute to increased downstream flood risks, by reducing natural flood storage capacity and increasing peak flows during heavy rainfall (Blanc *et al.*, 2012).

In the past decade, there has been increasing recognition of the need to provide flood alleviation by means other than permanent, hard-engineered defences. The use of temporary and demountable barriers will now be examined.

17.2.1 Temporary and Demountable Barriers

An advantage of these methods is their suitability for areas where, for example, tourism interests would militate against the construction of high walls obstructing river views. The towns of Bewdley (United Kingdom) and Givet (France) are such locations, which now benefit from the 'demountable' type of flood defences, originating from Germany: these are deployed only when flooding is forecast, leaving an uninterrupted riverside scene at other times (Environment Agency – Midlands Region, 2006; ArcelorMittal Commercial RPS S.a.r.l., no date). One example of a 'temporary' barrier method is a Swedish design termed the 'pallet barrier'; this is used to protect parts of the world heritage site of Ironbridge (United Kingdom). The original concept made use of standard wooden Europallets and waterproof polypropylene membranes, although more sophisticated versions with steel supports have subsequently been developed (Hydro Response Ltd, 2006). Another type of temporary defence employs highly robust air- or water-filled tubes, often described as 'mobile dams'; these conform to the terrain by friction and gravity, offering greater flexibility than the more rigid methods described above. These, and other, methods are discussed more fully elsewhere (e.g. Ogunyoye and van Heereveld, 2002; Jha *et al.*, 2012). It should be noted that the use of sandbags, although very common, is one of the least efficient methods, unless used in conjunction with impermeable membranes (as deployed by experts such as the armed forces) (Dhonau, 2009).

All of the above methods carry significant resourcing implications, including storage facilities for the equipment, transport to the relevant area when flooding is expected and availability of the human resources needed to deploy them. They are not, therefore, suitable for use in all instances.

In designing flood alleviation schemes, it is necessary to understand both the natural processes at work and how this relates to the vexed issue of 'how high is high enough?' in relation to the design standards to be employed. The concept of flood return periods will first be examined.

17.2.2 Flood Return Periods

> Knowledge of the past is a key for understanding present and future; this is especially true for climate history (Glaser and Stangl, 2004).

For many decades, the reference events for flood management in England and Wales were the extensive floods of 1947, often referred to by flood professionals as a 'one in a hundred year' event. Such phrasing, although intended merely as shorthand for the statistical probability of the recurrence of a flood of this magnitude, is, however, easily misunderstood by the general public, who interpret it as meaning that such floods occur on a predictable, cyclical basis. Expressing this 'return period' as a percentage chance of occurrence in a given year is less misleading. For example, the 1998 event on the River Leam (Leamington Spa, United Kingdom), in which the river reached its highest level since records began in 1735, was calculated to have a 0.57% chance of occurring in any given year (Horner and Walsh, 2000). It should also be noted that the return period of a rainfall event may differ from the return period of the flood to which it contributes: factors such as whether the soils of the river basin are in a dry or saturated condition prior to the inundation are relevant here.

One of the assumptions underpinning recurrence calculations is 'stationarity', meaning that the probability distribution is stable over time. The prospect of climate change is now introducing a major source of uncertainty into the picture, as extreme flood events, which have occurred only rarely in the past and for which detailed data are lacking, may now be expected to arise more often in some areas. Some existing flood defence structures have been designed to withstand a 1% flood, as it was understood in, say, the 1960s; if more extreme events do indeed become more commonplace, then overtopping can also be expected to occur more frequently.

As an example, the history of the town of Shrewsbury (Shropshire, United Kingdom) reveals long-standing flooding issues from the River Severn (Harding and Parker, 1974). Some of the more notable flood levels are marked on a plaque attached to the flood wall completed in 2004, with the highest being the winter flood of 1946 (5.43 m above normal river levels) (Figure 17.1). The events shown are, however, only those based on the available validated data: historic records show that a far more severe flood occurred in 1795, affecting the whole of the Severn catchment and damaging, or destroying, many bridges (University of Gloucester, no date), some of which dated from the 13th century.

Although reliable local data are not available for this event, it has been estimated that the level of the river in the centre of Shrewsbury reached 5.7 m (Environment Agency – Midlands Region, 2005). A flood alleviation scheme for one part of the town was completed in 2004; it is, however, designed to protect the relevant area from a 1% flood, not an event of the magnitude of 1795. The risks posed by infrequent extreme events must, inevitably, be balanced against the financial costs of constructing such schemes, and the Environment Agency's official publication covering the Shrewsbury scheme clearly states that:

> ... It should be borne in mind that a more severe flood could overtop the new defences (Environment Agency – Midlands Region, 2005).

Whether the residents of areas behind defences are aware of, still less accept, the existence of such a 'residual risk' is a moot point, as will be seen in the next section.

Figure 17.1 Plaque showing historic flood levels in Shrewsbury (UK) (photo courtesy of J.E. Lamond).

17.2.3 Design Standards

The engineering design of flood alleviation schemes defines the standard to which the structures will provide protection (the design flood level); a measure of 'freeboard' is then added, which is effectively a safety margin allowing for uncertainties, to produce a finished crest level (Rickard, no date). The general public, however, commonly perceive that 'flood defences' are capable of providing protection against any and all flood events. Where flood levels have exceeded a scheme's design standards, there have been accusations that the defences have 'failed', even though no breach or structural failure has actually occurred (e.g. Jarrow and Hebburn Gazette, 2012). Thus, the extreme floods that affected Carlisle (United Kingdom) in 2005 had an estimated return period of between 175 and 200 years (0.5% probability), but the local flood defences had been built to withstand a far less severe event, with a 1.33% average probability (DESURBS, no date) leading to widespread flooding of properties.

An increasing recognition of the need to manage water resources more holistically led to the EU Floods Directive of 2007; in anticipation of compliance with this, the UK Government published the policy review entitled *Making Space for Water* (Department for Environment, Food and Rural Affairs (DEFRA), 2005). This clearly articulated the premise that floods cannot be prevented, but flood risk can be managed. It should be noted, however, that this successful implementation of this strategy incorporates a need to manage the behaviours of the people living and working in the floodplain, as well as the fluvial environment itself. A key element of *Making Space for Water* is the intention to empower those at risk to take suitable action themselves (e.g. by investing in property-level protection measures) rather than being reliant solely on community-level schemes provided by the government (DEFRA, 2005). In this way, the 'residual risk' factor inherent in hard-engineered alleviation schemes can be addressed, at least in part. Raising awareness of flood risk amongst the population is an important aspect of this approach, but is not sufficient, of itself, to bring about behavioural change: an understanding of issues such as psychology and communication theory is also needed at policy-maker level. These are topics that are beyond the remit of the current volume but are discussed more fully elsewhere (e.g. Rose *et al.*, 2012).

Alongside hard-engineered alleviation measures, another aspect of alleviating flood risk is the provision of flood forecasting and warning systems, allowing the at-risk population to take appropriate action, where possible, before inundation commences.

17.2.4 Flood Forecasting and Warning

In order to forecast flood levels and, thus, be in a position to disseminate warnings with an adequate lead time, detailed data on both past events and prevailing conditions are needed. In the United Kingdom, telemetry systems and highly sophisticated computer models have been developed over many years, incorporating predicted rainfall data from the Met Office (Flood Forecasting Centre, no date). A detailed discussion of such systems is beyond the remit of this volume, but further information on the methods are available (e.g. Evans, 2011; Jha *et al.*, 2012).

This type of approach may not, of course, be appropriate or achievable in all locations, whether for reasons of lack of historical data, inadequate infrastructure, financing or governance issues, but early warning systems are a key element of flood risk management in terms of safeguarding life as well as property. It must, however, be borne in mind that forecasting is not an exact science: on occasion, there may be failures to give adequate warnings and there may also be instances of false alarms (flood warnings issued where no flooding subsequently occurs, sometimes described as 'crying wolf') (Parker *et al.*, 2007). Both situations can damage the relationship between the warning authority and the at-risk population they serve; hence adequate attention must be paid to issues of communication in this context (Orr and Twigger-Ross, 2009). There is currently a proposal for the UK warning services to move from deterministic to probabilistic forecasting, in order to facilitate risk-based decision-making, such as the need to evacuate properties or suspend transport links (Sene *et al.*, 2007).

Although 'hard' defence schemes are still maintained and new ones continue to be built in the United Kingdom (e.g. Elgin FAS; see Rickard, 2007; Moray Council, 2011) the strategy to be adopted in future years is more holistic in nature: for example, restoration of historic landscape features or changes in farming practices may offer a more sustainable long-term approach (Environment Agency, no date c). Some of the future options for flood risk management will now be examined.

17.3 The Future

> ... we must either invest more in sustainable approaches to flood and coastal management or learn to live with increased flooding (Evans *et al.*, 2004).

The anticipated climate changes associated with global warming, such as an increased frequency and intensity of rainfall events (UK Climate Impacts Programme, no date), will impact upon how new flood alleviation schemes are designed, as well as how the maintenance or refurbishment of existing structures should best be approached: should design crests be increased, and, if so, by how much? The *Foresight Future Flooding* report and the subsequent update (Evans *et al.*, 2004, 2008) represent a detailed, independent, examination of the issues as applying to the United Kingdom over the coming years. The future of the Thames Barrier, for example, has been subjected to intense scrutiny (e.g. Reeder and Tarrant, 2007; Environment Agency, 2008b), with the conclusion being that no major changes will be needed prior to 2030. Beyond that date, it has been estimated that additional flood storage could see the Barrier being effective up to 2100 (Environment Agency, 2008b).

Restoration of historic landscapes, such as reinstating peat bogs or blocking artificial drainage channels created for agricultural purposes, may offer additional mitigation of flood risk, together with ecological benefits in terms of improved wildlife habitats. Similarly, planting forests in floodplains may help to retard runoff rates by enhancing soil infiltration (Environment Agency, no date c), particularly if using endangered native wetland species such as the Black Poplar (*Populus nigra*, sp. *Betulifolia*). Sustainable drainage schemes, covered elsewhere in the volume, follow similar principles of working with water rather than attempting to constrain it.

An alternative suggested approach would allow humankind to live with flooding with minimal disruption, for example by designing houses that float to accommodate water-level fluctuations. One such scheme has been constructed at Maasbommel (Netherlands); 50 homes, known as 'amphibious housing', have hollow concrete bases, which permit the houses to rise and fall with changing water levels over a range of 5.5 m whilst linked to piles driven into the banks, which keep them in place. All utility links are flexible and the gangways for accessing the units from higher ground are also attached to piling (the grey columns seen on the left of Figure 17.2) such that life can continue in a normal fashion until the water levels subside (Warren, 2011). Other, more ambitious developments on similar lines are in development worldwide, envisaged to include 'water cities' and floating islands (e.g. see Figure 17.2).

By combining a number of the methods discussed here, as appropriate to different locations and circumstances, it may be possible to continue to live and work in the floodplains for the foreseeable future. A great deal will, however, depend upon the emissions scenario and, in particular, the effects exerted by emerging world economies such as China, India and Brazil.

17.4 Conclusions

In this chapter, river flood defences have been examined in the context of some of the practical and political changes that have arisen, together with some of the recent developments that move away from traditional 'hard-engineered' schemes. The concepts of return periods and design standards have been discussed, and the importance of considering the needs and perceptions of the at-risk population have been introduced, insofar as this impacts

Figure 17.2 Floating housing, Maasbommel (Netherlands).

upon the strategic choices to be made by policy-makers. A number of commonly applied methods for managing flood risk as well as some of the innovative nontraditional approaches have been examined.

The future of flood risk management as it relates to river flooding has been considered and the need to 'live with', rather than attempt to control, flooding has been emphasised. Sustainable, and affordable, approaches in the face of the challenges posed by climate change will be crucial to our progress in this area in the coming years.

References

ArcelorMittal Commercial RPS S.a.r.l. (no date) *Flood Defence – Givet, France*. ArcelorMittal Commercial RPS S.a.r.l., Luxembourg. Available at: http://www.arcelormittal.com/palancola/uploads/files/AMCRPS_CS%20Givet_GB_web.pdf.

Berendsen, H. (2007) *Flooding* [online], [Accessed 27/09/12]. Available at: http://www.geo.uu.nl/fg/palaeogeography/results/flooding.

Blanc, J., Arthur, S., Wright, G. and Beevers, L. (2012) *The Effect of Land Drainage on Flood Risk and Farming Practice*. Scotland's Centre of Expertise for Waters (CREW), Edinburgh.

Department for Environment, Food and Rural Affairs (DEFRA) (2005) *Making Space for Water: Taking Forward a New Government Strategy for Flood and Coastal Erosion Risk Management in England*. Department for Environment, Food and Rural Affairs, London.

DESURBS (no date) *Carlisle (UK) Flood 2005* [online], [Accessed 16 May 2012]. Available at: http://www.desurbs.net/Carlisle%20flood.html.

Dhonau, M. (2009) *Flood Defence Products* [online]. Gloucestershire County Council.

Environment Agency (2008a) *Severn Catchment Flood Management Plan – Draft Report [May 2008] – Executive Summary*. Environment Agency, Bristol.

Environment Agency (2008b) *Thames Gateway 2100 (TE2100)* [online], [Accessed 16 May 2012]. Available at: http://www.environment-agency.gov.uk/homeandleisure/floods/104695.aspx.

Environment Agency (2009) *Prepare Your Property for Flooding*. Environment Agency, Bristol.

Environment Agency (no date a) *Fluvial Design Guide* [online]. Available at: http://evidence.environment-agency.gov.uk/FCERM/en/FluvialDesignGuide/Fluvial_Design_Guide_Overview.aspx.

Environment Agency (no date b) *Thames Barrier Closures – Indicator Two* [online]. Available at: http://www.environment-agency.gov.uk/research/library/publications/41065.aspx.

Environment Agency (no date c) *Types of Flood Defences* [online], [Accessed 20/06/2012]. Available at: http://www.environment-agency.gov.uk/homeandleisure/floods/31738.aspx.

Environment Agency – Midlands Region (2005) *Frankwell Flood Alleviation Scheme Shrewsbury*. Environment Agency, Midlands Region, Solihull.

Environment Agency – Midlands Region (2006) *Creating a Better Place – Bewdley Flood Defences – The invisible defences*. Environment Agency, Midlands Region, Solihull.

Evans, K. (2011) Flood warning and incident management. In: Lamond, J., Booth, C., Hammond, F., Proverbs, D. (eds), *Flood Hazards: Impacts and Responses for the Built Environment*. CRC Press, Taylor & Francis Group, pp. 25–36.

Evans, E., Ashley, R., Hall, J., Penning-Rowsell, E., Saul, A., Sayers, P., Thorne, C. and Watkinson, A. (2004) *Foresight, Future Flooding*. London Office of Science and Technology, p. 366.

Evans, E.P., Simm, J.D., Thorne, C.R., Arnell, N.W., Ashley, R.M., Hess, T.M., Lane, S.N., Morris, J., Nicholls, R.J., Penning-Rowsell, E.C., Reynard, N.S., Saul, A.J., Tapsell, S.M., Watkinson, A.R. and Wheater, H.S. (2008) *An Update of the Foresight Future Flooding 2004 Qualitative Risk Analysis*. Cabinet Office, London.

Flood Forecasting Centre (no date) Homepage [online], [Accessed 2012]. Available at: http://www.ffc-environment-agency.metoffice.gov.uk/.

Glaser, R. and Stangl, H. (2004) Climate and floods in central Europe since AD 1000: data, methods, results and consequences. *Surveys in Geophysics*, **25** (5–6), 485–510.

Hansard (1940) *Land Drainage Grants. HC Deb 361* [online]. HM Government, London. Available at: http://hansard.millbanksystems.com/commons/1940/may/23/land-drainage-grants.

Hansard (2007) *Written Answers. 22 Oct 2007: Column 85W* [online]. Available at: http://www.publications.parliament.uk/pa/cm200607/cmhansrd/cm071022/text/71022w0020.htm.

Harding, D.M. and Parker, D.J. (1974) Flood Hazard at Shrewsbury, United Kingdom. In: White, G.F. (ed.) *Natural Hazards*. Oxford University Press, London.

Horner, M.W. and Walsh, P.D. (2000) Easter 1998 Floods. *Journal of Chartered Institute of Water and Environmental Management (CIWEM)*, **14** (December), 415–418.

Hydro Response Ltd (2006) *Flood Barriers from Geodesign Barriers*™ [online], [Accessed 11/09/2012]. Available at: http://www.hydroresponse.com/flood_barrier.htm.

Jarrow and Hebburn Gazette (2012) Residents' anger after £3m flood defence fails. *Jarrow and Hebburn Gazette*, Friday 29 June 2012, 10:51 pm.

Jha, A.K., Bloch, R. and Lamond, J. (eds) (2012) *Cities and Flooding – A Guide to Integrated Urban Flood Risk Management for the 21st Century*. World Bank, Washington DC, p. 637.

Moray Council (2011) *Contract Signing Heralds Start for Elgin Flood Scheme* [online], [Accessed 11/09/2012]. Available at: http://www.moray.gov.uk/moray_news/news_70428.html.

Ogunyoye, F. and van Heereveld, M. (2002) *Temporary and Demountable Flood Protection – Appendices to Interim Guidance – R&D Publication 130/2 Part 1*. Environment Agency, Bristol.

Orr, P. and Twigger-Ross, C. (2009) *Communicating Risk and Uncertainty in Flood Warnings: A Review of Defra/Environment Agency FCERM Literature – Science Project SC070060/SR2*. Environment Agency, Bristol.

Parker, D.J. and Harding, D.M. (1978) Planning for urban floods. *Disasters*, **2** (2), 47–57.

Parker, D.J., Tunstall, S.M. and McCarthy, S. (2007) New insights into the benefits of flood warnings: results from a household survey in England and Wales. *Environmental Hazards*, 7 (3), 193–210.

Reeder, T. and Tarrant, O. (2007) Future flood: risk management in London and along the tidal Thames. *Issues in Risk Science* [online], (7). Available at: http://www.abuhc.org/research/floods/Pages/publications.aspx.

Rickard, C.E. (2007) *Elgin Flood Alleviation Scheme Downstream Areas – Independent Review Final Report* [online], [Accessed 14/05/2008]. Available at: http://216.239.59.104/search?q=cache:kBHbuiKKmNUJ:www.moray.gov.uk/minutes/data/EN20070117/elginfloodalleviationscheme-downstreamapp.pdf+ELGIN+fas%2BCOSTS&hl=en&ct=clnk&cd=1&gl=uk&lr=lang_en.

Rickard, C.E. (no date) *Fluvial Design Guide – Chapter 9 – Floodwalls and Flood Embankments*. Environment Agency, Bristol.

Rose, C.B., Manktelow, K.I., Booth, C.A. and Proverbs, D.G. (2012) Psychological factors affecting flood coping strategies. In: Mambretti, S. (ed.), *Flood Risk Assessment and Management*. WiT Press, Southampton, pp. 105–114.

Sene, K., Huband, M., Chen, Y. and Darch, G. (2007) *Probabilistic Flood Forecasting Scoping Study (R&D Technical Report FD2901/TR)*. Department for Environment, Food and Rural Affairs – Flood Management Division, London.

Takeuchi, K. (2004) Dujangyan and Shingenteh: lessons from the World's longest living flood control works. In: *The Basis of Civilization – Water Science? (Proceedings of the UNESCO/IAMS/IWHA Symposium)*, Rome, December 2003, **286**, 2004. IAHS, pp. 195–200.

UK Climate Impacts Programme (no date) *UK Climate Projections – UK-Wide Key Findings* [online], [Accessed 04/04/12]. Available at: http://ukclimateprojections.defra.gov.uk/content/view/511/499/.

University of Gloucester (no date) *Historic Flooding in the Severn Catchment* [online], [Accessed 17/06/2012]. Available at: <http://insight.glos.ac.uk/tli/activities/co-fast/severnfloods/lowersevern/Pages/Bewdley.aspx>.

van Heerden, I.L. (2005) *Failure of Levee Systems Surrounding Greater New Orleans During Hurricane Katrina – Preliminary Assessment*. Committee on Homeland Security and Governmental Affairs, US Senate, Washington, DC.

Warren, R. (2011) Staying above water. *Financial Times*, March 18, 2011, 10:00 pm.

18

Coastal Flood Defences – Strategies for Protection in the United Kingdom

Trevor Goodhew

18.1 Introduction

It is perhaps not surprising that many towns and cities are positioned in areas at risk of coastal flooding because centres of commerce have historically grown around sites accessible to sea freight, typically adjacent to natural coastal basins or mouths of deep estuaries that have allowed the development of significant ports. Many such centres now face the concern of an increasing threat of flooding from rising sea levels and the coincidence of storms and surges. This chapter focuses on forms of coastal flood defence and approaches in the United Kingdom to address the threat of coastal erosion and flooding, introducing historic approaches and later presenting proposals for innovative modern solutions.

18.2 Holding Back the Sea

The traditional protective response to hold back the sea has been to build coastal walls as defences, but these are expensive to construct and maintain. They are also insensitive to the needs of the natural environment. A more sustainable approach to the natural environment has led to more thoughtful interventions with an awareness of the impact that these defences have upon the natural equilibrium of erosion and deposition. More complex strategies will be required and delivered in the future, by integrating the drivers of development with sustainable urban drainage and responses to rising sea levels. No doubt this will produce innovative adaptations responding to coastal, fluvial and pluvial flooding and, where feasible, may include navigable barriers to defend estuarine towns and cities from rising sea levels.

Strategies for dealing with flood risk have usually developed in response to particular significant flood events. This was the case for the 2007 floods across England, prompting

Water Resources in the Built Environment: Management Issues and Solutions, First Edition.
Edited by Colin A. Booth and Susanne M. Charlesworth.

the subsequent review by Sir Michael Pitt (2008) and the government's response in enacting the Floods and Water Management Act 2010. The consequence has been a change in policy and practice in defending against coastal flooding and in particular leading to:

- a proportionate risk-based approach to flood defence;
- expectation that the government will reduce its funding of schemes and expect those who benefit to pay more, including developers;
- the government reducing its operational roles, which have historically been exercised through the Environment Agency (EA), by passing more responsibility to local authorities (local, district or unitary authorities), which will carry these out in collaboration with the Environment Agency;
- increasing responsibility for the local communities, who will take a more proactive role; and
- continued growth in urban populations, many of which are located along the coastline.

The current UK Government policy on 'Floods and Coastal Erosion Risk Management' is set out in the document *Understanding the Risks, Empowering Communities, Building Resilience* (EA, 2011a). This strategy aims at managing the risk of flooding and coastal erosion and, where possible, improving the standards of protection. It puts 'sustainability at the heart of the actions' taken, whilst directing solutions and sources of funding towards local communities rather than to government-funded national activity. This document also sets out the scale of the challenge for 'coastal erosion risk management', with an approximate analysis of the task:

> There are 4,500 km of coast in England, of which 1,800 km is at risk of coastal erosion and only 340 km is presently defended. It is estimated that 200 properties are currently vulnerable to coastal erosion but by 2029, up to 2,000 residential properties, and 15 km of major road and railway may become vulnerable (Halcrow, 2009), indicating a more challenging future.

The 2010 Floods and Water Management Act sets out the responsibilities of local (district or unitary) authorities, which includes carrying out works to tackle coastal erosion and manage the risk from flooding. 'Coastal Risk Management Authorities' should carry out such works in collaboration with the EA. They have also been given powers to protect land against coastal erosion and to control third-party activities on the coast. This includes constructing private defences or removing beach material. A more strategic perspective is taken by 'Coastal Groups', comprising members from local authorities and other relevant organisations (such as Natural England, the EA, Network Rail and English Heritage) working in partnership. It is their responsibility to produce Shoreline Management Plans (SMPs), which assess the risks from coastal flooding and erosion and work out how to manage these risks over the short-, medium- and long-term perspectives.

Responding to climate change provides opportunities for creative urban development, whilst protecting the natural environment. At the same time, pressures from:

- growing urban populations;
- consequential development; and
- increased flood risk, from more extreme weather patterns and sea level rise due to global warming are driving future policies and practice.

There will need to be recognition that some locations should service human habitation as a priority whilst other locations place enhancement of the natural environment as the priority. This could be as much a response to the historic location of urban developments as it is a need to balance population growth, the economy and the environment that is social, economic and environmental sustainability.

With a local approach and closer involvement of stakeholders there is the potential for local groups to initiate more innovative and energetic approaches to coastal defence. A recent publication proclaimed the strategies of 'retreat, defend, attack', emphasising the opportunity to develop coastal communities further by building out into the sea (Building Futures, 2010). Whilst this may make sense for the towns and cities concerned it is an approach that could be at odds with contemporary understanding and sympathy for sustainable coastal management and acceptance of a more naturalistic approach. Therefore, care needs to be taken to ensure that the sum of the impacts does not detract from the aims for sustainable futures. More complex solutions may also develop to deliver a more appropriate balance of outcomes to the benefit of all parties.

18.3 The Nature and Complexity of the Coastal Processes

The coastline is in constant change from the processes of erosion and deposition. Wave action,principally erodes soils and rocks, transports the eroded material and then deposits it back on the coast to form beaches. It is the creation and movement of those coastal sediments that is fundamental to the stability of the coastline. This natural process has an equilibrium level of activity and the consequence is a naturally changing coastline. Rates of erosion and deposition depend upon the geology and form of the coastline, as well as its exposure to wave action. The outcome from defending or protecting part of that coastline is a reduction in the body of eroded material available for deposition. The consequence is that defending one part of the coastline will most likely detrimentally affect other parts by reducing the protection provided by naturally deposited material (French, 1997). It is for this reason that in recent years there has been more reluctance to construct unnatural defences.

Landslides are a major component contributing to coastal erosion. Principally, this occurs where undercutting by the sea causes oversteepening of the slope profile and collapse, but this is not the sole cause of coastal landslides. For instance, in the case of the Holbeck Hall Hotel landslide, the sea wall at the foot of the failed cliff remained buried, but apparently undamaged (see the Case Study below). The control of coastal landslides requires an understanding of the physical characteristics of the soils and rocks, the lithostructure and the hydrogeological context landward of the slope, as well as considering the seaward influences. There is a common correlation between landslides and periods of intense rainfall. A dramatic increase in landslide activity after the very wet autumn and winter of 2000 (in the United Kingdom) caused considerable damage to the road and rail network that was incredibly costly in terms of both the damage done to the infrastructure and the delays to travellers who were caught up in the disruption (Forster and Culshaw, 2004).

This case demonstrates the complexity of coastal erosion processes where it is not only action by the sea that causes instability and collapse. The frontage had been completely defended with a vertical wall and the cliffs had been re-graded with drainage installed. Nevertheless, the cliff slope became unstable from heavy rainfall, negating efforts to protect the toe of the slope from erosion by wave action.

Case Study 18.1 Holbeck Hall, Scarborough, June 1993

The Holbeck landslide, south of Scarborough in North Yorkshire, attracted considerable interest when it destroyed the four-star Holbeck Hall Hotel between the nights of 3 June and 5 June 1993. The hotel stood on the cliffs at Scarborough with views over a lawn and rose garden to the North Sea (British Geological Survey (BGS), 2012). The undercliff below the hotel was owned by Scarborough Borough Council. Over the years, natural erosion of the coastline led to relatively minor landslips in the area. The Council engaged engineers to investigate them and remedial works were carried out in 1989, but in 1993 there was a massive landslip. The lawn and rose garden fell into the sea, and the ground under the seaward wing of the hotel collapsed so badly that the whole building had to be demolished (Hopkins, 2000). In total one million tonnes of rock and sediment had been displaced in this landslip (French, 2001).

A rotational landslide involving about one million tonnes of glacial till cut back the 60 m high cliff by 70 m. It flowed across the beach to form a semicircular promontory 200 m wide projecting 135 m outward from the foot of the cliff. The likely cause of the landslide was a combination of rainfall of 140 mm in the two months before the slide took place, issues related to the drainage of the slope, pore water pressure build-up in the slope and the geology (BGS, 2012).

The first signs of movement on the cliff were seen six weeks before the main failure, when cracks developed in the tarmac surface of footpaths running across the cliffs. These were filled to stop ingress of water to the cliff, but when the cracks reopened, shortly before the main failure, the Council closed the cliff paths below the hotel. At this time a small part of the hotel garden was also observed to have suffered a minor movement (BGS, 2012). Its owners claimed damages from the Council and the Judge, John Hick QC, held the Council liable for a breach of a 'measured duty of care'. However, the Court of Appeal held the Council not liable because it owed no duty of care in relation to a defect that could only have been discovered by extensive geological investigation (Hopkins, 2000) (see Figure 18.1;).

Figure 18.1 The Top left corner shows the landscaped remains of the Holbeck Hall landslip, South Bay, Scarborough. Copyright Peter Church.

18.4 Coastal Flood Defences

Coasts and estuaries are natural locations for harbours and ports for sea-going transport and trade, and are often highly populated areas. The coast is also seen as a place for leisure, where the view, the beach and the sea itself offer opportunity for holiday, recreation, rest and recuperation. Sea defences have principally been constructed for the purpose of protecting financial or commercial activity and gain.

Historically, early defences were not to defend property but were constructed as a means of claiming land from the sea, for the purposes of agriculture or development. Earliest evidence of this dates back to Romano Britain. As ports and harbours became increasingly important to economic development, land claim provided space for expansion of ports as well as associated activities. Later, the fashionable amenity and recreation associated with taking the sea air and bathing encouraged developers to modify the natural environment with defences and promenades. The consequence of the taking of land by creating a static coastline with defences in this way leads to a need to maintain these structures so as to retain an artificial position of the coastline (Goodhead and Johnson, 1996). It follows, therefore, that there are two forms of coastal defence: firstly, *flood* or *sea defences* are structures that have been created solely to stop land from being inundated by water from the sea and, secondly, *coastal protection* structures prevent erosion of the land from being scoured and lost to the sea. At times one structure may serve both of these purposes (French, 2001).

There are four main types of hard approach to sea defences (French, 2001):

- *Defensive walls*. These forms include embankments, revetments rubble mound, reflecting concrete wave return walls and vertical walls; the latter is often associated with a promenade. The most fundamental problem with defensive walls is that they reflect the wave energy back to the sea, which can lead to increasing erosion when it meets incoming waves and forms a standing wave. This increase in scour can undermine the wall itself. Figure 18.2 shows an example of a defensive sea wall constructed from rock, as a protection to the toe of the sandstone cliff.
- *Groynes and jetties*. These are structures 'normal' to the shore, meaning that they run perpendicular to the line of the shore and hence intrude into the sea from the shore. Their effect is to interrupt the usual natural movements of sediments along the shore, thus retaining beach materials that would otherwise be lost from the particular stretch of shore being protected. They can be constructed in a variety of forms and materials, including timber, masonry and concrete. Jetties tend to be longer structures than groynes, attached to the shore but extending beyond the surf zone and are often used to protect the mouths of ports and harbours. They can only be effective in areas where there is longshore drift but also have the effect of starving downshore stretches of beach of sediment and creating more loss of land along those stretches. Figure 18.3 shows a stone jetty off the coast of South Wales.
- *Cliff stabilisation*. Cliff strength and stability is dependent upon not just the rocks and soils of which the cliff comprises but also the geological structure, amount of weathering and degree of wave attack. It is the complexity of factors that makes stabilising cliffs challenging and requires an understanding of the mechanisms of failure. Measures used might include regrading of the cliff slope, protection of the toes, rock anchoring or other measures to secure cliff material and drainage to reduce pore water pressure on the soils and rocks. Beaches can be fed or nourished with fresh material to help protect cliffs from wave action.

Figure 18.2 Defensive sea wall, near Cadiz (Spain).

- *Offshore structures – breakwaters and sills.* The purpose of offshore structures is to provide a barrier that reduces wave energy and, thus, reduces erosion rates of beaches and cliffs. They are usually constructed close to the shore as either submerged or emergent structures rising out of the sea and hence have the effect of dissipating wave energy away from the beach area. They are in effect a replication of natural wave dissipating structures like coral reefs and offshore sand banks. Although first constructed in the modern era in the 1930s they did not become common until the 1980s. Their use can be dated back to the ancient Egyptians. Most frequently their form is of a rubble mound or rock structure, though cast concrete units that can interlock are also used. Part of their effect in reducing wave energy is to cause more sediment to be deposited between the offshore structure and the shore, effectively moving the beach profile out towards the sea.

Man-made, artificial, static coastlines can be problematic. They can create circumstances with an inability to respond to sea-level change in the longer term; cessation of beach/dune interactions; cessation of sediment input and instability in fronting beaches. These disrupt the normal natural equilibrium of erosion and sedimentation, which can cause unwelcome instability in these normal processes to the detriment of the line of the coast. Such defences can also change the natural drainage of the foreshore and land behind the defence, causing fluvial flows to back up and not be able to discharge to the sea without causing flooding, or if the defences are 'overtopped' by the sea thus trapping saline waters inside the defended areas.

Figure 18.3 Stone jetty, Pembrokeshire (South Wales).

Defences have historically comprised hard construction using natural materials, soils and rocks or man-made masonry, concrete or timer structures. More recently, over the last few decades there has been more realisation and interest in the use of 'softer' defences utilising more natural methods of protection. Nature's own processes form defences against the sea and wave action. The inundation of New Orleans in the United States during Hurricane Katrina has been blamed as much on the disruption to sediment deposition from fluvial flows into the sea as it has by the failure of inadequate man-made levees. Work there has now begun on recreating the seaward sand banks as a means of naturally removing energy from braking waves, to help protect the city.

18.5 The Strategy of Managed Retreat

Given the expense of constructing defences, the appraisal of projects to maintain defences has frequently led to consideration that there should be no further investment to 'hold the line'. Decisions have been made to manage retreat and allow the sea to take, or retake, lands that have been protected in the past. One example is at Porlock Bay, where the breach of a ridge during a severe storm allowed the sea to reclaim land owned by the National Trust. However, such strategies will become increasingly controversial as homes, land and properties are lost to the sea.

Case Study 18.2 Porlock Bay, Somerset

Porlock Bay is situated on the north coast of Somerset, within the Bristol Channel. The site comprises a shingle ridge and associated saltmarsh hinterland extending for a distance of approximately 4 km along the west Somerset coast, immediately north of Porlock village. The Porlock shingle ridge was formed as the sea level rose during the middle part of the Holocene epoch, from shingle eroded from head deposits that masked the sea cliffs to the west after the last glacial period. This major source of coarse sediment has long since disappeared, leaving only a relatively insignificant input of sediment from occasional cliff falls. The inputs of sediment to the beach ridge from this modern source are too small to sustain the earlier beach profile and the increase in the length of the ridge as it continues to roll back, in a lengthening curve, into Porlock Bay. This means that the ridge has been growing steadily thinner ever since it was formed, a condition exacerbated by the further reduction in shingle inputs caused by the construction of groynes at Gore Point at the western extremity of the ridge. The modern ridge was therefore unable to withstand recent extreme storm events and a breach opened during the storm of October 1996, which flooded the low lying marsh hinterland (see Figure 18.4). Rapid evolution of the beach following the breach is providing a unique opportunity to study the development of a coarse sediment barrier system in an open coast location (English Nature, 2002).

Bossington in Porlock Bay is one of the National Trust's earliest and best examples of how, by working with partners and stakeholders, we have worked with the natural process of sea-level rise to extend existing and create new habitats as a result of the boulder bank breaching in the 1990s. The salt marshes and lagoons developed in the area behind the boulder bank, where the River Horner meets the sea, have received SSSI designation. The biggest challenge for the Trust and its partners now is how to respond as sea levels continue to rise and pose a greater threat to natural habitats and the local village of Bossington a little further upstream. They are undertaking detailed studies to enable them to plan for the future (National Trust, 2008).

Figure 18.4 Shows flooding behind the shingle barrier at Porlock Bay. Copyright Martin Bodman.

18.6 Flooding Challenges from Climate Change and Sea-Level Rise

The world's climate is constantly changing and in recent years there has been a realization that not only is the sea level rising but also there is increased storminess and extremes of rainfall. This is expected to present significant challenges in managing the coastline in the future. In response, the UK Climate Impacts Programme (UKCIP) has been investigating and theorising over likely scenarios for the coming century. Estimates for sea-level rise have been the basis for guidance on allowances for shoreline management set out in the Department for Environment, Food and Rural Affairs (DEFRA) and EA reporting on the UKCIP02 theories (produced in 2002). The UKCIP02 climate change scenarios were generated from a climate model developed by the Hadley Centre and reflect scientists' best understanding of how the climate system operates, though UKCIP02 has been superseded by the UK Climate Projections 2009 (UKCP09), which is now administered through the EA. There will no doubt be future reports updating future expectations.

The UK government arranged a Foresight Project (2004) to provide more information on future flood and coastal defences in the United Kingdom (between 2030 and 2100), considering flooding from rivers and the sea and the risks of coastal erosion. They used alternative theories to assess the possible scale and nature of future risks and to assess options for responding to those risks. These theories took account of different social and economic visions of the United Kingdom and different amounts of climate change. Future works of UKCIP may include more detailed climate predictions and make specific efforts to predict possible changes in extreme conditions. DEFRA intend to provide further guidance specific to SMP studies and designing coastal defences when the results of these and other research studies are available (DEFRA, 2011a).

Climate Change and Sea-Level Rise

The climate is changing and this is likely to have an impact on flooding and coastal erosion. Sea levels are rising and winter rainfall may become more intense. Changes in weather patterns and, in particular, more torrential rainfall is likely to increase flood risk from surface water and ordinary watercourses as well as rivers. Today, around 490 000 properties face a 1 in 75 chance (in a given year) of flooding from rivers and the sea. In reviewing future risk, *Investing for the Future* (EA, 2009a) suggested that if overall investment remains at 2009 levels (in 'cash terms') and if there is no additional development in the areas at risk, by 2035 there will be an additional 350 000 properties, 280 000 of them residential, in areas with a 1 in 75 or greater annual chance of being flooded. Rising sea levels mean that waves and storm surges could cause greater coastal erosion. Changes to the currents acting on the coast could also lead to changes in the movement of coastal sediments, affecting both coastal deposition and erosion. This could expose new risks from coastal flooding, lead to a greater risk of coastal defences failing and increase the need for maintenance work on defences and more extensive warning systems (EA, 2011a).

As our coastlines experience sea-level rise, the natural state of dynamic equilibrium of erosion and deposition will require beaches and nearshore profiles to naturally relocate upwards and landwards. The inability of the coastline profile to adjust to changing levels

Case Study 18.3 Thames Estuary Tidal Flood Risk and the Thames Barrier and Thames 21

As a result of the 1953 flood, a system of flood defences was constructed. The most iconic element of this is the Thames Barrier, which has been operational since its completion in 1982. The Thames Barrier has been closed 116 times (up to September 2009) to prevent flooding. There are also around 400 smaller barriers and movable flood gates downstream of the Thames Barrier and over 300 km of river walls and embankments stretching into Essex and Kent, which have been raised by 2 metres to give additional protection from storm surges. Upstream of the Thames Barrier river walls are still necessary to prevent the normal range of high tides from flooding parts of inner and central London. This system of tidal flood defences made allowance for a sea-level rise and London is therefore protected to a very high level. It is estimated that the level of protection will reduce down to a standard of 1 in 1000 years (0.1% chance per year) by 2030 and this will continue to decline if no further measures are taken (Lavery and Donovan, 2005).

The EA undertook the Thames Estuary 2100 (TE2100) project (EA, 2009c). Developing the TE2100 plan involved extensive consultation with stakeholders and forms the basis of an implementation plan under development, to be approved by the government. Consistent with government guidance on climate change, this plan considers a 1 metre rise in sea level over 100 years, and presents a set of incremental adaptation measures as actions for the short (to 2034), medium (to 2069) and long term (from 2070):

- Shorter-term decisions are nested within a longer-term framework that explicitly identifies key thresholds and options for dealing with much larger extents of change. (For example, 10 indicators for change will be formally monitored to identify if or when a switch to alternative options may be needed.)
- The plan allows for flexibility on the timing of introduction of different options and interventions, and the ability of the plan to change between options, based on the monitoring programme.
- Detailed guidance is provided on how the recommendations contained in the plan should be applied in the event that more extreme change is realised; for example, if it becomes necessary to divert to an alternative adaptation pathway.

This guidance also shows how lead times for major interventions need to take account of any such changes, and is underpinned both by the identification of key decision points and by the inclusion of the monitoring and review cycle (Innocenti and Albrito, 2011).

The present system of flood risk management for tidal flooding can continue to provide an acceptable level of risk management up to 2030 without major alterations. Beyond 2030 those actions identified are as follows.

2010–2035

- Work with local authorities and the construction industry to ensure that existing and new development is safe through spatial planning and local resilience measures.
- Prepare joint riverside strategies establishing a shared vision for the riverside.
- Continue to maintain, enhance, improve or replace existing flood management systems.
- Work with Local Authorities and communities on the future use of the Thames Barrier in managing fluvial flooding in West London.

- Continue flood forecasting and emergency planning activities.
- Commence the creation of a new intertidal habitat in the Lower estuary, which is being lost as sea levels rise. (Large areas of currently undeveloped land such as Rainham/ Wennington Marshes, Erith Marshes and Dartford/Crayford Marshes could be used as Strategic Flood Storage areas to use as emergency storm surge flood storage.

2035–2070

- Maintain, improve or replace the walls, embankments, barriers and gates along the Estuary.
- Work with Local Authorities and communities on enhancing and revitalising the Thames riverside.
- Continue flood forecasting and emergency planning activities.
- Continue replacing areas of intertidal habitats as sea levels continue to rise.
- Decide on and construct the option to manage increasing flood risk for the end of the century and beyond.

2070–2100

- End of the century option operational (see 2035–2070).
- Further raising and adaptation of defences where required to keep new Barrier closures to within operational arrangements.
- Continue programme of maintenance replacement and repair of upstream and downstream defences.
- Continue flood forecasting and emergency planning activities.

(EA, 2009c)

of sea can often result in the loss of beach sediment and more intense erosion of unprotected beaches (French, 2001). The best way to defend and manage the coast is to do it in a way that most closely mimics nature.

There is a clear expectation that climate change will have a major impact on the threat of tidal flooding, as the rising sea level steadily reduces the level of protection that a defence can offer (Lavery and Donovan, 2005). However, predictions for future rates of sea-level rise vary considerably depending on the models and assumptions used. The Thames Estuary Project (TE2100) has considered sea-level rises, which may be derived from climate change ranging from 0.9 m (EA, 2009c) to 4 m (a scenario that is described as 'High++ Level', where all conceivable sea-level rise contributions up to 2100 occur). This uncertainty leads to a need for a flexible, emergent strategy as a response, which is the aim of the current London Plan policies. The plan looks ahead to 2030, when there are limited differences between predictions and so, until then, existing flood risk management options can continue to provide appropriate risk management for tidal flooding. There is more variation in the projections beyond 2030. Planning for those later changes now will improve chances of successfully managing more extreme future situations. This time is necessary to plan for and agree improvements and replacements to London's tidal defences, which include the Thames Barrier, since they will present very significant funding and engineering challenges (EA, 2009c).

18.7 Changing Populations, Land Management and Development

More people and properties may well be placed at risk from the consequences of a flood or coastal erosion incidents because of increasing population and growing urbanised settlement. This is particularly the case where there is a coincidence between development and areas at risk of flooding or erosions. It is also significant that developments outside the flood plain also raise flood risk, through increased runoff from impermeable surfaces and consequential faster entry to watercourses through new drainage systems. Where, further downstream, fluvial river flows meet tidal waters, this can exacerbate flooding from the sea and is therefore another factor amongst the complex systems of flood processes to be managed.

The population of England is predicted to increase by 10 million by 2030 (Hughes, 2009), which will undoubtedly lead to increasing demand for homes, businesses and infrastructures. Without appropriate consideration for flood and coastal erosion risks this development will put more people and property in areas at risk of flooding. With recent housing developments in the Thames Gateway and proposals for infrastructure such as a new London Airport in the Thames Estuary itself, there will be increasing pressure to place more people and properties in locations where flooding and erosion could be a significant risk. It is therefore essential that spatial planning alongside sound engineering ensures that new developments take flood and coastal erosion risk fully into account, and where possible incorporate measures to reduce those risks over the development's lifetimes. It is also just as important that appropriate flood forecasts and warnings are provided to enable individuals and communities to respond effectively. Even if the likelihood of flooding were to be reduced, the consequences may still increase as the value of property and contents continues to rise with inflation and as the value of possessions and household contents and fittings rise with general increases in wealth over time. The achievement of wider environmental and social objectives and other benefits, alongside flood risk and coastal erosion

Proposals included in *Understanding the Risks, Empowering Communities, Building Resilience, The National Flood and Coastal Erosion Risk Management Strategy for England* (DEFRA, 2011b) involve working with natural processes that can include taking action to manage flood and coastal erosion risk by protecting, restoring the natural function of catchments, rivers, floodplains and coasts. This could, for example, involve using farmland to store floodwater temporarily, reinstating washlands and wetlands to store flood water away from high-risk areas or allowing cliffs to erode to provide sediment that may be deposited elsewhere. Other techniques include protecting and restoring natural river, estuarine and coastal systems and features. The maintenance and restoration of a range of ecosystem services, or natural functions of the environment, can provide valuable additional benefits including:

- water quality improvements through reductions in runoff and diffuse pollution;
- water resource provision through aquifer recharge;
- mitigation of and adaptation to climate change through measures such as wetland creation and coastal and fluvial realignment and
- the provision of urban biodiversity and amenity green spaces through sustainable drainage systems (SUDS).

risk management, will require the use of measures that will work with natural processes wherever possible, whilst also being based on partnerships with local communities (DEFRA, 2011a). Such strategies are set out in the document *The National Flood and Coastal Erosion Risk Management Strategies for England* (DEFRA, 2011b).

18.8 Important National Infrastructure

In the summer of 2007 the UK floods created a situation where water, sewerage and electrical services were disrupted on a scale that was close to the point of creating a national disaster. This highlighted the need to make sure that essential infrastructure, necessary for the continuation of safe living (such as water and electricity supply services), are suitably resilient to flooding and coastal erosion. The Cabinet Office has published a framework for achieving this: *The Strategic Framework and Policy Statement on Improving the Resilience of Critical Infrastructure to Disruption from Natural Hazards* (Cabinet Office, 2010), which establishes an interim standard for resilience to flooding that critical infrastructure should meet. This framework emphasises the need for infrastructure providers to maintain overall service provision and to take account of the importance of specific sites within infrastructure networks. The Cabinet Office has also published guidance on improving the resilience of critical infrastructure and essential services in *Keeping the Country Running: Natural Hazards and Infrastructure* (Cabinet Office, 2011).

18.9 Proportionate Risk-Based Approaches to Flood Risk Management

The National Flood and Coastal Erosion Risk Management Strategy for England (DEFRA, 2011b) states:

> ... it is not feasible technically, economically or environmentally, to totally prevent flooding and coastal erosion. Therefore, a risk-based management approach which targets limited resources to those areas where they can have the greatest effect is appropriate. Such risk management approaches consider both the probability over time of a flood or coastal erosion happening and the consequences that might arise if it did.

This can be done, for example, by assessing the average annual damages that arise from floods or coastal erosion. This requires a clear understanding of the flooding process and systems, such as the 'sources', 'pathways' and 'receptors' as well as the consequences of risk, to enable them to be addressed as appropriate to manage all of the factors that combine to create risk. Further details on this approach are available in *Guidelines for Environmental Risk Assessment and Management* (DEFRA, 2011c).

Such an approach involves using a tiered assessment, that is starting at a high, screening level and in stages becoming more detailed to address the risks identified. It seeks to make risk management more straightforward, removing unnecessary barriers while ensuring that legal and government policy requirements are met. All aspects of risk management should be carried out in a proportionate way that reflects the size and complexity of the risk and society's ability to manage it. Investment in managing risk, and who pays for it, should reflect the benefits that result (DEFRA, 2011b).

18.10 Beneficiaries Should be Encouraged to Invest in Risk Management

Sir Michael Pitt (2008) suggested better aligning those that benefit with those that pay would create a more efficient and responsive system to managing flood and coastal erosion risks. To achieve this he recommended that 'Government should develop a scheme that allows and encourages local communities to invest in flood risk management measures.' He also said that developers, in potentially increasing local flood risk, should 'make a full contribution towards both the costs of building and maintaining the necessary defences'. In taking this recommendation forward, the government has made clear that they 'cannot continue all of the work that the Environment Agency has historically done at the tax-payer's expense. Government investment in flood and coastal erosion risk management is significant, but we need to ensure that we get best value for money.'

Changing climate, increasing populations and the subsequent increasingly unaffordable risk management measures demand more innovative responses to flood and coastal erosion risk management. These approaches need to be more integrated and responsive to the holistic nature of natural and man-made systems.

18.11 A View to the Future – Strategies Requiring Innovations and New Solutions

It may well turn out that the future will see a wider variety of types of response to the coast and the need to protect people, properties and investments as appropriate. However, more extreme scenarios may well prompt more extreme responses. The tradition of defending at all costs, to retain lands won and protected, is receding into history as managed retreat is adopted as a solution (EA, 2009b). There is also an appetite to create natural barriers out into the sea as well as to 'win land' and construct in the sea itself. Such examples as the Cardiff Barrier, proposals for an airport in the Thames Estuary and examples abroad, such as Hong Kong airport, as well as 'The Palm' and 'The World' off Dubai, demonstrate this desire.

A collaborative think tank between the Royal Institute of British Architects and the Institution of Civil Engineers has produced *Facing up to Rising Sea Levels – Retreat Defend*

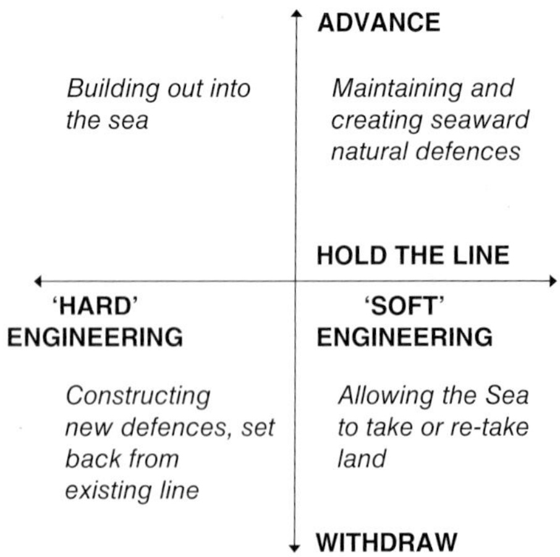

Figure 18.5 Mapping the strategic options.

Attack, which identifies a range of responses for urban developments on the coast (Building Futures, 2010). In considering the types of options in response to threats from coastal erosion and flooding, Figure 18.5 shows two key dimensions, which can be used to describe those options. They relate firstly to whether engineered options are hard or soft and, secondly, to whether the approach is to allow the sea to take land or whether the response is to build out into the sea, with holding the line as an intermediary strategy. Consideration of these strategies, alongside developing practice and policy, would lead one to realise that responses to coastal flood risk are increasingly combining all available solutions in complex holistic approaches.

18.12 Conclusions

Coastal erosion is a complex process and early attempts by communities to drain land or protect it from the sea have frequently led to unintended consequences and problems elsewhere along the coastline. Hard engineered structures can take a variety of forms to modify erosion and deposition patterns and comprise both onshore and offshore structures. It has been apparent for some years that the solution requires schemes that respect and work with natural processes. Such solutions prove to be far more sustainable and beneficial, not least because they can accommodate improvements and benefits for economic and social elements. Increasing flood risk has led to a realisation of the importance and risk posed to essential infrastructure and the need to develop policies and practices to ensure its resilience to coastal floods.

It has become apparent through a better understanding of climate change, and the likely impact that this is having on the environment and the world, that sea-level rise will become an exceedingly serious threat to many shore-based communities and cities. Responding to that challenge is beginning to prompt creative and innovative solutions, even consideration of building out into the sea to accommodate growing populations and increasing urbanisation. This widening of potential options will allow the consideration of schemes that can bring many simultaneous benefits and improvements to both urban developments and the natural environment.

Increasing risk leads to increasing costs to manage coastal flood risk, to which the response has been the targeting of funds and resources through a risk-based approach and a move towards expecting those benefiting to pay for protection. Communities are engaging in raising funds and in developing schemes as stakeholders rather than relying totally on government funds. Climate change and sea-level rise are characterised by great uncertainty and a broad range of forecasts against which to plan. The scale of responses, like the replacement of London's sea defences, requires early identification of options, long planning lead times and 'space' to build community and political commitment to expensive and, necessarily, innovative solutions. Those pressures from population growth, alongside changes to the climate, will further exacerbate the challenges faced at the coast. Visions of the future now include strategies of 'retreat, defend and attack'.

References

British Geological Survey (BGS) (2012) *Holbeck Hall Landslide, Scarborough* [online], [Accessed 1 July 2012]. Available at: http://www.bgs.ac.uk/landslides/holbeckHall.html.

Building Futures (2010) *Facing up to Rising Sea Levels Retreat? Defend? Attack?* [online], [Accessed 9 July 2012]. Available at: http://www.buildingfutures.org.uk/projects/building-futures/facing-up/.

Cabinet Office (2010) *Strategic Framework and Policy Statement on Improving the Resilience of Critical Infrastructure to Disruption from Natural Hazards*. The Cabinet Office, Natural Hazards Team, London.

Cabinet Office (2011). *Keeping the Country Running: Natural Hazards and Infrastructure*. The Cabinet Office, London.

Department for Environment, Food and Rural Affairs (DEFRA) (2011a) *Shoreline Management Plan Guidance Volume 1: Aims and Requirements* [online], [Accessed 30 June 2012]. Available at: http://www.defra.gov.uk/publications/files/pb11726-smpg-vol1-060308.pdf.

Department for Environment, Food and Rural Affairs (DEFRA) (2011b) *Understanding the Risks, Empowering Communities, Building Resilience, The National Flood and Coastal Erosion Risk Management Strategy for England*. The Stationery Office, London.

Department for Environment, Food and Rural Affairs (DEFRA) (2011c) *Guidelines for Environmental Assessment and Management*. The Stationery Office, London.

English Nature (2002) *Citation: Porlock Ridge and Saltmarsh* [online], [Accessed 2 September 2012]. Available at: http://www.english-nature.org.uk/citation/citation_photo/2000380.pdf.

Environment Agency (EA) (2009a) *Investing in the Future – Flood and Coastal Risk Management in England – A Long Term Investment Strategy* http:// [online], [Accessed 8 June 2012]. Available at: www.environment-agency.gov.uk/research/library/publications/108673.aspx.

Environment Agency (EA) (2009b) *Securing Community Support* [online], [Accessed 8 July 2012]. Available at: http://publications.environment-agency.gov.uk/PDF/GEHO0609BQEE-E-E.pdf.

Environment Agency (EA) (2009c) *Thames Estuary 2100 Managing Flood Risk Through London* [online], [Accessed 2 September 2012]. Available at: http://www.cityoflondon.gov.uk/services/environment-and-planning/planning/planning-policy/local-development-framework/Documents/Thames%20Estuary%202100%20Plan%20Chapters%201–4.pdf.

Environment Agency (EA) (2011a) *Understanding the Risks, Empowering Communities, Building Resilience – National Flood and Coastal Erosion Risk Management Strategy for England, Summary Strategy* [online], [Accessed 30 June 2012]. Available at: http://publications.environment-agency.gov.uk/dispay.php?name=GEHO0711BTZE-E-E.

Environment Agency (EA) (2011b) *Dunwich Heath and Minsmere – National FCERM Strategy Case Study* [online], [Accessed 30 June 2012]. Available from: http://publications.environment-agency.gov.uk/dispay.php?name=GEHO0711BTZQ-E-E.

French, P.W. (1997) *Coastal and Estuarine Management*. Routledge, Abingdon.

French, P.W. (2001) *Coastal Defences, Process, Problems and Solutions*. Routledge, London.

Forster, A. and Culshaw, M. (2004). Feature: implications of climate change for hazardous ground conditions in the UK. *Geology Today*, 20(2), 61–66.

Goodhead, T. and Johnson, D. (1996). *Coastal Recreation and Management: The Sustainable Development of Marine Leisure*. E. and F.N. Spon Publishing, London.

Halcrow Group Ltd (2009) *Receptors Vulnerable to Coastal Erosion*. Environment Agency.

Hopkins, C.A. (2000). Slipping into uncertainty. *The Cambridge Law Journal*, 59, 421–471.

Hughes, M. (2009) *Social Trends*. Office of National Statistics, Newport.

Innocenti, D. and Albrito, P. (2011) Reducing the risks posed by natural hazards and climate change: the need for a participatory dialogue between the scientific community and policy makers. *Environmental Science and Policy*, 14, 730–733.

Lavery, S. and Donovan, B. (2005) Flood risk management in the Thames Estuary looking ahead 100 years. *Philosophical Transactions of the Royal Society A*, 363, 1455–1474.

National Trust (2008) *Shifting Shores* [online], [Accessed 2 September 2012]. Available at: http://www.nationaltrust.org.uk/servlet/file/store5/item349149/version1/shiftingshores%20–%20sw.pdf.

Pitt, M. (2008) *Learning Lessons from the 2007 Floods*. The Pitt Review, London [online], [Accessed 1 September 2012]. Available at:http://webarchive.nationalarchives.gov.uk/20100807034701/http:/archive.cabinetoffice.gov.uk/pittreview/thepittreview/final_report.html.

19

The Costs of Flooding on Households

Rotimi Joseph, David G. Proverbs, Jessica E. Lamond and Peter Wassell

19.1 Introduction

The occurrence of flood events is a major environmental risk. This is clear from the scale of destruction which was brought about by major floods over the past decade in New Orleans, United States (in 2005), Jakarta, Indonesia, and the United Kingdom (in 2007), Pakistan (in 2010), Australia, Japan, Colombia and Bangkok (in 2011). According to Fay *et al.* (2009), floods currently account for half of the fatalities across the world arising from natural disasters. However, many of the costs of flooding extend beyond immediate fatalities, imposing major damage and disruption to housing, living conditions, infrastructure and the local economy. In the United Kingdom, between 1998 and 2000, there was a significant increase in the costs of insurance payout due to flooding. This increased from £500 million (in 1998) to £1 billion (in 2000) within two years interval (Environment Agency, 2009b). This increase is partly due to inflation but more so an increase in the frequency and intensity of flood events coupled with development and urbanisation (Brath *et al.*, 2006; Hadjimitsis, 2010; Pitt, 2008; Jha *et al.*, 2012). It is estimated that the annual economic and financial costs of damage to residential and commercial property from flooding in England could rise from £1 billion to £4 billion by 2035 (Environment Agency, 2009b).

Costs associated with flooding can be categorised as tangible and intangible, as well as direct and indirect costs (Joseph *et al.*, 2011b, 2011c; Lamond, 2012). The tangible costs of flooding are costs that can be estimated using data from actual markets, which include reinstatement costs of flood-damaged property, emergency services and cost of providing alternative accommodation. Conversely, the intangible costs are mainly social impact costs that are difficult or, in most cases, impossible to estimate by using actual market data. These include loss of life, loss of irreplaceable items and emotional stress as a result of flood event.

Water Resources in the Built Environment: Management Issues and Solutions, First Edition.
Edited by Colin A. Booth and Susanne M. Charlesworth.

Costs of flooding (tangible and intangible) can further be categorised as economic and financial costs (Lamond, 2012). This division depends on whether the costs are considered on a large geographical scale, such as the national level (i.e. economic costs to the nation) or at individual or business level (i.e. financial costs to households or businesses). Financial costs capture the losses incurred to one particular business or individual as a result of flooding (and is broadly related to insurance claims) whereas economic costs refer to opportunity costs (North West Development Agency, 2009). More precisely, flooding results in economic costs only if it decreases the overall welfare of society (i.e. if flood impacts cause a decrease in net consumption that cannot be offset by an increase elsewhere). For example, if flooding causes one shop's goods to no longer be available yet consumers can readily go to an alternative shop to purchase the same goods then there is no net change in consumption. Businesses in the United Kingdom experienced a variety of flood impacts, ranging from direct physical impacts to indirect effects (to supply chains, for example); damage to premises, equipment and fittings; loss of stock; and reduced customer visits and sales as well as disruption to business activities (Ingirige and Wedawatta, 2011). For instance, in the Australia flooding of 2011, it was reported that the biggest short-term impact in terms of dollars was the disruption to coal exports, and some statistics set that at $400 million dollars a week (Skyes, 2011). The long-term financial costs are related to the future expected damage and may be expressed via the risk that insurance coverage will not be renewed or subject to high policy excess (Lamond, 2012). Alternatively, the economic long-term costs relate to expected future damage, maybe through the increase in the cost of flood protection provision by the government, but may also be incurred if capacity is permanently reduced or business is redirected in the long term. According to the Environment Agency (2009a) one in six properties (commercial and residential) in the United Kingdom are estimated to be at risk of flooding and 500 000 of these are said to be at 'significant risk' (Environment Agency, 2009a). It is estimated by the Association of British Insurers (ABI) (2009) that this figure will rise to 835 000 properties by 2035. The consequence of this revelation is that both the economic and financial costs of flooding will be on the increase.

In order to place the costs of flooding into a wider context, a brief discussion on the economic costs of flooding is presented. Of course, the real cost of flooding is not only economic but also includes other nonfinancial losses such as fatalities, injuries, moral damages, historical and cultural losses, environmental losses and societal disruptions. However, the focus of this chapter is on the financial costs of flooding as relates to individuals and businesses in the United Kingdom. This is discussed under the themes of tangible and intangible costs of flooding in the subsequent sections.

19.2 Economic Costs of Flooding

The economic costs of flooding represent those costs that are borne by the national economy; these include overall damage to businesses, that is premises that have been adversely affected by flood water, the loss of stock and loss of sales, due to temporary relocation of community members and the wages of employees who cannot get to work (North West Development Agency, 2009). The economic costs of flooding include the wider social costs. For instance, the total economic costs of the summer 2007 floods in the United Kingdom were estimated at about £3.2 billion in 2007 prices, within a possible range of between £2.5 billion and £3.8 billion (Chatterton *et al.*, 2010). It was suggested by Penning-Rowsell *et al.* (2005) that the direct economic costs arising from a flood event and their indirect consequences are assessed during flood protection investment appraisals rather than the financial losses to individuals and organisations. This is based on the theory that the UK economy is generally an open, competitive, service-based economy with many product substitution choices available to consumers and producers. Any financial loss to a business would not, therefore, affect net spending or economic activity because it would be offset by an increase in trade with alternative businesses.

Jha *et al.* (2012) suggested that in assessing the economic impacts of flooding, care must be taken to adopt both local and national perspectives. This is because disasters have a large impact on those directly affected but a much smaller effect on the national economy. Some local impacts, such as the effect on the tourist trade, may be balanced by growth in trade elsewhere in the country. Conversely, the scale of flooding determines the effect of flooding on the national economy. Further, the location of flooding relative to the location of critical economic assets determines the economic impact of flooding on a nation's economy. For instance, Gurenko and Lester (2004) estimated that, on average, the direct cumulative costs of natural disasters in India account for up to 12% of central government revenues. This can have a significant impact on the national economy, resulting in important infrastructure spending being delayed or cancelled.

19.3 Financial Costs of Flooding

Financial costs are borne by individuals and organisations; examples include costs of replacing damaged or lost possessions, costs of repairing damaged properties and alternative accommodation costs. The costs associated with flooding can be divided into tangible and intangible costs, with further division into direct and indirect as well as short-term and long-term costs. The relationship between these cost perspectives and the specific period when they are incurred during flood events is presented in Table 19.1. Having a full knowledge of the different categories of costs of flooding is an important factor in decision-making on spending on flood risk management both at household and governmental levels. For instance, as shown in Table 19.1, the long-term costs of flooding, both tangible and intangible, came about as a result of indirect impacts of flooding. The financial costs of flooding are considered in more detail below under the broad themes of tangible and intangible costs.

19.3.1 Tangible Costs of Flooding

The tangible costs of flooding on households are costs that can be estimated by using actual market data; these costs are both direct and indirect in nature as well as short and long term in nature (Table 19.1). Examples of tangible cost of flooding are expenditure necessary to cope with the immediate impact of inundation and bringing life back to normality. A majority of the direct costs of flooding are covered under the domestic home insurance policy in the United Kingdom for those homeowners who are fully insured (Lamond, 2012), while the indirect tangible costs are extra expenditures incurred by households as a result of a flood event, which in most cases are not recovered via insurance. The short- and long-term categories of the costs of flooding depend on whether the costs are incurred immediately after the flood events (short term) or incurred long after the flood events (long term).

Table 19.1 Categories of flood costs.

	Tangible costs		Intangible costs	
	Direct costs	Indirect costs	Direct costs	Indirect costs
Short-term costs	Building reinstatement costs	Extra expenditure on feeding	Loss of life	Inability to move house immediately after flood event
	Contents replacement costs	Extra travel expenditure	Physical injuries	
	Clean-up costs	Loss of income	Loss of irreplaceable personal belongings	
	Alternative accommodations costs	Loss of utility	Stress of dealing with insurer, builder and loss adjusters	
Long-term costs	Costs of treating rot and damp	Inability to renew property insurance policy		Deterioration of physical and mental health (depression, anxiety, etc.); physical and psychological trauma
		Increase in insurance premium		Worrying; – about future flood – about loss of borrowing power
		Loss of property value		Loss of community spirit
		Cost of providing flood mitigating measures		Stains between families

The short-term direct tangible financial costs of flooding are costs associated with dealing with the immediate aftermath of a flood incident. These costs include costs of reinstating flood-damaged buildings to pre-flood conditions, contents replacement costs, clean-up costs and costs of provision of alternative accommodation if there is a need for evacuation. These costs are typically borne by the insurer for fully insured property owners in the United Kingdom. For instance, many people were evacuated from flooded areas during the summer 2007 events in the United Kingdom. According to the Pitt review, approximately 14 500 households were provided with alternative accommodation (Pitt, 2008) and the Environment Agency estimated the costs of providing alternative accommodation by the UK insurance industry to be just under £100 million (Chatterton *et al.*, 2010). While these costs are not being borne directly by households, the consequence of flood events always brings about indirect financial costs on households.

Indirect tangible impacts of flooding comprise damage, which occurs as a further consequence of the flood and the disruptions of economic and social activities as a result of power outages and in some cases damage to infrastructures (Joseph *et al.*, 2011c). This damage can affect households who may not have been affected directly by the flood. Such short-term indirect tangible financial costs include extra travel costs (e.g. to school or the work place) due to damage to the road network or the collapse of bridges (Watts, 2009/2010) and loss of utilities (Joseph *et al.*, 2011c; Lamond, 2012).

The loss of utilities such as water supply and electricity brings about extra expenditure on households. For example, 140 000 houses were without clean water for up to 17 days in Gloucester during the summer 2007 flood event in the United Kingdom. In the electricity sector, supply companies bore only 6% of total costs, whereas consumers incurred 94% of total economic costs due to loss of value associated with disruption of supply (Chatterton *et al.*, 2010). These costs are important to households because in many cases these costs are not recoverable through insurance claims. The occurrence of a flood event can bring about long-term indirect tangible financial costs, such as the inability to renew property insurance policy; the potential for an increase in an insurance premium; the loss of property value; and loss of borrowing powers.

In the United Kingdom, the Environment Agency estimate that every £1 currently invested in new and improved flood risk management assets reduces the long-term economic and financial costs of flooding and coastal erosion damages by around £8 (Environment Agency, 2009b). However, despite this flood protection spending, households are still experiencing long-term direct tangible costs such as costs of treating rot and damp, which resulted from the effect of being previously flooded. These costs in most cases may be declined by insurer, or householders may have to engage services of building surveyors/structural engineering to prepare a detailed report to support their insurance claim against such occurrences. Further, there is a long-term indirect financial cost of flooding such as the costs of providing property-level flood protection measures by homeowners.

The average cost of adopting temporary/permanent resistance flood protection measures such as installation of door and window boards, airbricks, service duct covers, waterproof doors and windows for residential properties in the United Kingdom ranges between £2000 to £10 000 for a single property (Bowker, 2007). However, installing resilience measures could cost as much as £30 000 depending on the house type and timing of the works, that is whether it was installed as part of planned renovation works or during a flood recovery period (Thurston *et al.*, 2008; Wassell *et al.*, 2009; Joseph *et al.*, 2011a). These costs may be higher than the stated amount if the installation was undertaken in isolation.

19.3.2 Intangible Costs of Flooding

Intangible damages arise from adverse social and environmental effects caused by flooding, including factors such as loss of life and both physical and stress-related symptoms, for example loss of sleep, anxiety, a reduced immune system response and increased susceptibility to certain illnesses (Environment Agency and Department for Environment, Food and Rural Affairs (DEFRA), 2004). The intangible costs of flooding are difficult to quantify and ethically problematic to monetise; for instance in extreme cases flooding may cause loss of life. The intangible costs of flooding can be divided into short- and long-term costs and also with further classification into direct and indirect costs. A majority of the intangible costs of flooding are subjective and depend on the individual in question. The direct short-term costs include loss of life, physical injuries, loss of irreplaceable personal belongings and the stress of dealing with insurers, builders and loss adjusters (Lamond, 2012), while the indirect short-term costs include an inability to move house immediately after a flood event.

The number of deaths associated with flooding is closely related to the life-threatening characteristic of floods: rapidly rising water, objects carried by the rapidly flowing water, the behaviour and the vulnerability of the victims. However, when looking at risk trade-offs that people make with regard to their health, economists often consider the value of a statistical life (VSL). The VSL is the value that an individual places on a marginal change in their likelihood of death. Economists often estimate the VSL by looking at the risks that people are voluntarily willing to take and how much they must be paid for taking them (Mankiw, 2012). Conceivably, the value of life can be estimated through the use of insurance models using the amount of sum insured people are willing to pay to insure their lives against death. Other methods of valuation of human life, according to Jha *et al.* (2012), are the expected future earnings, the economic contribution of an individual and the replacement cost of the investment in health care, education and social welfare that the state has provided.

Flood-related injuries may occur as individuals attempt to escape from objects being carried by fast-flowing waters or as a result of the collapse of buildings or other structures (Du *et al.*, 2010). Injuries can be relatively minor and self-treated, such as cuts and abrasions, or may be more serious (Ahern and Sari, 2006). Flood-related injuries may occur in the pre-onset, onset and post-onset phases of the flood events (Few *et al.*, 2004). In the pre-onset and onset phases, injuries may be sustained when individuals are attempting to remove themselves, their family or valued possessions from the approaching waters (Ahern *et al.*, 2005). The post-onset injuries are likely to occur in the aftermath of a flood disaster as residents return to their homes and businesses and begin the clean-up process (World Health Organization (WHO), 2002; Few *et al.*, 2004; Ahern and Sari, 2006). Few *et al.* (2004) noted that a lack of coordinated monitoring of injuries related to flooding, including from clean-up activities, meant that it was difficult to assess the true cost of injuries sustained due to flood events. However, the cost can be estimated by, for example, finding the number of visits made to the doctors following flood events.

Minimal information is presently available on the frequency of nonfatal flood injuries, as they are mostly not routinely reported or identified as flood-related. Further, in the United Kingdom the availability of flood warning systems, which inform people of the potential flood event and allow both the authorities and the households to take necessary precautionary measures before the flood water comes, could be seen as a contributing factor for the reduction in the number of reported flood-related injuries.

Loss of personal belongings of sentimental value is one of the flood impacts that in most cases affect households both in the short and long term. Items such as family photographs, which cannot be retrieved or repaired, can lead to the development of health-related problems such as depression. Research shows that on average women are placed under a particular strain in the months following the flood event, as the stresses of managing the recovery process often appears to fall on them. Many of them end up taking on the responsibility of project-managing the repairs to the house (Enarson and Fordham, 2001), thereby going through the stress of managing the insurance claim and the repair work. This can be attributed to the fact that more women may work part time or are at home with children and are, therefore, prone to suffer more from the disruption of the home environment. Further, their presence at home may encourage them to take up the responsibility of project-managing the repair works; therefore, they will be expected to supervise workmen, receive deliveries and make phone calls to the insurance company (Joseph *et al.*, 2011b) and in some cases deal with letting agents (Whittle and Medd, 2012).

In the United Kingdom, guidance on economic appraisal for flood and coastal risk management is provided by the Environment Agency (2010). This includes a section on the 'nonmonetary impact on households', in which it is stated that 'impacts of flooding on households, such as increased stress, health damage and loss of memorabilia can be far more important than the direct material damages to their homes and their contents'. Green *et al.* (1985) also suggested that the most significant intangible impact of flooding is the effect of flooding on the health of the people. Given the possible acute and chronic long-term health effects that can arise from a flooding incident, the costs attributed to these effects could be a major factor in flood risk decision-making, particularly at a smaller spatial scale.

There has been a dearth of research in establishing the long-term intangible cost of flooding on households; this is due to the fact that these costs are difficult to value owing to the subjective nature of some of the impacts. However, a recent study jointly commissioned by the Environment Agency and DEFRA (2004) adopted the stated preference method of valuation to infer the willingness to pay value from respondents. The results of this research shows that more than 60% of flooded and at-risk respondents expressed a willingness to pay (WTP) to avoid the health impacts associated with flooding. Based on the analysis of the survey, the authors recommended that the value of £200 (in 2004 prices) per household per year be taken as representing the benefits of reduced health impacts at a household level as a consequence of a significant reduction in the risk of flooding. This can be seen as a positive development with regards to the valuation of the long-term intangible impact of flooding on households.

19.4 Conclusions

The costs of flooding can be examined from a variety of perspectives (tangible and intangible) and at a variety of spatial and temporal scales. The important issue is to use the appropriate temporal and spatial scales and the correct perspective for the given decision when establishing the full costs of flooding. The categorisation of the costs of flooding depends on factors, such as the ease of monetising the impact, when the impacts are felt and for how long (short or long term) and whether the impacts are felt directly or indirectly. However, it is recognised that the impacts on the lives and livelihoods of those caught up in flood events are difficult to quantify in monetary terms. For many people affected worldwide, floods are a personal tragedy from which full recovery may be very slow or not possible; therefore, the costs of flooding to such people may be unquantifiable. Financial costs of flooding are borne by private individuals and businesses, but the overall economic impact of flooding is the damage to businesses, that is premises that have been adversely affected by flood water, the loss of stock and loss of sales due to temporary relocation of community members.

The tangible financial costs of flooding on households are huge and include both direct and indirect costs; these are of the utmost importance in the assessment of the full economic costs of flooding. However, it is recognised that for fully insured homeowners in the United Kingdom, the majority of the financial costs of flooding are covered under their domestic insurance policy. However, the intangible impacts of flooding can be more costly and they are not ordinarily insurable. These impacts include psychological effects, loss of irreplaceable items of personal possession, significant disruption caused to households through the occurrence of a flood event, living in properties being repaired, having to deal with recovery with little or no help and the amount of time and effort involved in organising the household recovery and dealing with insurers, loss adjusters and building contractors.

While there are empirical data for the assessment of the tangible financial costs of flooding on households, assessing the full financial intangible costs of flooding on households has received less attention within the flood research community. This is due to the fact that they are difficult to value owing to the subjective nature of some of the impacts. However, the jointly commissioned Environment Agency and DEFRA (2004) research concluded that the sum of £200 per household per year should be taken as representing the benefits of reduced health impacts as a consequence of a significant reduction in the risk of flooding. This can be seen as a positive move in the assessment of intangible flood impacts on households. However, the value of £200 may be somewhat questionable when compared to the value of tangible impacts, bearing in mind that the intangible impacts are said to be more devastating than tangible impacts. It is, therefore, recommended that further research is needed to address the assessment of the full costs of the impacts of flooding on households, as this will enhance households' understanding of the costs associated with flooding, thereby helping inform households to make informed decisions about how to manage flood risk. Conceivably, if the full costs of flooding were fully accounted for then it would have a higher priority in government spending both in the United Kingdom and elsewhere.

References

Ahern, M. and Sari, K. (2006) The health impact of floods. In: Few, R. and Matthies, F. (eds), *Flood Hazards and Health: Responding to Present and Future Risks*. Earthscan, London, pp. 28–53.

Ahern, M., Kovats, R.S., Wilkinson, P., Few, R. and Matthies, F. (2005) Global health impacts of floods: epidemiologic evidence. *Epidemiologic Reviews*, **27**, 36–46.

Association of British Insurers (ABI) (2009) *Delivering on Our ClimateWise Commitment: Second Year Report*. Association of British Insurers (ABI), London.

Bowker, P. (2007) *Flood Resistance and Resilience Solutions an R&D Scoping Study*. Department for Food and Rural Affairs (DEFRA).

Brath, A., Montanari, A. and Moretti, G. (2006) Assessing the effect on flood frequency of land use change via hydrological simulation. *Journal of Hydrology*, **324**, 141–153.

Chatterton, J., Viavattene, C., Morris, J., Penning-Rowsell, E. and Tapsell, S.M. (2010) *Delivering Benefits Through Evidence: The Costs of the Summer 2007 Floods in England*. Environment Agency, Bristol.

Du, W., Fitzgerald, G., Clark, M. and Hou, X. (2010) Health impacts of floods. *Orehospital and Disaster Medicine*, **25**, 265–272.

Enarson, E. and Fordham, M. (2001) Lines that divide, ties that bind: race, class, and gender in women's flood recovery in US and UK. *Australian Journal of Emergency Management*, **15**, 43–52.

Environment Agency (2009a) *Flooding in England: A National Assessment of Flood Risk*. Environment Agency, Bristol.

Environment Agency (2009b) *Investing for the Future: Flood and Coastal Risk Management in England. A Long-Term Investment Strategy*. Environment Agency, Bristol.

Environment Agency (2010) *Flood and Coastal Erosion Risk Management appraisal guidance (FCERM–AG)*. Environment Agency, Bristol.

Environment Agency and DEFRA (2004) *The Appraisal of Human Related Intangible Impacts of Flooding*. Joint DEFRA/EA R&D technical report FD2005/TR, DEFRA/EA.

Fay, M., Block, R., Carrington, T. and Ebinger, J. (2009) *Adapting to Climate Change in Europe and Central Asia*. World Bank. Available at: http://www.worldbank.org/eca/climate/ECA_CCA_Full_Report.pdf.

Few, R., Ahern, M., Matthies, F. and Kovats, S. (2004) *Floods, Health and Climate Change: A Strategic Review*. Norwich.

Green, C.H., Emery, P.J., Penning-Rowsell, E.C. and Parker, D.J. (1985) *The Health Effects of Flooding: A Survey at Uphill, Avon*. Flood Hazard Research Centre, Middlesex Polytechnic, Enfield.

Gurenko, E. and Lester, R. (2004) *Rapid Onset Natural Disasters: The Role of Financing in Effective Risk Management*. The World Bank, Washington, DC.

Hadjimitsis, D.G. (2010) Brief communication: determination of urban growth in catchment areas in Cyprus using multi-temporal remotely sensed data: risk assessment study. *Natural Hazards Earth System Science*, **10**, 2235–2240.

Ingirige, B. and Wedawatta, G. (2011) Impacts of flood hazards on small and medium sized companies. In: Lamond, J.E., Booth, C.A., Hammond, F.N. and Proverbs, D.G. (eds), *Flood Hazards: Impacts and Responses for the Built Environment.* Taylor and Francis Group, London.

Jha, A.K., Bloch, R. and Lamond, J. (2012) *Cities and Flooding: A Guide to Integrated Urban Flood Risk Management for the 21st Century.* The World Bank, Washington, DC.

Joseph, R., Proverbs, D., Lamond, J. and Wassell, P. (2011a) An analysis of the costs of resilient reinstatement of flood affected properties: a case study of the 2009 flood event in Cockermouth. *Structural Survey*, **9**, 279–293.

Joseph, R., Proverbs, D., Lamond, J. and Wassell, P. (2011b) A critical synthesis of the intangible impacts of flooding on households. In: *International Conference in Building Resilience: Interdisciplinary Approaches to Disaster Risk Reduction and the Development of Sustainable Communities*, Sri Lanka.

Joseph, R., Proverbs, D., Lamond, J. and Wassell, P. (2011c) A critical synthesis of the tangible impacts of flooding on households. In: *27th ARCOM Annual Conference 5–7th September 2011.* University of the West of England, Bristol, UK. ARCOM.

Lamond, J.E. (2012) Financial implications of flooding and the risk of flooding on households. In: Lamond, J.E., Booth, C.A, Hammond, F.N. and Proverbs, D.G. (eds), *Flood Hazards: Impacts and Responses for the Built Environment.* Taylor and Francis Group, London, pp. 317–326.

Mankiw, N.G. (2012) *Principles of Economics.* 6th Edition. South-Western, Cengage Learning, Mason.

North West Development Agency (2009) *Economic Impacts of Increased Flood Risk Associated with Climate Change in the North West (Final Report).* London.

Penning-Rowsell, E., Johnson, C., Tunstall, S., Tapsell, S., Morris, J., Chatterton, J. and Green, C. (2005) *The Benefits of Flood and Coastal Risk Management: A Manual of Assessment Techniques*, Flood Hazard Research Centre. Middlesex University Press.

Pitt, M. (2008) *The Pitt Review – Learning Lessons from the 2007 Floods.* HM Government Cabinet Office.

Sykes, E. (2011) *Counting the Cost of the Floods*, 612. ABC, Brisbane. Available at: http://www.abc.net.au/local/stories/2011/01/04/3106141.htm.

Thurston, N., Finlinson, B., Breakspear, R., Williams, N., Shaw, J. and Chatterton, J. (2008) *Developing the Evidence Base for Flood Resistance and Resilience. Joint Defra/EA Flood and Coastal Erosion Risk Management R&D.* DEFRA, London.

Wassell, P., Ayton-Robinson, R., Robinson, D., Joseph, R., Hack, K., Butler, D., Salkeld, I. and Twomey, J. (2009) *Resilient Reinstatement: The Costs of Flood Resilient Reinstatement of Domestic Properties.* Association of British Insurers, London.

Watts, A. (2009/2010) Cumbria flooding: are we witnessing the seeds of change? *The Damage Management Magazine: Recovery*, **10**, 6–7.

Whittle, R. and Medd, W. (2012) Living with flood understanding residents' expericences of recovery. In: Lamond, J.E., Booth, C.A., Hammond, F.N. and Proverbs, D.G. (eds), *Flood Hazards: Impacts and Responses for the Built Environment.* Taylor and Francis Group, London, pp. 281–290.

World Health Organization (WHO) (2002) *Flooding: Health Effects and Preventative Measures.* World Health Organization (WHO).

20

The Role of Market-Based Flood Insurance in Maintaining Communities at Risk of Flooding: A SWOT Analysis

Jessica E. Lamond

20.1 Introduction

Managing the risk of flooding to human settlements can involve multiple strategies, which can include flood prevention, flood avoidance, damage reduction and recovery planning. Among these strategies, flood insurance has the specific function of facilitating recovery after a flood event by providing finance for reinstatement and replacement. Massive investment has been seen over the years in flood risk management infrastructure to prevent flooding and this has proved effective in providing protection to people and property. However, despite this, the annual losses from flooding continue to grow and this trend is expected to continue. Reported losses from flooding since 1950 (Figure 20.1) show there has been a growing trend in losses and that in 2011 the highest ever loss was reported. Over the past two decades the average annual reported loss has amounted to $20 billion dollars, which is likely to be an underestimate of the true cost of flooding as many minor events go unreported or are not classed as disasters.

Much of the increase in monetary flood losses is attributable to growth in the value of assets at risk from flooding, due to global patterns of urbanisation and increasing wealth (Jha *et al.*, 2011). These factors are predicted to continue, creating even greater concentrations of assets at risk and in need of protection; therefore, there is increasing recognition that it will not be possible to prevent flood damage losses rising even further in the future. Approaches such as 'making space for water' in the United Kingdom (Penning-Rowsell and Wilson, 2006), '*room* for the rivers' in Germany (Lange and Garrelts, 2007), 'living with flood' in the Netherlands (Bruen and Gebre, 2001), amongst others, reflect the shift in flood risk management thinking away from hard engineering and towards other measures. Flood insurance can be seen as an alternative to massively increased investment in harder flood risk management measures, as it has the potential to mitigate the effects of flooding and maintain communities in the floodplain. More usually it is seen as an additional measure

Water Resources in the Built Environment: Management Issues and Solutions, First Edition.
Edited by Colin A. Booth and Susanne M. Charlesworth.

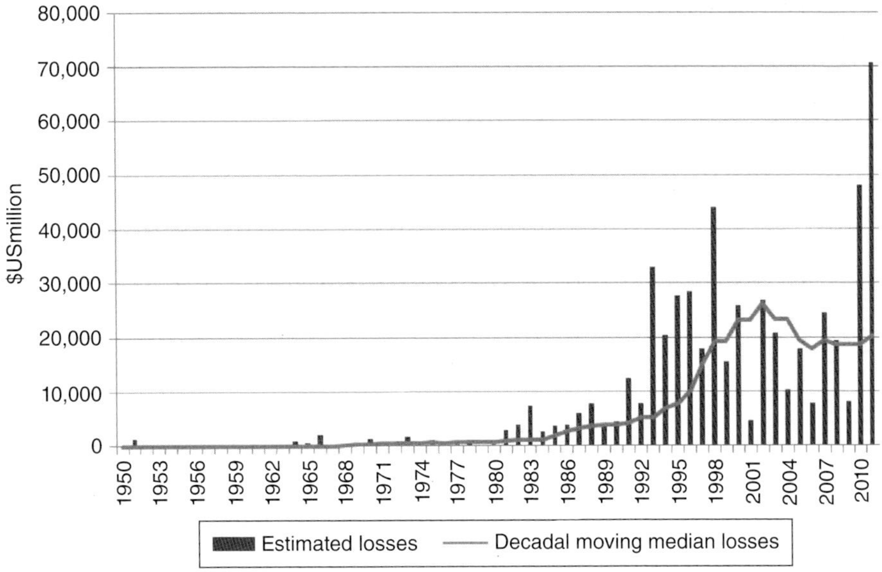

Figure 20.1 Global reported losses from flood events with moving median (data sourced from Emdat.be).

to absorb residual risk after the implementation of flood damage reduction measures. As flood events become more costly and flood losses accelerate globally, the need for insurance has grown, and governments are increasingly considering the role of market-based insurance schemes as part of their flood risk management policy.

Flood insurance is regarded, by policyholders, as a strategy of risk avoidance. Insurance can never prevent a flood from happening but it allows floodplain occupants to avoid much of the financial risk from flooding by providing funds to rebuild and replace assets after a flood event. Flood insurance has proved of great benefit to insured flooded households, as those without insurance can suffer great financial hardship after a flood (Welsh Consumer Council, 1992; Fordham and Ketteridge, 1995). Insurance can be a successful and effective way of managing flood risk and can be particularly attractive in areas at risk of low-frequency high-impact events.

Coverage of flood risk by insurance is somewhat difficult to measure as it is often bundled with other perils, and different types of flooding may be included under a range of alternative covers. However, an indicative estimate is that insurance covers 40% of high income country losses, falling to 10% in middle-income countries and <5% in low-income countries (Jha *et al.*, 2011). The issue is not limited to a developed/developing world dichotomy: on a country-by-country basis it varies within the developed world from almost 100% in countries such as Spain and Switzerland to <1% in the Netherlands. In the developing world, while many countries see minimal to nonexistent coverage, flood reinsurer Swiss Re has estimated that Indonesian flood insurance cover may be as high as 20% (Gaschen *et al.*, 1998). It appears that the propensity to buy insurance is dependent on many external and internal influences. Availability and cost of insurance is a crucial limiting factor but take-up of cover is also subject to provision of disaster relief, income levels, general risk awareness and attitudes to collective and individual risk (Lamond and Proverbs, 2009).

It is prudent to note, although from a policyholders perspective insurance is risk avoidance, in reality, from a broader perspective, risk is not avoided but passed on, spread out or possibly delayed depending on the insurance model. Insurance models are diverse throughout the globe and each insurer or scheme will have very specific features that cater to local risk profiles.

Table 20.1 shows a general categorisation of models that considers two important dimensions thought to affect take-up and viability of flood insurance (Lamond and Penning-Rowsell, 2011). Policies are categorised from bundled, that is flood insurance included into all household policies, through optional add-on or optional separate cover, through to no cover or minimal cover. The other dimension is market-based versus state-backed provision, relating to the eventual underwriting of flood risk and the security of provision in the face of possible catastrophic events.

Within this chapter the focus is on the more common market-based schemes regardless of their delivery on dimension one. Common features of market schemes are that private insurance companies provide cover directly to property owners or occupiers. Policy terms, typically, cover direct damage but may also include cover for disruption, alternative accommodation and other losses. There is usually a need for some offsetting of risk, to cover the extreme losses experienced in major events, and this is often achieved by reinsurance via international reinsurance markets. Risk is, therefore, spread out across local policyholders and international risk markets with a greater or lesser degree of cross-subsidy, depending on the precision of actuarial pricing.

Perceived attractions of market- or private company-based insurance include reduction of the need for national governments to plan or set aside funds for catastrophe finance and the improved efficiency and lower costs often anticipated with a move from public to private sector provision. However, it is important to recognise the limitations to the protection that insurance can offer and to balance the advantages and disadvantages of reliance on a market based system now and in the uncertain future. Therefore, this chapter examines the

Table 20.1 Common types of flood insurance models.

	Bundled into general household policy	Add-on or bundled with other specific perils	Separate policy	No/minimal cover available
Private	United Kingdom Hungary Brazil Australia Norway	Germany Indonesia Israel China Czech Republic Ecuador Italy Japan Philippines Poland Portugal South Africa Taiwan	Australia Austria Belgium	Canada (surface water) Denmark (sea flooding covered) Turkey (some might be covered under earthquake risk) Netherlands (some pluvial cover)
State-backed or operated	Spain China Switzerland	France Indonesia New Zealand (compulsory)	United States	

strengths and weaknesses of market-based insurance with reference to extant market-based systems of flood insurance. It also considers the opportunities and threats for future flood risk insurance, as identified from the strengths and weaknesses and by insurers and reinsurers that are scoping the future of catastrophe risk finance, to provide food for thought about the possible interventions that can be most successful in reducing future risk.

20.2 Strengths

20.2.1 Secure Financing for Recovery

Market-based insurance, in common with state-backed schemes, exists primarily to offset the financial risk of damage to property caused by flooding. In the United Kingdom, for example, flood insurance is bundled into standard household polices and the majority of households will have some level of cover. An estimated 95% of UK property will have buildings cover, whereas 75% will be covered for movable contents (Gaschen *et al.*, 1998). Studies have shown UK homeowner satisfaction with the eventual reinstatement of their properties after flooding is high (Pitt, 2008; Samwinga, 2009), albeit there is often substantial delay after a major flood event (Association of British Insurers (ABI), 2009) and households experience stress in dealing with the disruption while reinstatement is carried out (Whittle and Medd, 2011). In Germany, households with insurance receive money more quickly and recovered better than those receiving state payouts (Thieken *et al.*, 2006). Market-based schemes often cover a wider range of losses, such as alternative accommodation costs, whereas state-backed schemes may be limited to physical damage and may have an upper limit for cover (Burby, 2001; Michel-Kerjan, 2001; Consorcio de Compensacion de Seguros, 2008; Abbott, 2008). There is some merit in delegating flood insurance to the market experienced in handling claims with processing systems in place for other perils, rather than set up parallel government systems in departments that may not be as well equipped. Secure finance for reinstatement from a commercial company also allows for the professionalisation of damage management companies and, potentially, faster and better reinstatement.

Involvement of the market in supporting some of the financial losses from flooding alleviates the pressure on national and local governments to raise or set aside funds through taxation or loans. For example, in the United Kingdom, central government has responsibility for repairing damage to major national infrastructure and reimburses responsible authorities for some emergency and disruption costs. Other losses are expected to be managed by local authorities, private companies and individuals. Business can benefit from the purchase of flood insurance and this is particularly important for small- and medium-sized companies with limited geographical diversity, as these are most likely to be hardest hit by localised flood events (Bhattacharya *et al.*, 2011) and often fail to recover after the event (Crichton, 2006). Local government and public organisations can also consider the advantages of insurance, for example, in the UK floods of 2007, where ~12% of total local authority losses were covered by market insurance and other losses were covered by self-insurance funds or local taxation (Audit Commission, 2007). Local authorities and housing associations can insure housing stock through large specialist companies and are not subject to the ABI statement of principles. They should, therefore, be priced on the basis of assessed risk and the experience of local authorities seeking to reinsure, since claiming in 2007, has shown that premiums are increasing. Many more are now moving towards self-insurance and mutual funds. Individuals without insurance in the United Kingdom have to rely on their own resources, perhaps using loans or support from their networks or hope funds will be set up to assist those uninsured. Market-based flood insurance therefore reduces the liability of UK Government in the event of a flood to a great extent. In theory, this frees government resources to invest in preventative risk management measures and other priorities.

20.2.2 Maintenance of Communities via Peace of Mind, Property Value

Security of finance for reinstatement via insurance allows occupation of dwellings that might otherwise be seen as undesirable, due to the periodic need to invest in repairing flood damage. This security also allows property to be transferred at prices close to, or equal to, the market value of comparable property not at risk (Shilling *et al.*, 1989; Lamond *et al.*, 2010), thus securing the long-term viability of floodplain property. Owners therefore have the incentive to maintain their property and invest in their community, leading to the avoidance of property blight.

Desirability of maintaining populations in areas at risk of flooding is a debatable issue. In some cases, governments take the view that a useful option for reducing flood risk is to relocate settlements out of the floodplain. This has formed part of policy and practice in the United States (Federal Emergency Management Association, 2008), Canada (Babcock and Mitchell, 1980), France (Bruen and Gebre, 2001), China and Brazil (Jha *et al.*, 2011). In these places property has been bought at market rates, or exchanged for similar or even superior accommodation elsewhere, to allow residents to relocate to areas not at risk. Such extremes can sometimes be justified in areas of frequent and severe flooding, but it is worth noting populations at risk from flood do not usually want to move from their settled communities (Lamond, 2006). Equally, many properties at risk from flooding are at risk of shallow flooding at a low return frequency and the advantages of a floodplain location could outweigh the expected future damage costs, given adequate warning systems and emergency planning processes. Relocation policies are only likely to apply to those settlements or properties at risk from very severe or very frequent flooding and, therefore, the ability to maintain communities in the floodplain will be important for the foreseeable future.

20.2.3 Risk-Based Pricing Generates Awareness of Risk and Incentives to Reduce Risk

Market-based insurance is more likely than other risk financing mechanisms to reflect risk in the setting of premiums. Due to the pressures of competition, a market-based scheme, particularly one with add-on or stand-alone policies, will be less likely to sustain cross-subsidy from areas not at risk to those at risk or to sustain underwriting losses in the long term (Dlugolecki *et al.*, 2009). Therefore, premiums for market-based insurance are a signal of flood risk to occupiers or owners of property at risk of flooding. Insurance has the potential to generate raised, and realistic, flood risk awareness, which could translate into improved preparedness and is a necessary, though not sufficient, condition for flood adaptation (Lamond and Proverbs, 2009; Rose *et al.*, 2011). However, this advantage may only be realised if insurance for flooding is mandatory or regularly sought. Where residents are completely unaware of living in an area at risk they may not seek insurance and, thus, not experience high premium quotes. Therefore, high premium can be seen as one way of reinforcing a message of flood risk awareness but is not the best or preferred means of communicating risk to a population.

Conversely, as an incentive to reduce risk, market-based insurance has the advantage of the potential to provide a tangible, ongoing financial incentive to reduce risk via adaptation or preparedness. Where this has been applied, for example in the UK commercial property market (Lamond *et al.*, 2009a), there is some evidence that insurers can influence the behaviour of property stakeholders at risk.

20.2.4 Ability to Offset Risk Internationally

A major advantage of market-based insurance, over national government schemes and disaster assistance is the ability to offset risk in one geographical location with risk elsewhere in the world via reinsurance or a large international portfolio (Dlugolecki *et al.*, 2009). For most weather systems causing flood events the geographical scope is limited and therefore, for reinsurers, the worldwide spread of losses makes their portfolio of risk, relative to received premium income, more sustainable. Reinsurers need to hold expensive reserves is lower than nationally based insurers would be for the same risk; therefore, insurance costs are lower and larger risks may be accepted.

Such means are available to national governments by going out to the reinsurance markets, but these would be seen as market-backed schemes by definition. For national governments, market-backed financing is an expensive option that is justifiable if the probable maximum loss is beyond the capacity of the tax base to recover quickly (Jha *et al.*, 2012).

20.2.5 Ability to Offset Risk across Perils

The national and international insurance market has a similar ability to offset the risk of flooding against other perils. The French scheme is an example of this on a national scale (Michel-Kerjan, 2001). However, the potential for the market to use international expertise, regional models and capitalise on economies of scale is much greater. The United Kingdom, for example, has a higher degree of flood risk and very low earthquake risk. The ability of the UK government to spread risk across hazard is, therefore, quite low. Conversely, in other countries, such as Japan, flood risk is not considered important against the much higher risk of earthquakes (Consorcio de Compensacion de Seguros, 2008). International profiles of risk are naturally much more diverse and the potential to reduce costs by offsetting risk is much greater.

20.2.6 Risk Assessment

The ability to predict, with any degree of reliability, where the next event will take place is quite low, but global predictions of damage are more reliable. Future climate risks are also much more predictable at a global or regional scale than at detailed national levels. It is also likely that increased flood losses may coincide with reduction in claims from other perils, such as drought. The expertise of the global insurance market lies exactly in quantifying and exploiting patterns of global risk. Investment to do these calculations at the national scale may be less reliable, outside the capacity of small nations, and an inefficient allocation of investment, which could be spent in short-term prediction and emergency planning.

20.3 Weaknesses

As the above section outlines, there are many advantages to the provision of market-based insurance for populations at risk from flooding. However, most communities at risk from flooding are very poorly covered by insurance and those covered by market-based schemes tend to have a lower coverage rate than other options. This is linked to the many weaknesses of the market-based insurance mechanism as detailed below.

20.3.1 Low Coverage

Schemes based on competitive market provision of insurance have been shown to have lower coverage rates than state-based schemes (Lamond and Penning-Rowsell, 2011). Unless flood insurance is bundled into standard policies, as it is in the United Kingdom, the element of electing to buy market insurance in advance of a flood event means coverage tends to be low. In contrast, disaster assistance, as a financing mechanism, will usually cover flood affected property on a more widespread basis. State-backed insurance is more likely to be compulsory, available, affordable and cross-subsidised, resulting in higher coverage levels.

20.3.2 Potential for Insolvency

Failure of an insurance company may be seen as a potential weakness in market-based insurance that would leave policyholders without cover and liable for their own damages. In practice, although individual companies may become insolvent, the risk to the policyholder is very low. Most insurers diversify risk internally, reinsure or subscribe to national schemes with guaranteed payouts (Dlugolecki *et al.*, 2009). Financial regulations are continually updated to ensure that risk of insolvency is minimised. An insurer withdrawing from the market leading to lack of availability of cover is much more likely.

20.3.3 Availability and Affordability of Insurance

There are some flood insurance markets that are viewed as commercially uninsurable, for example the United States (Burby, 2001; Federal Emergency Management Agency, 2011), France (Michel-Kerjan, 2001) and Canada (Swiss Re, 2010). Provision of a market-based system relies on private companies identifying a profit opportunity and seeing this opportunity as currently exploitable, in the context of alternative profitable options within their strategic priorities and capital base. Insurance companies are ultimately answerable to their shareholders; they also have a duty of care to their policyholders but no responsibility to a potential customer base. Commercial reality may lead to the decision of an insurer not to enter a market or to withdraw from it, making flood insurance unavailable to purchase on a market basis. This has happened in the United States on a large scale (Mills *et al.*, 2005) and in the United Kingdom on an individual insurer basis (Stevenson, 2002). Further, the price of market insurance, where it is provided on an actuarial basis, may be very high, as in some parts of Hungary (Vari *et al.*, 2003), or policyholders may be subject to high excesses. If excesses are set high then the policyholder will be responsible for providing large sums of capital for repair, as part of the insurance contract, and could be seen as effectively uninsured.

These effects are much more likely to occur in areas at high flood risk. For example, in Hungary, Germany and Australia residents in the highest risk areas are not covered by insurance (Vari *et al.*, 2003; Chen *et al.*, 2004; Thieken *et al.*, 2006). A stark weakness of a market-based system is that it may become the case that those most in need of insurance cannot find a policy or cannot afford to buy it.

20.3.4 Adverse Selection

As indicated above, the provision of flood insurance is reliant on the balance of risk and premium recoverable from the population at risk. Some degree of cross-subsidisation is inevitable because the level of information about the exact hazard and associated risk is difficult to assess at a detailed local level. As a result, it is often the case that local populations have a greater understanding of their individual risk than an insurance company can hope to have. This results in the phenomenon of adverse selection, where only those who are at very high risk choose to insure but the insurer underestimates their risk based on averaging across a wider population with varied risk profiles (Clark *et al.*, 2002). The insurer is, therefore, much more exposed to risk than they expect and are vulnerable to higher claims and possible insolvency.

The potential for adverse selection in actuarially priced insurance markets leads to the need for insurers to gain higher levels of risk and hazard information than is needed in cross-subsidised or mandatory schemes. This may result in higher premiums, lower attractiveness for commercial insurers and, ultimately, may make a market uninsurable.

20.3.5 Moral Hazard

The phenomenon of moral hazard is a familiar insurance paradox; whereby the act of providing insurance increases the level of risk in a population. Moral hazard exists because insurance cover provides a disincentive for individuals to undertake other risk management activities. Evidence suggests that moral hazard is greatest for low-probability, high-impact events like flooding, where the cost of mitigation activity is real and current but the potential consequences of flooding are possible, regarded as unlikely and covered by insurance (Kunreuther and Pauly, 2005). In the flood risk context, moral hazard is particularly apparent in the disincentive to protect property from flood damage. Research has shown adaptation to flood risk is generally low and insurance cover is implicated in the tendency to maintain property in a vulnerable state in the United Kingdom (Lamond *et al.*, 2009b) and in the United States (Federal Emergency Management Agency, 2011), resulting in large repetitive claims.

20.3.6 Tendency to be Inequitable

The question of whether it is the responsibility of society in general or floodplain residents in particular to provide funds for their own protection is an open one and beyond the direct scope of this chapter. However, factors affecting the judgement for a particular community at risk may include the degree of choice in location of residence, the socioeconomic constitution of the population at risk, cultural and historical norms and policy with respect to the desirability, or not, of supporting the continued occupation of the floodplain. Social justice with respect to climate change is a thorny topic and crosses national boundaries, but equity with respect to who pays for different flood protection measures can also divide national opinion. In the United Kingdom, for example, there appears to be a concentration of poorer residents in coastal areas at risk (Walker and Chalmers, 2004; Zsamboky *et al.*, 2011). If rising sea levels place more of these areas at risk or at higher risk in the future and the decision not to increase or maintain hard defences is taken, it may not be reasonable to expect these communities to pay the increased risk-based premium. Market-based insurance schemes may struggle to continue to cross-subsidise high-risk households. Equally, market-based insurance is often more costly than equivalent state-provided cover (World Bank and the United Nations, 2010) and has been shown to be, on average, more expensive for households on lower incomes (Treby *et al.*, 2006). Thus, market-based insurance often disadvantages the poor whereas, in theory, state systems can be adjusted to protect the poorest and most vulnerable.

20.4 Threats

20.4.1 Climate Change

Climate change and associated expected higher extreme weather claims have the potential to put increased strain on the supply of reinstatement capital via insurance or aid. Climate change threatens the liquidity of all insurance models, but there are particular threats specific to market-based models that are related to the lack of a strategic view of flood recovery. As it remains the case that the majority of market-based insurance policies are renewed annually and can, therefore, be speedily cancelled; in an era of concerns about a changing climate the stability of privately supplied insurance cover is questionable. Climate change is likely to bring higher levels of insurance claims in the future and result in greater uncertainty. For flood insurance, this threatens the insurability of markets if insurers and reinsurers assess the maximum loss potential to be too great to cover (Dlugolecki *et al.*, 2009). As required capital reserves rise globally to protect the solvency of the market, it is possible that areas at highest risk will be sacrificed to more profitable lower-risk markets. Market-based insurance has no social directive to insure the most vulnerable as state-backed schemes may have. Increased uncertainty may also lead to insurers and reinsurers prioritising other perils and not pursuing viable flood insurance markets. In the past, new money has entered the market in response to increased demand after extreme events; the increased premium levels yield a profit-rich opportunity for global money lenders. This may not continue to be the case if global demand rises dramatically (Dlugolecki *et al.*, 2009).

20.4.2 Improved Risk Information

Better quality and more detailed and accurate risk information is likely to lead to more actuarially based pricing. This could lower the demand for flood insurance as it becomes unaffordable for those who are most at risk and need it most. This has been seen to be the case in some European countries, and in Australia, where insurance in high-risk areas is so expensive that insurers find very low uptake. The increased ability to discriminate between properties in terms of risk may effectively end the insurability of very high risk property that is now being cross-subsidised due to the inaccuracies of current information systems. This will inevitably reduce cover levels. The paradox is that good information is necessary for cover to be viable and yet very precise risk assessment may leave the most vulnerable uninsured. In a market-based system, it may be argued that these properties should not be receiving the benefit of cover, whereas in a state-backed scheme there may be more support for cross-subsidisation.

20.5 Opportunities

As has been identified, there is a large and growing risk of flooding on a worldwide basis and much of it is not covered by insurance. There is, therefore, a potential for higher levels of cover to protect more assets and property and give financial security to more people.

20.5.1 Potential for Growth Leading to Better Offsetting

The potential for the insurance market to grow in the future could be seen as an opportunity for insurers and reinsurers to generate larger profits (Munich Re, 2004). However, market competition should dictate that a massively increased risk portfolio will bring lower premium rates on average. The ability to offset risk across a wider geographical range, economies of scale in information systems and processing may also be achieved. The risk of flooding can be used to balance against other perils and, therefore, increased levels of flood cover may yield benefits in other disaster insurance areas.

20.5.2 Climate Change

Climate change brings greater uncertainty and is predicted to result in an increase in intense rainfall events, making patterns of risk less predictable. For flood insurance, this may give the opportunity to spread the risk across an even greater number of policies, whereby the action of adverse selection is minimised and more markets could become insurable.

20.5.3 Potential for Innovative Products that Encourage Adaptation

Evidence suggests that insurance markets are not currently very adept at encouraging adaptation to flood risk (Kunreuther, 2006; Lamond *et al.*, 2009b; Federal Emergency Management Agency, 2011). The potential for insurance to develop innovative models and products that incentivise adaptation is a major future opportunity warranting investment in research, product development and market research. It is possible this may be achieved through improved risk-based pricing, but other alternatives may include longer insurance contracts enabling better investment returns for the insurer.

20.5.4 Potential to Influence Adaptation and Mitigation Policy

The Gentlemen's Agreement between the government and the insurance industry in the United Kingdom is an example of the ability of a partnership approach to flood risk management that could form a model for the industry as a whole. Despite issues identified with incentives for individual households to adapt, the UK system has allowed the insurance industry to enter into and help to shape the debate in flood defence and adaptation policy. This has been somewhat successful in raising standards and keeping flood risk high on the national agenda. Involvement of the market in insurance can also provide a platform for insurance industry involvement in carbon abatement policy and the drive towards greenhouse gas (GHG) emission targets.

20.6 Conclusions

Market-based insurance can be a force for positive maintenance of communities at risk from flooding. Consideration of the amount of insurance paid after major flood events and the number of people affected suggests that the importance of insurance in flood risk management should not be underestimated. However, flood insurance is just one of many flood risk management measures and may be seen, within an integrated approach, as a solution to manage residual risk, remaining after appropriate hard engineered and damage limitation schemes have been implemented.

This chapter has considered the strengths and weaknesses of market-based flood insurance, one of the existing models of catastrophe financing and insurance commonly observed. The analysis considers the role of insurance in protecting property at risk but also in encouraging adaptation and long-term reduction of risk. It has also considered some opportunities and threats that may emerge in the future. A multiple perspective approach of policyholders, insurers and societal good has been adopted.

In terms of protection, there are many advantages to the market-based system but these are, in some sense, offset by shortcomings in the lack of a mandate to protect the most vulnerable by cross-subsidisation. Hybrid schemes that combine market-based insurance with government or donor support as a last resort may overcome these issues, but research is needed to establish whether this would provide the best value for tax payers.

It is also apparent that proactive market-based insurance has the potential to not only protect against financial losses associated with climate change but also to change behaviours. The use of premium discounts for flood adaptation and mitigation of losses by individuals has the potential to increase take-up of such practices and reduce global risk. However, this is an area where further research is needed to ensure such schemes are cost-efficient and produce the expected outcomes.

The provision of insurance is threatened by the potential growth in claims associated with climate change. Market-based cover is seen as the system most likely to capitalise cover, given the relative size of global rather than national money markets. The appetite for continued provision is, however, a potential issue. Clearly, an opportunity exists to leverage mitigation and adaptation policy to reduce risk for the insurance industry. Such activities might also benefit populations at risk through the encouragement of climate sensitive policy.

Flood insurance could form an important part of future flood risk management strategy; therefore, an objective participatory analysis of multiple stakeholder goals and outcomes on an international basis could assist the development of insurance models and partnerships fit for this purpose.

References

Abbott, E.B. (2008) Floods, flood insurance, litigation, politics – and catastrophe: The National Flood Insurance Program. *Sea Grant Law and Policy Journal*, **1**, 129–155.

Association of British Insurers (ABI) (2009) *Resilient Reinstatement – The Cost of Flood Resilient Reinstatement of Domestic roperties*. Research Paper. Association of British Insurers, London.

Audit Commission (2007) *Staying Afloat, Financing Emergencies*. Public Services, National Report. Audit Commission, London.

Babcock, M. and Mitchell, B. (1980) The impact of flood hazard on residential property values in Galt (Cambridge) Ontario. *Water Resources Bulletin*, **16**, 532–537.

Bhattacharya, N., Proverbs, D. and Lamond, J. (2011) Identifying the causes of the vulnerability of UK commercial property to flood damage: a synthesis of research. In: *International Conference in Building Resilience, Interdisciplinary Approaches to Disaster Risk Reduction and the Development of Sustainable Communities*, Kandalama, Sri Lanka.

Bruen, M. and Gebre, F.A. (2001) Worldwide public perception of flood risk in urban areas and its consequences for hydrological design in Ireland. In: *Irish National Hydrological Seminar*, Tullamore.

Burby, R. J. (2001) Flood insurance and floodplain management: the US experience. *Global Environmental Change Part B: Environmental Hazards*, **3**, 111–122.

Chen, K., McAneny, J., Blong, R., Leigh, R., Hunter, L. and MacGill, C. (2004) Defining area at risk and its effect in catastrophe loss estimation: a dasymetric mapping approach. *Applied Geography*, **24**, 97–117.

Clark, M.J., Priest, S.J., Treby, E.J. and Crichton, D. (2002) *Insurance and UK Floods: A Strategic Reassessment. Report of the Tsunami Project*. Universities of Southampton, Bournemouth and Middlesex.

Consorcio de Compensacion de Seguros (2008) *Natural Catastrophies Insurance Cover. A Diversity of Systems*. CCS, Madrid.

Crichton, D. (2006) *The Effects of Insurance Changes and Climate Change on Small and Medium Size Enterprises (SMEs) in Britain*. AXA Insurance.

Dlugolecki, A. et al. (2009) Coping with Climate Change: Risks and Opportunities for Insurers. Chartered Insurance Institute, London.

Federal Emergency Management Agency (2011) *History of the National Flood Insurance Program and Reform Efforts. Roundtable on Flood Risk Management*. GFDRR/ The World Bank, Washington, DC.

Federal Emergency Management Association (2008) *Mitigation Best Practices Search*. FEMA, Washington, DC.

Fordham, M. and Ketteridge, A.M. (1995) *Flood Disasters – Dividing the Community*. Emergency Planning, Lancaster, UK.

Gaschen, S., Hausmann, P., Menzinger, I. and Schaad, W. (1998) *Floods – An Insurable Risk?* A Market Survey. Swiss Re, Zurich.

Jha, A., Lamond, J., Bloch, R., Bhattacharya, N., Lopez, A., Papachristodoulou, N., Bird, A., Proverbs, D., Davies, J. and Barker, R. (2011) *Five Feet High and Rising – Cities and Flooding in the 21st Century*. Policy Research Working Paper 5648. The World Bank, Washington, DC.

Jha, A., Lamond, J. and Bloch, R. (2012) *Cities and Flooding: A Guide to Integrated Urban Flood Risk Management for the 21st Century*. GFDRR/World Bank, Washington, DC.

Kunreuther, H. (2006) Disaster mitigation and insurance: learning from Katrina. *The Annals of the American Academy of Political and Social Science*, **604**, 208–227.

Kunreuther, H. and Pauly, M. (2005) Insurance decision-making and market behavior. *Foundations and Trends in Microeconomics*, **1**, 63–137.

Lamond, J. (2006) Elephants or ostriches, does the collective memory of flood events fade with time? In: *First World Conference for Accelerating Excellence in the Built Environment*, Birmingham.

Lamond, J. and Penning-Rowsell, E. (2011) *A Review of the International Approaches to Flood Insurance*. Knowledge for Climate.

Lamond, J. and Proverbs, D. (2009) Resilience to flooding: learning the lessons from an international comparison of the barriers to implementation. *Urban Design and Planning*, **162**, 63–70.

Lamond, J., Dhonau, M., Rose, C. and Proverbs, D. (2009a) Overcoming the barriers to installing property level flood protection – an overview of successful case studies. In: *Road Map Towards a Flood Resilient Urban Environment*. Paris, Urban Flood Management Cost Action Network C22.

Lamond, J., Proverbs, D. and Hammond, F. (2009b) Accessibility of flood risk insurance in the UK – confusion, competition and complacency. *Journal of Risk Research*, **12**, 825–840.

Lamond, J., Proverbs, D. and Hammond, F. (2010) The impact of flooding on the price of residential property: a transactional analysis for the UK. *Housing Studies*, **25**, 335–356.

Lange, H. and Garrelts, H. (2007) Risk management at the science–policy interface: two contrasting cases in the field of flood protection in Germany. *Journal of Environmental Policy and Planning*, **9**, 263–279.

Michel-Kerjan, E. (2001) *Insurance Against Natural Disasters: Do the French Have the Answer? Strengths and Limitations*. Working Paper. Laboratoire d'économétrie, Ecole Polytechnique, Paris.

Mills, E., Roth, R.R.J. and LeComte, E. (2005) *Availability and Affordability of Insurance under Climate change. A Growing Challenge of the US*. CERES.

Munich Re (2004) *Megacities – Megarisks Trends and Challenges for Insurance and Risk Management*. Münchener Rückversicherungs–Gesellschaft, Munich.

Penning-Rowsell, E. and Wilson, T. (2006) Gauging the impact of natural hazards: the pattern and cost of emergency response during flood events. *Transactions of the Institute of British Geographers*, **31**, 99–115.

Pitt, M. (2008) *The Pitt Review: Learning Lessons from the 2007 Floods*. Cabinet Office, London.

Rose, C.B., Proverbs, D.G., Manktelow, K. and Booth, C.A. (2011) Improving community resilience: education, empowerment or encouragement. In: Lamond, J.E., Booth, C.A., Hammond, F.N. and Proverbs, D.G. (eds), *Flood HazardsImpacts and Responses for the Built Environment*,. Taylor CRC Press, New York, pp. 305–316.

Samwinga, V. (2009) Homeowner satisfaction and service quality in the repair of UK flood-damaged domestic property. *School of Engineering and the Built Environment*. Unpublished PhD thesis, University of Wolverhampton, Wolverhampton.

Shilling, J.D., Sirmans, C.F. and Benjamin, J.D. (1989) Flood insurance, wealth redistribution and urban property values. *Journal of Urban Economics*, **26**, 43–53.

Stevenson, R. (2002) Esure axes insurance for property at risk of flooding. *Money Marketing*. 20 June 2002 edition.

Swiss Re (2010) *Making Flood Insurable for Canadian Homeowners*. Swiss Re/Institute for Catastrophic Loss Reduction.

Thieken, A.H., Petrow, T., Kreibich, H. and Merz, B. (2006) Insurability and mitigation of flood losses in private households in Germany. *Risk Analysis*, **26**, 1–13.

Treby, E.J., Clark, M.J. and Priest, S.J. (2006) Confronting flood risk: implications for insurance and risk transfer. *Journal of Environmental Management*, **81**, 351–359.

Vari, A., Linnerooth-Bayer, J. and Ferencz, Z. (2003) Stakeholder views on flood risk management in Hungary's upper Tisza basin. *Risk Analysis*, **23**, 587–600.

Walker, G. and Chalmers, H. (2004) Delivering a better environment, understanding environmental inequalties. *Sustain*, **5**, 33–34.

Welsh Consumer Council (1992) *In Deep Water – A Study of Consumer Problems in Towyn and Kinmel Bay after the 1990 Floods*. Welsh Consumer Council, Cardiff.

Whittle, R. and Medd, W. (2011) Living with flood: understanding residents' experience of recovery. In: Lamond, J.E., Booth, C.A., Hammond, F.N. and Proverbs, D.G. (eds), *Flood Hazards, Impacts and Responses for the Built Environment*. Taylor CRC Press, New York, pp. 281–289.

World Bank and the United Nations (2010) *Natural Hazards, Unnatural Disasters*. The International Bank for Reconstruction and Development, Washington, DC.

Zsamboky, M., Fernanddez-Bilbao, A., Smith, D. Knight, J. and Allan, J. (2011) *Impacts of Climate Change on Disadvantaged UK Coastal Communities*. Joseph Rowntree Foundation.

21

Holistic Property-Level Flood Protection

David W. Beddoes and Colin A. Booth

21.1 Introduction

The UK floods of 2000 were the result of the wettest autumn in England since records began in 1766. Four people lost their lives and the aftermath left ~10 000 flooded properties in over 700 locations, resulting in widespread disruption and damage estimated to cost > £1000 million (Environment Agency (EA), 2001). In 2005, the equivalent of two months average rainfall fell in one night at Carlisle (northwest England) causing flooding that resulted in the loss of three lives and flood damage to ~2000 properties at a cost of £250 million (Association of British Insurers (ABI), 2005a). The UK summer of 2007 was the wettest period during summer months for 250 years (Balmforth, 2008) and extensive flooding meant 13 people lost their lives and damage to ~48 000 homes, with insurance payouts costing > £3000 million (ABI, 2007).

In England and Wales, the threat of flooding affects ~10% of the population and ~2 million homes, with the UK government predicting these figures will most probably double over the next 25 years (Harman *et al.*, 2002; Department for Environment, Food and Rural Affairs (DEFRA), 2008a, 2009). Faced with such disturbing prospects, the government has decided that improvements in traditional engineering infrastructure will not be sufficient to address the future risk and a change in policy is necessary (Ashby *et al.*, 2005; Thorne *et al.*, 2006). In fact, the 2004 *Foresight Study* of future flooding and coastal defence recommended that a range of new measures are required to address the effects of the increasing threat posed by climate change (Evans *et al.*, 2004). As a result, the UK government policy to address flood and coastal risk management, *Making Space for Water*, in effect informs the owners of flooded homes that it is now their own problem that they must manage accordingly (DEFRA, 2005). The government has accepted that large-scale engineering projects aimed at protecting whole communities are no longer sustainable and recommends that homeowners install property-level resistance or resilience measures (Treby *et al.*, 2006).

Water Resources in the Built Environment: Management Issues and Solutions, First Edition.
Edited by Colin A. Booth and Susanne M. Charlesworth.

Home insurance has always been readily available to UK homeowners to mitigate the risk of flooding, under the terms of a voluntary agreement between the government and the insurance industry known as the *Statement of Principles* (see Chapter 20). Understandably, the government strategy of *Making Space for Water*, combined with a real-term reduction in government spending on community flood protection is leading to the demise of the longstanding agreement, as the deal has always ensured flood insurance cover to UK homes provided that the government continues to spend and improve flood defences (ABI, 2005b, 2007). The insurance industry has campaigned for more government money to manage flood risk and mitigate potential losses but the UK government is now unwilling to invest more money in avenues, which it views as actually doing nothing to solve the problem (Crossman *et al.*, 2006). Imminent loss of home insurance, particularly if the *Statement of Principles* ends in July 2013, may finally encourage homeowners to invest in flood protection measures (Lamond *et al.*, 2009).

Homeowners are in need of advice with regard to the application, effectiveness and relative merits of available products and systems. In this chapter a progressive, methodical approach to flood protection will refer to individual flood depths and durations, in order to provide the homeowner with an explanation, so they may better understand how available systems may work for their particular home. The proffered systems incorporate innovative and affordable new products that are passive in nature and, therefore, acceptable to insurers.

21.2 Flood Resistance

Flood resistance measures aim to keep floodwater outside a property or to minimise floodwater ingress. Resistance can involve a range of approaches (Table 21.1) that are often expensive and may also be prevented by the local planning department (e.g. building bunds, boundary walls and fences around a property). In particular, the planning officer will consult the EA and require a flood risk assessment to be carried out and if, as a result, the works are seen to deflect floodwater on to unfortunate neighbours then this will lead to a refusal of permission (Colins, 2009). Similarly, local authorities are choosing not to invest in demountable barriers (as deployed along the River Severn at Bewdley, Worcestershire) due to residents complaining about deflection of floodwater on to downstream properties and the large costs involved in abortive deployment when expected floodwater levels do not arrive.

Property-level flood resistance, involving construction or adaptation of a building, must be installed as a complete package, as one small entry point will render a whole suite of measures ineffective (Department of Transport, Local Government and the Regions Development and Flood Risk (DTLR), 2002; DEFRA, 2008b). An important feature of resistance measures is that they can be incorporated into the building and, as such, are passive in nature or, conversely, temporary items that need deployment by the homeowner. For example, a flood-proof external door is a passive resistance product whereas a removable door aperture guard must be deployed at the appropriate time. As well as floodwater ingress through apertures in the building envelope, building elements themselves are often permeable (e.g. a typical external masonry wall may leak as much as 400 litres/hour/m^2 when subjected to a one metre head of floodwater (Department for Communities and Local Government (DCLG), 2007)). Furthermore, due to structural considerations, there is a limit to the height of floodwater that a typical masonry wall can support (DTLR, 2002; Bowker, 2007). This is because the hydrostatic pressure exerted onto the building structure

Table 21.1 Flood resistant measures at the property level.

Temporary measures	Comments
Door aperture guards	Require storage, accessibility, checks on surrounds and edge seals and deployment by homeowner
Airbrick covers	As above
Flood resistant skirt around property	For detached property, extensive groundworks, needs maintenance, power and homeowner to operate
Permanent measures	
Low bund walls, fences and flood gates	Require groundworks, access to property may be a problem and may transfer flood risk downstream to neighbour
Raise door thresholds	Limited affect and access to property may be a problem (particularly for disabled)
External storm porch	As above
External render to walls	Expensive, needs regular maintenance and may trap moisture within walls/impede drying out
Automatic below-ground rising flood doors	Expensive, extensive groundworks, special maintenance and need power to function
External flood resistant doors	Expensive, need maintenance to seals, frame to brickwork joins and durable locking mechanisms
Automatic airbricks	For timber floors; timber floor only suitable for very short duration and low-level flood as floodwater enters space below floors
Periscope airbricks	As above
Nonreturn valves on drainage	Need yearly maintenance

can cause damage at apertures in the building envelope, particularly where there are no cross–walls or the original construction failed to incorporate sufficient ties within the walls themselves (Kelman and Spence, 2004). Furthermore, a great many Victorian buildings (late 19th century), although appearing of substantial construction, will only feature slender dividing walls between adjacent properties of ~100 mm width.

Public safety in a flood event is of paramount importance (Fundter, 2008). For this reason, the floodwater differential height of 600 mm is the acknowledged maximum that should be resisted prior to intentionally letting floodwater inundate a property (US Army Corps of Engineers (USACE), 1988). However, homeowners instinctively prefer to keep floodwater out, which means they favour resistance products over resilience (Harries, 2007). This is because resistance products are easier to understand and provide a cost beneficial package with minimal disruption during installation. At the very least, as a flood event unfolds, resistance measures can buy precious time for the homeowner to move valuables, sentimental items and furniture to a higher level.

21.3 Flood Resilience

Property-level flood resilience involves designing or adapting a building in such a way that when floodwater inundates it causes minimal or no permanent damage to a property (Table 21.2) and concomitantly facilitates easier cleaning and rapid drying (e.g. ceramic tiled walls are not affected by floodwater and can be easily cleaned and dried (DEFRA, 2008b)). Resilience measures are permanent and passive in nature but are typically

Table 21.2 Flood resilient measures at property level.

Measures	Comments
Internal tanking	Expensive, specialist work and requires maintenance when sump/pumps installed
Solid floors with tiled finish, rendered walls with ceramic tile finish	Property will be inundated with water – not favoured by homeowners as must accept threat of water ingress
Removable internal doors	Need deployment by homeowner and storage
Raise services and appliances (boiler)	Essential when water entry strategy employed
Sump/pump installed in floor	For collection of water ingress, needs power supply
Raise kitchen appliances	Constant reminder of threat of floodwater ingress – not favoured by homeowner
Resilient or sacrificial kitchen units	As above

expensive to install. However, they can bring individual benefits when carried out separately (ABI, 2007) and are most cost-effective as part of reinstatement work after a flood event (Proverbs and Lamond, 2008). Insurance companies are becoming receptive to resilient reinstatement as they realise such measures can reduce the costs of future claims (Sims *et al.*, 2008). Adaption of resilient measures can also aid early rehabilitation after flood events. However, the majority of homeowners do not like homes having resilient features (e.g. tiled walls, tiled floors, kitchen equipment raised on plinths) and are uneasy with the concept of allowing floodwater into their homes (Harries, 2007).

21.4 Current Situation Facing Homeowners

The installation property-level flood protection measures, even in homes that have already experienced flooding, remains very low (McCarthy *et al.*, 2008). There is little doubt most of the general public are bewildered with regard to flood protection of their homes. This is because homeowners tend not to understand the terms resistance and resilience and, in general, cannot identify products as examples of each. Coupled with concerns over the costs involved, this has resulted in minimal or no action taken by flooded homeowners (Norwich Union, 2008). The confusion has been added to by the handling of recent government grant aid whereby funds allocated for design of protection measures were capped at unrealistic levels, which are altogether inadequate for the extensive work involved.

Proffering homeowners with separate lists of resistance and resilience products, and associated methods, appears to be of minimal worth as this has been done repeatedly and fails to solve the flood protection dilemma (Bowker, 2007). The general public remains baffled over how best to protect their homes against flooding and cannot be expected to simply choose from a list of available options themselves (Broadbent, 2004). In the past, very often soon after a flood event, when unfortunate homeowners are suffused with leaflets and advice on all types of products and special offers, they have been advised by contractors promoting only their own products, which are peddled in some cases by 'double glazing salesmen', leading to a lack of confidence in all products (Bowker, 2007). For instance, many aperture guards are readily promoted with a standard 'kitemark' that homeowners understandably believe is the indicator for products that will solve their water ingress issues.

Aperture guards alone are not the promised panacea they appear to be as their deployment does nothing to stop floodwater ingress through the permeable structure of the building itself. Then, of course, there are the problems of actual deployment (e.g. Are they accessible at short notice? Who deploys when people are at work? Are the edge seals and fixing clips intact? Are they fitted correctly?). The flood protection industry and the EA have spent considerable time and money adapting specialised test tanks in order to progress aperture guards to 'kitemark' status on a product that has a multitude of inherent issues. The costs associated with 'kitemark' status are sizeable and chiefly funded by larger companies, who aggressively promote their own products and take sales from smaller manufacturers who cannot afford to progress their own, sometimes, exemplar and first-rate products.

The 'Achilles heel' of aperture guards will always be the necessity for deployment and, unfortunately, the insurance industry is unlikely to recognise any product that requires deployment because insurers are in the business of risk and they simply have no guarantee they will be both robust and deployed in time. As the UK government has decided not to spend any further public funds on approaches that do not actually solve the problem it would, therefore, be wise for them to consider how homeowners should be advised to spend their limited money and, perhaps, only promote passive measures recognised by insurers. The insurers, looking to their own ends, are now favouring resilience on the grounds that it is passive in nature and it reduces the reinstatement costs associated with future flooding of the property, this being particularly attractive to them when funded by the homeowner as part of reinstatement (ABI, 2012). However, homeowners are understandably not willing to suffer the anxiety of living in a home furnished with flood resilient features that portray constant reminders of being in a flood prone area and the ever-present threat that eventually their precious home will be inundated with floodwater (Tapsell and Tunstall, 2008).

21.5 Holistic Solutions

For the benefit of the homeowner, a successful system of flood protection will always be of bespoke design and consist of a carefully considered combination of both resistance and resilience. Therefore, it would be far more industrious for governments, insurance and flood repair companies to stop referring to flood resistance and resilience products as separate entities and, in future, to simply refer to the 'flood defended home' (FDH), regardless of how the measures used to achieve the aim on behalf of the homeowner are categorised. The FDH must feature products and systems that are easily understood by the homeowner, whilst being both functional and aesthetically acceptable, and which will also provide the necessary confidence for the insurers to offer reduced premiums and/or excesses and, along with a government grant, provide the much needed financial incentive for take-up. In order to meet these criteria, the FDH needs to feature initial innovative passive resistance combined with innovative resilient products to mitigate residual risk.

An initial desktop study should be guided by EA flood maps (EA, 2011) and data to determine probable floodwater heights and expected durations of flood events, together with details of any previous works on flood protection. The types of external walls and floor construction will be needed and whether the building is detached or semi-detached. As an alternative to considering resistance and resilience methods separately the mitigation measures are allowed for by reference to the available grades for a 'flood defended home' (Table 21.3).

All grades of the 'flood defended home' should feature initial passive resistance in the form of a newly developed three-stage hydrophobic chemical treatment for masonry walls

Table 21.3 The 'flood defended home'.

Level		Flood severity	Measures	DIY costs
Increasing severity and additional measures	**Flash**	<300 mm for <6 h[14]	Innovative hydrophobic treatment for external masonry and waterproof coating below DPC[1] A pair of innovative safety flood doors[2] Replacement of patio/French doors with safety flood door[2] Self-closing backflow valve to main sewer[3] Wastewater nonreturn valves[4] Puddle-sucker pump[5]	£1500
	Low(13)	<300 mm for >6 h[15]	Additional requirements: Sump/pump installation[6] Battery back-up or small generator for pumps (Note: generator must be mounted externally)[7]	£2000
	Medium	<600 mm for <6 h	Additional requirements: Maintainable perimeter floor drain[8] Toilet panseals[9]	£2500
	High	<600 mm for >6 h	Additional requirements: Floor membrane needed to transfer water to maintainable perimeter floor drains. Wall membrane may also be necessary for external walls[10]	£3500
	Deep	>600 mm – water enters property	Additional requirements: Dado Wallboard[11]	£6500
	Newbuild	<600 mm	Wall cavity used to manage water ingress[12]	£3500

Attachment of property (semi or detached) must be considered and upgraded to the next level where necessary.
Fill weep holes/holes below DPC with cement mortar 1:3.
Floors need to be solid with tiles and/or carpet tiled finish. Timber floors only suitable for grade 1 < 4 h maximum duration and must feature automatic airbricks.
All grades to feature automatic self-closing backflow valves on sewers.
Toilet panseals and sink waste no-return valves (NRVs) optional but become essential when flood level exceeds 600 mm.
All depths are differential depths.
Battery back-up or small generator needed for extended time periods and further periods over 48 h may need further measures.
(Note: all internal combustion generators must be externally mounted and never run inside a property.)

(1) Special mortar. http://www.drainangel.co.uk
(2) Safety flood door. http://www.drainangel.co.uk
(3) Foul (sewage) mains NRV. http://shop.revetment.uk.com/home/foul-sewage-mains-nrv
(4) Waste water NRV. http://shop.revetment.uk.com/home/waste-water-nrv
(5) Puddle-sucker pump. http://www.allpumpsdirect.co.uk/simo-pump
(6) Safeguard sentry pump system. http://www.safeguardeurope.com/products/basement-sump-system.php
(7) Hyundai generators. http://www.seddondirect.co.uk/moredetail.asp?productID=528
(8) Safeguard Aquadrain. http://www.safeguardeurope.com/products/aquadrain.php
(9) Toilet panseal. http://www.pansealuk.com/toilet-flood-defence.asp
(10) Safeguard membranes. http://www.safeguardeurope.com/products/cavity_drainage.php
(11) Dado Wallboard. http://www.drainangel.co.uk
(12) Cavity wall flood protection. http://www.drainangel.co.uk
(13) Low flood level < 300 mm (Bowker, 2007).
(14) Flash floods <6 h. http://www.srh.noaa.gov/mrx/hydro/flooddef.php
(15) Short duration flood (Penning-Rowsell *et al.*, 2003)

together with a waterproof breathable coating applied below a dampproof course and novel safety flood doors. Apertures in external walls should be limited, wherever possible (e.g. patio or French doors replaced by a masonry wall with a window sill set >600 mm), or convert to a single flood door. In all situations, the external resistance of the property must be increased to limit the amount of floodwater that can enter, this being applicable even if full internal resilience measures are necessary. It will always be better to initially eliminate water ingress completely as the actual flood event may not turn out to be as severe as expected and it is unacceptable to have to completely clean and dry a building when the actual ingress level was low (<300 mm deep). This approach provides the homeowner with a measure of hope and some security, softening the blow if floodwater has eventually to be allowed into the property, when it rises to 600 mm depth. Homeowners always express the opinion that to simply open the door, show no resistance and let floodwater into their home is totally intolerable. Therefore, the latest design of flood doors incorporates self-sealing, together with an automatic feature to allow floodwater into the building when the critical depth of 600 mm is reached. To guard against backflow from sewers a self-closing valve is installed on the main sewer together with waste water nonreturn valves on any small-diameter low-level connections. Finally a 'puddle-sucker' pump is supplied in order to manage any water ingress that breaches defences by entry through party walls of adjoining properties (these pumps for use on all flat surfaces will pump down to 3 mm of water with a pumping rate of 170 litres per minute). The above measures will satisfy the requirements of a flash-level FDH (low-level floodwater for short duration), provided that the property has solid floors and irrespective of whether adjacent homes receive similar treatment.

At the next stage, for a low-level FDH (low-level floodwater for long duration), the increased time of exposure will require additional measures in the form of resilience. Add-ons are a floor sump/pump system that the homeowner can discretely direct water into when, over time, it seeps through the initial defences. Already having the 'puddle-sucker', the homeowner will now be equipped with two extraction pumps and the portable 'puddle-sucker' may be moved next to any specific source of ingress, such as through a party wall. Battery back-up or preferably a small portable generator is always a preferred extra but will be essential for all flood periods in excess of 12 hours (note that internal combustion generators must always be mounted externally).

A medium-level FDH (elevated floodwater for short duration) will require maintainable perimeter floor drains that are linked to the sump/pump system as much higher rates of water ingress will result from differential flood depths as they approach 600 mm. Toilet pan seals should also be fitted as water rises to these higher levels.

Measures for a high-level FDH (elevated floodwater for long duration) will be floor and wall membranes in order to manage any water ingress by safely directing it into the sump/pump system. The wall membranes will be essential for solid external walls where the flood heights of 600 mm and longer durations will eventually allow water to permeate across the width of the wall itself.

A deep-level FDH (floodwater enters property) is the stage at which floodwater must be allowed to inundate the property (differential depth >600 mm) due to structural considerations and more importantly the safety of occupants, as rescuers cannot lead or carry their charges in water when it surrounds the property and is over 750 mm deep. The safety flood doors begin to let floodwater into the property at 600 mm differential depth to equalise pressure on the walls and also to act as a signal to the occupiers that it is time to abandon the property. Further measures such as the innovative Dado Wallboard system then facilitate rapid cleaning and drying (Beddoes and Booth, 2012). Finally, new properties at newbuild-level FDH should again use the initial external treated masonry resistance and safety flood doors together with new innovative measures that make use of the current features of cavity walls to manage water ingress and direct it to sump/pump systems for evacuation (UK Patent Number GB 1021323.9).

21.6 Conclusions

Traditionally there have been two separate approaches used to address property-level flood protection, resistance and resilience. Both methods have associated uptake issues and so a holistic solution has been proposed that adopts a series of collective measures. The resultant combination uses innovation in both masonry treatments and flood doors to provide passive, functional and aesthetically acceptable initial defence for all properties. Where necessary, this protection is then reinforced by further measures that effectively manage possible water ingress that results from deeper or longer flood periods.

Governments, insurance and flood protection companies must promote holistic flood solutions and cease mystifying homeowners by referring to resistance and resilience approaches in isolation. The concept of developing a series of scenario levels for a 'flood defended home' that uses a combination of measures acceptable to the homeowner must be promoted. The government should be promoting new measures that are easy for the homeowner to understand and passive in nature, so that the insurers will consider their installation when deciding on premiums.

Research has revealed the significance of masonry treatments and innovation will now enable the take-up of flood protection. Successive government campaigns have already shifted the responsibility for flood protection on to the homeowner, who has shouldered that responsibility and now has the desire to act. New passive flood doors combined with a chemical treatment that does not change the look of a property, reduction in insurance costs as a result of passive products and a realistic level of government grant are the drivers needed.

References

Ashby, R.M., Balmfort, D.J., Saul, A.J. and Blanksby, J.D. (2005) Flooding in the future – predicting climate change, risks and responses in urban areas. *Water Science and Technology*, **52**, 265–273.

Association of British Insurers (ABI) (2005a) *Financial Risks of Climate Change*. Association of British Insurers, London.

Association of British Insurers (ABI) (2005b) *Revisiting the Partnership: Five Years on from Autumn 2000. October 2005 Anniversary Report*. Association of British Insurers, London.

Association of British Insurers (ABI) (2007) *Summer Floods 2007: Learning the Lessons*. Association of British Insurers, London.

Association of British Insurers (ABI) (2012) *Flood Resilience and Resistance Factsheet for Insurers and Loss Adjusters*. Association of British Insurers/The Chartered Institute of Loss Adjusters, London.

Balmforth, D. (2008) Editorial. *Journal of Flood Risk Management*, **1**, 1–2.

Beddoes, D.W. and Booth, C.A. (2012) Factors for the design of novel property-level flood resilient products: the Dado Wallboard. In: Proverbs, D.G., Mambretti, S., Brebbia, C.A. and de Wrachien, D. (eds), *Flood Recovery Innovation and Response III*. WIT Press, Southampton, pp. 205–219.

Bowker, P. (2007) *Flood Resistance and Resilience Solutions: An R&D Scoping Study*. R&D Technical Report. DEFRA, London.

Broadbent, C. (2004) Improving the flood resistance of domestic property. *Structural Survey*, **22**, 79–83.

Colins, C. (2009) *The Environment Agency South West Regional Flood Defence Committee Meeting January 2009. Environment Agency* [cited 30 January 2009]. Personal communication.

Crossman, M., Richardson, D. and Milne, J. (2006) A partnership approach to managing flood risk. *Proceedings of the Institution of Civil Engineers: Civil Engineering*, **159**, 41–45.

Department for Communities and Local Government (DCLG) (2007) *Improving the Flood Performance of New Buildings. Flood Resilient Construction*. DCLG, London.

Department for Environment, Food and Rural Affairs (DEFRA) (2005) *Making Space for Water: Developing a New Government Strategy for Flood and Coastal Erosion Risk Management in England: A Delivery Plan*. DEFRA, London.

Department for Environment, Food and Rural Affairs (DEFRA) (2008a) *Developing the Evidence Base for Flood Resistance and Resilience: R&D Technical Report FD2607/TR1*. DEFRA, London.

Department for Environment, Food and Rural Affairs (DEFRA) (2008b) *Consultation on Policy Options for Promoting Property-Level Flood Protection and Resilience*. DEFRA, London.

Department for Environment, Food and Rural Affairs (DEFRA) (2009) *Adapting to Climate Change. UK Climate Predictions June 2009*. DEFRA, London.

Department of Transport, Local Government and the Regions Development and Flood Risk (DTLR) (2002). *Preparing for Floods – Interim Guidance for Improving the Flood Resistance of Domestic and Small Business Properties*. TSO, London.

Environment Agency (2001) *Lessons Learned: Autumn 2000 Floods*. TSO, London.

Environment Agency (2011) *Environment Agency Flood Map* [online], [Accessed 2 January 2012]. Available at http://maps.environment-agency.gov.uk/wiyby/wiybyController.

Evans, E., Ashley, R., Hall, J., Penning-Rowsell, E., Saul, A., Sayers, P., Thorne, C. and Watkinson, A. (2004) *Foresight Future Flooding. Scientific Summary: Volume II – Managing Future Risks*. Office of Science and Technology, London.

Fundter, D. (2008) Health impacts of large-scale floods: government decision making and resilience of the citizens. *Prehospital and Disaster Medicine*, **23**, 70–73.

Harman, J., Bramley, M.E. and Funnel, M. (2002) Sustainable flood defence in England and Wales. *Proceedings of the Institution of Civil Engineers: Civil Engineering*, **150**, 3–9.

Harries, T. (2007) *Householder Responses to Flood Risk: The Consequences of the Search for Ontological Security*. Unpublished PhD thesis, Middlesex University.

Kelman, I. and Spence, R. (2004) An overview of flood actions on buildings. *Engineering Geology*, **73**, 297–309.

Lamond, J.E., Proverbs, D.G. and Hammond, F.N. (2009) Accessibility of flood risk insurance in the UK: confusion, competition and complacency. *Journal of Risk Research*, **12**, 825–841.

McCarthy, S., Penning-Rowsell, E. and Tunsall, S. (2008) Public attitudes to 'community-based' small-scale reduction measures in England. In: Bolsher, L. (ed.), *Hazards and the Built Environment – Attaining Built-in Resilience*. Taylor and Francis Group, pp. 150–179.

Norwich Union (2008) *Homeowners Fear Future Flooding but Fail to Take Measures to Protect Their Properties* [online], [Accessed 14 February 2012]. Available at: http://www.norwichunion.com/media-centre/story/3981/homeowners-fear-future-flooding-but-fail-to-take-measures-to-protect-their-properties/.

Penning-Rowsell, E.C., Johnson, C., Tunstal, S., Tapsell, S., Morris, J., Chatterton, J., Coker, A. and Green, C. (2003) *The Benefits of Flood and Coastal Defence: Techniques and Data for 2003*. Flood Hazard Research Centre, Middlesex University.

Proverbs, D. and Lamond, J. (2008) The barriers to resilient reinstatement of flood damaged homes in International group for Research and Information on Post-disaster Construction. In: *4th International i-Rec Conference 2008 on Building Resilience: Achieving Effective Post-Disaster Reconstruction*, April 30–May 2, 2008, Christchurch, New Zealand.

Sims, R., Medd, W., Mort, M., Watson, N., Walker, G. and Twigger-Ross, C. (2008) *Perspectives on Resilience from Households in Hull – Response to Defra Consultation on Policy Options for*

Promoting Property-Level Flood Protection and Resilience. Lancaster University and Collingwood Environmental Planning.

Tapsell, S. and Tunstall, S. (2008) 'I wish I'd never heard of Banbury': the relationship between 'place' and the health impacts from flooding. *Health and Place*, **14**, 133–145.

Thorne, C., Evans, E. and Penning-Rowsell, E. (eds) (2006) *Future Flooding and Coastal Erosion Risks*. Thomas Telford, London.

Treby, E., Clarke, M. and Preist, S. (2006) Confronting flood risk: implications for insurance and risk transfer. *Journal of Environmental Management*, **81**, 351–359.

US Army Corps of Engineers (USACE) (1988) *Tests of Materials and Systems for Flood Proofing Structures*. US Army Corps of Engineers, Washington, DC, p. 89.

Section 6 Flood Solutions in the Urban Landscape

22

Sustainable Drainage Systems – Features and Designs

Simon Watkins and Susanne M. Charlesworth

22.1 Introduction

Sustainable drainage systems (SUDS) primarily provide flood attenuation, for example, Wilson *et al.* (no date, p.5) state that 'the primary and overriding function is to provide effective drainage'. In doing so, they also improve water quality, but what is generally relegated to third place in the 'triangle' of benefits is that of amenity and biodiversity. Once hydraulic considerations have been taken account of, which may possibly lead to the sacrifice of more aesthetically pleasing features in the short term, nonetheless, 'Unlike conventional drainage systems, most SUDS features should be visible' (Bray, no date) providing 'Useful or pleasurable spaces'. This is a *volte face* from the conventional approach whereby water in urban areas is hidden underground. This chapter considers the evolution of this attitude towards water and provides examples of the successful integration of water into the city environs by the suitable design of sustainable drainage systems.

22.2 Water and Landscape – One and Indivisible

The Victorians achieved extraordinary things with water. In the face of the twin challenges of rapid population expansion and industrial urbanisation, a pressing need developed for high-capacity systems of water supply and treatment. Similar approaches were exported or developed independently across the globe as other areas faced similar challenges. In the United Kingdom, by a combination of philanthropy, public subscription and corporate vision, the infrastructure that would provide vastly increased urban areas with sufficient clean water and to discharge from those areas the surplus they did not need was put in place, and with it the notion of the management of water as a single problem with one overarching solution: drains. Whilst the solutions created by the Victorian engineers was

Water Resources in the Built Environment: Management Issues and Solutions, First Edition.
Edited by Colin A. Booth and Susanne M. Charlesworth.

magnificent, the legacy of putting water underground has created a collective mental block out of which we have only recently begun to emerge.

This perception of water as problematic is one of the hurdles that must be overcome by the design of open SUDS systems. Visible measures whose mechanical responsibilities include retaining, treating or conveying water are often viewed with an unnatural paranoia so that they must be more than functional, more than safe and maintained with absolute ease. Safety of open water is understandably of public concern, but the fear of water in the public realm is also in part due to society's longstanding separation from it in the urban landscape. However, this antipathy towards open water in the urban environment is determined more by a cultural, geographical and historic context than by a natural disinclination, as demonstrated by the range of approaches taken to the integration of drainage and the public realm globally.

In 2005 a formerly culverted drainage channel in Seoul, South Korea, the Cheong Gye Cheon, was uncovered and converted into a linear water park in the heart of the city (see http://webarchive.nationalarchives.gov.uk/20110118095356/http:/www.cabe.org.uk/case-studies/cheonggyecheon-restoration-project). This surprising public space is all the more remarkable when one considers that what is now living, sculptural and planted space, teeming with visitors, replaced a 16 m wide raised highway, and still sits between busy roads lining the edges of this high-rise city street. The stream receives pumped drainage from subways in the city and from other sources. In this context there is apparently no perceived conflict between public amenity and this drainage function, and the many additional measured benefits include reduction in the peak temperatures of the surrounding areas, the reintroduction of wildlife into the city and even a rebalancing of travel modes used in the city towards public transport resulting from the removal of the former highway (Lee, 2006).

The example of the Cheong Gye Cheon serves to illustrate what is possible if assumptions about public safety and segregation of functions can be reconsidered. It departs from the principles of sustainable drainage in that its benefits are largely dependent upon a continual stream flow; hence it relies upon diverting and pumping water from a variety of sources. Sustainable drainage systems collect and manage water passively and the resulting environments are ideally as close to self-sustaining as possible. In the remainder of this chapter a number of designs will be considered from the perspective of the *management train*: the way in which a series of techniques and means of conveyance may be linked to deal with surface water at varying degrees of proximity to its source. The means of *prevention* of runoff through vegetated surfaces will briefly be considered and also systems that enable runoff to be minimised through *source control*, such as permeable paving and underground storage. Examples of *site control* in the form of rain gardens, swales and settling ponds are described. Reed beds, wetland systems and detention basins installed as a means of *regional control* are also reviewed. Finally, an example of a comprehensive management train applied to a small-scale housing area is described.

The discussion focuses on the functionality, maintenance and aesthetic appeal of each example. Functionality considers the range of functions supported by a system – technical functions such as the contribution made by the system to the management of water – and other ecosystem functions such as wildlife habitat, urban cooling, etc. Maintenance is related to functionality and to aesthetics but deserves separate mention as a key component of good design. Aesthetic appeal concerns the contribution to the visual environment and the likely public perception of systems, particularly with regard to safety.

Examples are largely taken from the urban context, but where good examples from peri-urban environments could be applied in some measure to urban situations, these are included. The aim of this discussion will be to demonstrate that a re-identification of water as not only a desirable but a fundamental component of the urban environment is possible and in many cases has already been achieved.

22.2.1 Prevention: Vegetated Surfaces

The surest way to reduce runoff is to maximise areas of vegetated surfacing. Left to its own devices, nature will rapidly colonise unmaintained surfaces with a succession of plants and other organisms, with the result that water is naturally detained in the surface of the landscape. In areas of poor natural drainage, this can lead to localised flooding as part of normal ecosystem functioning. However, when used as a tool in drainage, vegetated surfaces can be designed to receive and absorb an appropriate quantity of water, retaining sufficient for the needs of the surface and by doing so reducing the quantity transmitted without necessarily flooding (unless intended), as well as trapping minor pollutants prior to onward flow.

Mature street trees clearly play an important role in reducing flooding in urban areas (Charlesworth, 2010), not only through the extensive surface area presented, but through the volume of root zone in which water is retained, absorbed and taken upwards to be transpired from the above ground part of the tree. Newly planted trees require generations to become effective in this way; hence in design terms one of the most valuable contributions to sustainable drainage is site planning around existing mature trees. Surprisingly, examples of such a simple strategy are less common than may be expected, although the Angel Building in London (2010) (Figure 22.1) demonstrates how setting back a development to accommodate existing mature trees not only complements the character of the street but provides an opportunity to create a useful multifunctional space in front of the building for the benefit of its users.

Figure 22.1 Angel Building Islington.

Figure 22.2 Whittington Park changing rooms.

Installing vegetated surfaces on new structures is perceived as technically complex, though there are a growing number of precedents ranging in sophistication from simple sedum-covered roofs to fully clothed green walls. As a rule the further removed from a natural environment the more technically demanding the project will be to ensure ongoing maintenance is minimised. An ideal approach to green roofs simply replaces the flat roof surface with a substrate supporting locally appropriate grassland flora. The roof of the changing rooms in Whittington Park, Islington (Figure 22.2), is one example where the use of native grasses and wildflowers on the roof enhances the attractiveness of the building, supports insect and bird life, reduces discharge of rainwater to the drainage system and provides insulation to the interior. This particular example is drained via a collection gutter directly to the positive system; however, the quantity of runoff in a rainfall event is significantly reduced and it is easy to envisage how this approach could form the first of a number of techniques employed in a SUDS management train addressing building runoff.

22.2.2 Source Control: Permeable Paving

There are a number of options available for marrying the durability of hard surfaces with sustainable drainage design. Porous grass grid systems, available either in plastic or concrete, can provide the appearance of a grass surface whilst accommodating traffic. The most appropriate locations for such systems, especially where installed permanently, are in contexts where the surface is used by relatively light traffic such as overflow car parks and restricted access lanes (Figure 22.3). Visually preferable to tarmac or large expanses of gravel, these systems also potentially encourage careful driving simply by appearing different from the normal carriageway surface.

Permeable block paving, however, represents a far more durable, low-maintenance and flexible alternative to conventional hard surfacing. A wide range of products from major

Figure 22.3 Bus access, Blythe Valley.

Figure 22.4 Permeable block paving.

manufacturers are suitable for either foot or vehicle trafficked areas, many of which utilise recycled materials in the block construction. Paving installed in 2012 at Coventry University, United Kingom (Figure 22.4), includes permeable blocks, laid on a double thickness of clean angular stone acting as a laminar soakaway. During a three month

Figure 22.5 Porous resin-bound surfacing, Rugby.

period of unusually heavy rainfall coinciding with the completion of this scheme no ponding occurred at any point on the surface of these areas. As an alternative to increased depth hardcore, block paving may also be combined with underground storage with control and overflow structures, allowing simple conveyance to subsequent stages of a management train.

One disadvantage of permeable block paving is that the 'nibs' that extend from the sides of each block, thereby maintaining the spaces between them, may be regarded as visually intrusive (although this varies depending upon manufacturer). However, permeable monolithic paving such as porous asphalt dressed with permeable resin-bound surfacing avoid this problem. Also, by combining permeability with the potential for tree planting, these surfaces also combine source control with prevention (Figure 22.5). Easy to maintain, the method of construction, which involves rolling over the hardcore subbase, means that the surface can accommodate undulating ground more easily than block paving. Around mature trees, a subsurface mesh is necessary to protect the root zone from excessive pressure during construction and when in use.

22.2.3 Site Control: Rain Gardens, Swales and Settling Ponds

Vegetated and permeable surfaces offer a means to reduce the volume and pollutant burden on the drainage system without giving rise to public concerns over maintenance and the health and safety aspects of water in the urban environment. In considering more visible SUDS techniques, it is important to understand this aspect of design.

22.2.3.1 Rain gardens

A rain garden is a small depression or enclosure, positioned to receive runoff directly from an adjacent source and filled usually with ornamental plants selected for their tolerance of both wet and dry conditions. Potentially it is one of the most eye-catching and attractive component of a SUDS management train: a visible indication that a radically different approach is at work. The three examples discussed here illustrate a range of different contexts in which the technique may play a part.

(a) Green Street: Infiltration Planters, Portland, Oregon

These features have been installed in a variety of contexts in the city. They comprise a contained, vegetated basin either in the form of a 'bump-out' that extends into the street or a low planter box between the sidewalk and the street. Cuts in the kerb allow runoff to flow into the stormwater facility where the plants and soil encourage infiltration. Overflow is directed to the adjacent positive (stormwater/sewer pipe) system. The planting is a mixture of ornamental grasses and herbaceous plants that are tolerant of both wet and dry conditions. Together with microbes in the soil, the plants help break down pollutants from the adjacent road surfaces. Hence, water quality is addressed through biofiltration and uptake prior to onward discharge.

The City of Portland publishes fact sheets on the design and implementation of these features (see http://www.portlandonline.com/bes/index.cfm?c=46962&a=188636). Given their public context, safety and structural matters are key issues determining their proximity to foundations, property boundaries, level and provision of overflow. Periodic maintenance includes inspection of the structural components, removal of sediments, clearance of debris and replacement of any failed plants. In general, it has been determined that the combination of 'grey and green' drainage (i.e. SUDS and pipe infrastructure) is less expensive to implement than a purely pipe-based system (see http://www.portlandonline.com/bes/index.cfm?c=47203). Additionally, the green infrastructure approach provides many more benefits for liveability than hidden piped drainage. The success and track record of this approach since implementation must surely provide additional incentive to trial the use of similar systems in any urban context.

(b) Ashby Grove, Islington, London

An illustrative example of a simple management train, this garden accepts runoff from approximately 30 m^2 of roof from an apartment block, designed by Bob Bray of Robert Bray Associates for Islington Borough Council (Figure 22.6). Planting includes Crocosmia, Acanthus, Alchemilla, Geranium, Hemerocallis and Miscanthus varieties – plants that may be found in any domestic garden or local park, and not normally associated with any water features, standing or otherwise. These cover an area of approximately 3 m^2 and are contained in a shallow valley at the head of which is a short surface drain of granite setts. This in turn is connected to the downpipe of the roof gutter. An overflow to the positive drainage network sits at the top of a gabion mattress ~30 cm above the lowest point. At the request of the residents a control pipe was also included near the base of the depression, which sets a maximum standing water level of a few cm, although the combination of infiltration and uptake by plants would naturally leave very little standing water for any length of time.

A flash of colour at the end of an otherwise monochrome street, in the growing season the plants are evidently healthy, benefitting from the increased moisture available in

Figure 22.6 Ashby Grove rain garden.

comparison with the surrounding flat ground. They also gain nutrients from the flow from the roof and downpipe and also from the retention of moisture and nutrients in the bowl created for the purpose. In winter, some of these plants would remain visible, ensuring the feature maintains a degree of attractiveness throughout the year.

This simple example illustrates how a single-stage management train can add visual interest to a street scene with minimal intervention and in a small space. The small scale of this rain garden contrasts with the large, repeated apartment blocks. However, rolled out in similar situations, this approach could radically enhance the visual environment of urban streets whilst reducing runoff and particulate pollution to receiving systems.

(c) London Wetland Centre

An indication of how a vegetated detention pond might work as a step in a series of SUDS components is illustrated by the chain of bog gardens in the permanent London Wetland Centre's 'Sustainability Garden' exhibit, designed by Nigel Dunnett of the University of Sheffield and The Landscape Agency (Figures 22.7 and 22.8). The core of the system comprises four lined raised beds each planted with a different selection of herbaceous perennials and grasses, including, amongst them, Calamagrostis, variegated Phragmites, Helianthus, Lythrum, Penstemon and varieties of Primula as well as other species. The water source is the green roof of an adjacent small building, whilst the final outflow is to a wetland. The individual beds are linked by hollowed timber rills, which very visibly illustrate the way in which water is passed through the system, being intercepted and utilised at each stage on its journey

Figure 22.7 London Wetland Centre 1.

Figure 22.8 London Wetland Centre 2.

to final outflow by vegetation. Hence, even without infiltration, the net result is a decrease in the quantity of water being discharged compared with a positive system attached to the same roof. The presence of the green (sedum) roof adds runoff prevention to the management train.

Whilst not located in a heavily used urban area, this very attractive example progresses both the design and public awareness of SUDS systems. The gardenesque nature of the design is particularly appropriate in the domestic setting or any other controlled environment. The principle of multistage rain gardens may also be adapted for fully public areas, incorporating more robust ground-level structures and allowing for infiltration similar to the Ashby Grove scheme.

22.2.3.2 Swales and Settling Ponds

Swales are an increasingly common means of surface-level conveyance in large-scale infrastructure developments, often leading to settling ponds prior to onward discharge. Their form is frequently linear, as dictated by the spatial constraints and form of the highway corridor as well as gradients; however, where possible, opportunities should be taken to deviate from a strict linear route in order to encourage oxygenation, reduce flow rates and thereby limit erosion and nutrient loss, enhance habitat diversity and improve visual appeal (Figure 22.9). A settling pond enables particulates to be removed from the water column and, if constructed with sufficiently low gradients, the colonisation by marginal plants including Phragmites will assist with biofiltration of pollutants. Occasional periodic removal of accumulated silts is necessary in order to maintain the functionality of these features.

Figure 22.9 Industrial estate, Milton Keynes.

22.2.4 Regional Control: Intermediary Wetlands and Detention Basins

Typical of early SUDS systems are those that collect surface water via a conventional drainage network to a series of receiving bodies adjacent to the site of a development. Managing water in this way is design efficient – since it involves a relatively simple set of decisions and drawings – though more energy and land intensive than utilising source and site control. However, the resulting wetland environments can provide significant additional benefits to people and wildlife. Systems involving regional control typically incorporate swales, settling ponds, detention basins and sometimes reed bed filtration ponds.

(a) Blythe Valley Business Park

Located between Solihull and the M42 in the English Midlands, this site lies adjacent to a series of former marl pits associated with a mediaeval manor. These were utilised as attenuation basins in the drainage scheme, which includes an extensive chain of reed beds and other waterbodies linked by ditches and swales and also fed by positive drainage from the estate.

The whole system includes elements of on-site prevention (in the form of soft landscape and a small area of grass grid paving), conveyance via swales and regional control of drainage gathered from each developed zone of the estate. The scheme was designed by landscape architect David Singleton of Munro+Whitten and engineer Paul Tinley of Cameron Taylor Bedford and was completed in 1998 based on a masterplan by architects Stephen George & Partners. A major factor influencing the design was the requirement to avoid impact on water levels or quality in the Blythe Brook – a Site of Special Scientific Interest relating to its rarity as a lowland clay-based river.

Two sequences comprise attenuation basins effectively acting as settling ponds (Figure 22.10), flowing via control structures into shallow reed beds (Figure 22.11), thence

Figure 22.10 Blythe Valley, balancing pond.

Figure 22.11 Blythe Valley, reed bed.

Figure 22.12 Blythe Valley Park, wetland.

to wetlands (Figure 22.12) and finally outflowing to local watercourses. During construction the various ponds were lined (although the underlying clay soils were sufficiently heavy to avoid this as a requirement); hence the main impact upon quantity is through detention rather than infiltration. Detention in the initial basins also enables a constant a flow to be maintained within the reed beds. This is achieved by the use of a 'throttle pipe' at the outflow of the attenuation basin preceding the first bed in each chain. An even flow across the breadth and length of each reed bed is affected by the use of a single row of gabion baskets, which also detain a low head of water in a stilling pool immediately upstream of the reed bed. A very shallow gradient is applied to the base of the reed bed from inlet to outfall, which is via a castellated concrete sill – the castellations allowing for batons to be inserted to raise the water level in the bed in the event of siltation or excessive vegetative growth. The effect of the reed beds can be seen by visually comparing the water surface in the initial attenuation basins with that of the wetlands lower in the chain, which is noticeably clearer. In fact, annual Biological Monitoring Working Party (BMWP) testing at seven monitoring points carried out by the ecological consultancy DSA Environment + Design Ltd confirms that the quality of water leaving the system consistently exceeds that of the receiving watercourses, according to biological indicators. The continued health of the receiving watercourses is also confirmed by the Environment Agency's own quality monitoring stations downstream (Singleton, personal communication).

At the time of design the UK legislative requirement for green field development was that drainage should accommodate stormwater runoff from a rainfall event of the maximum intensity likely to occur within 100 years at the outfall of the system. (This requirement has been upgraded in the light of climate change so that an additional 20% volume must be added above the 1:100 year event for a new development.) Although half of the planned business park is yet to be developed, all plots have been fitted with site drainage; hence the current operation of the system may be considered to be broadly representative of its function in the fully developed state. The size of the ponds is such that the capacity of the system has not been exceeded since completion (Singleton, personal communication). Siltation, which may be a concern in systems incorporating reed beds, has not occurred, and only minimal clearance of bulrushes, leaf fall and encroaching vegetation has been necessary. The whole site benefits from a proactive Landscape Management Plan designed to facilitate the organic development of the drainage system and the landscape in general.

One benefit of managing surface water storage 'regionally' on this site (as opposed to directly on each development plot) is that the resulting water features also benefit wildlife through the creation of relatively undisturbed habitat and provide a public amenity that would be less attractive were it solely based within the business park. These spaces are known to be popular both with staff and members of the public. Further interest is added in the form of interpretation panels each highlighting a different aspect of the ecology of the site (Figure 22.13) and in this way the positive contribution of sustainable drainage to the environment is conveyed to a wider audience.

(b) Brooklands Meadows Park, Milton Keynes

The challenge of sustainable drainage for large-scale development plots is multiplied where the ultimate land use is residential. The means of drainage to private dwellings sold as individual units may not be as easily controlled as for offices or other large drained land uses. Further, residential streets are often small in scale and a complex set of local regulatory requirements on highway design, materials and public spaces may apply. For these reasons, regional control is suitable for major residential development, as it enables a single design solution to be applied to the drainage of relatively large areas. Lying between Milton Keynes and the M1 motorway, the Brooklands development is an example of the use of public space to manage drainage from the surrounding streets.

Unlike the Blythe Valley example above, the management train does not include prevention or site control, instead simply comprising a large-scale detention basin on the line of a

Figure 22.13 Blythe Valley, site interpretation.

modified existing water course. This is fed by positive drainage from the adjacent areas, the maximum water level being controlled overflow via a culvert. As an online detention basin, the park is treated as a reservoir; however, for the majority of the time, most of the park is dry. In these conditions, the smaller-scale landforms and wetlands (Figure 22.14) designed by the author (Simon Watkins) on behalf of Roger Griffiths Associates Ltd provide visual and wildlife amenity. The permanent waterbodies in the bottom of the basin include both on- and off-line ponds, providing a variety of microhabitats and acting as settling ponds between significant rainfall events. Stagnation in the on-line ponds is avoided by the exchange of water, which occurs during each period of high flow.

22.2.5 The Comprehensive Management Train

Given the highly developed techniques of sustainable drainage, the existence of high-quality precedents and the well-rehearsed benefits, it is perhaps surprising that few examples have been constructed utilising more than two or three components in a simple management train. However, the recently constructed Springhill co-housing project in Stroud represents one of the most comprehensive systems in the United Kingdom, demonstrating the potential for efficient, safe management of surface water in a public environment.

Designed by Bob Bray in consultation with the residents of the project, this scheme employs permeable surfacing, swales, rills, ponds and a wet–dry detention basin, all in close proximity of houses. A high degree of customisation by residents of the various components was enabled and encouraged, resulting in a diverse and visually interesting environment and ensuring that the approach is both understood and welcomed by those living in the immediate environs. The site lies on a steep hillside, with vehicular access from

Figure 22.14 Brooklands Meadows Park online ponds.

a road lying above. To protect the system from incoming flows and associated pollutants, an interceptor drain straddles the entrance driveway, fitted with a wildlife-friendly silt trap.

The first stage of the on-site drainage proper consists of a permeable paving system comprising block paving over storage crates and forming the car parking area. This accepts runoff from the highest placed buildings and the driveway, and is linked via a control structure to an outfall pipe. An overflow swale lies above this pipe to accommodate excessive rainfall (Figure 22.15). The outfall from the crates emerges from a brick section of an otherwise permeable retaining wall, dropping to a gravel splash bed. This outfalls to a vegetated swale (Figure 22.16), which itself connects to a rill running along the upper edge of the pedestrian street between the houses (Figure 22.17). The other side of this street also incorporates a drain in the form of a rill. This forms the boundaries of the private dwellings and in order to protect the buildings from sudden excesses of surface water runoff, includes a higher lip on the building side of the rill. Gravel has been placed in much of this rill, which is decorated by plants of the householders' choice. The rill on this side of the street overflows into an underground box storage via silt traps, the remainder outfalling directly to a wet–dry detention basin in the form of a bowl-shaped lawn (Figure 22.18). This is a communal area used for play and other recreational activities for the majority of the time, when it remains dry. The upper rill also overflows to this detention basin via a formal pond (Figure 22.19) and a channel under the street. The basin itself overflows via a control pipe set into a gabion mattress to an open channel running to a final outflow. A third rill serves the lowest part of the street and outfalls to a shallow ditch running within the boundary of the site. Both this and the open channel flow from the basin outfall to the site of a natural spring located at the lowest point of the site.

The initiation, design and implementation of this drainage and landscape scheme were undoubtedly aided by the unusual context of the project. The co-housing concept is one in which the individual householders cooperate as a community to manage the building and upkeep the whole site. There is a degree of correlation between people who are sufficiently

Figure 22.15 Springhill overflow swale.

Figure 22.16 Springhill vegetated swale.

Figure 22.17 Springhill vegetated rill.

Figure 22.18 Springhill detention basin.

Figure 22.19 Springhill formal pond.

community oriented to choose this kind of living environment and those who are open to innovative environmental solutions. Nonetheless, the scheme illustrates what can be achieved physically even – or possibly especially – where a degree of control over the appearance and management of a system is left with those whom it serves. The single drawback that has occurred has been a lack of attention given by the residents to regular clearing of silt traps; however, the designer considers this could be addressed through the use of more open structures.

Typically, urban areas contribute 150–300 litres per second per hectare to onward flows. With its combination of permeable surfacing and underground storage (source control), vegetated swales and rills and local detention basin (prevention and site control), this system maintains the overall discharge to 5 litres per second per hectare – the equivalent of a greenfield rate.

22.3 Conclusions

Sustainable drainage features need to be well designed hydraulically in order to function effectively. However, beyond efficient attenuation of the flood peak, individual devices and trains can provide the 'useful and pleasurable spaces' mentioned by Bray (no date) in which the landscape they are designed into is enriched. This chapter has illustrated one of the many benefits of sustainable drainage, that of providing aesthetically pleasing spaces, but it is also one of the most neglected areas of the sustainable drainage triangle. The case studies detailed here show the added value of integrating green and hard infrastructure, the way they can be designed together and the way they can provide a more sustainable and attractive means of attenuating the storm peak, addressing water quality issues and providing amenity and biodiversity in urban areas.

References

Bray, R. (no date) *Presentation to Cambridge City Council: The Art of SUDS Design – Integrating Amenity and Biodiversity into the Management of Rainfall*. Available at: http://www.cambridge.gov.uk/public/docs/Art-of-SUDS-by-Bob-Bray-RBray-and-Associates.pdf.

Charlesworth, S.M. (2010) A review of the adaptation and mitigation of Global Climate Change using Sustainable Drainage in cities. *Journal of Water and Climate Change*, **1** (3), 165–180.

Lee, I.-K. (2006) Cheong Gye Cheon Restoration Project – a revolution in Seoul. In: ICLEI Conference. Available at: http://worldcongress2006.iclei.org/uploads/media/K_LEEInKeun_Seoul_-_River_Project.pdf.

Wilson, S., Bray, R., Neesam, S., Bunn, S. and Flanagan, E. (no date) *Sustainable Drainage, Cambridge Design and Adoption Guide*. Cambridge City Council. Available at: http://www.cambridge.gov.uk/public/docs/SUDS-Design-and-Adoption-Guide.pdf.

Acknowledgements

With thanks to: Douglas McNab, Sustainability Officer, Islington Council; Emily Hauth, Sustainable Stormwater Management, City of Portland; Deborah Fennell, Property Manager, Liberty Property Trust UK Ltd; David Singleton, Director, DSA Envrionment+Design Ltd; Stuart Hodgkiss, THDA Ltd, Consulting Engineers; and Bob Bray, Robert Bray Associates Ltd.

23

Drainage Benefits of Porous, Permeable and Pervious Paving

Miklas Scholz, Susanne M. Charlesworth and Steve J. Coupe

23.1 Introduction

A porous or permeable paving system (hereinafter both referred to as PPS, although differentiation between the terms 'porous' and 'permeable' with respect to paving are given later in this chapter) is generally taken to be a 'hard' sustainable drainage system (SUDS) device constructed from concrete, bricks and aggregate, although there are 'soft' PPS that utilise vegetation PPS (VPPS). However, the fundamental purpose of all PPSs is to allow stormwater to infiltrate slowly, to be stored in its structure and then allowed to be conveyed elsewhere in association with the SUDS (or storm sewerage network) to groundwater or to a receiving watercourse (Charlesworth and Warwick, 2012). A PPS can be used for both pedestrian and vehicular traffic. Whatever their application, PPSs conform to the SUDS triangle (see Chapter 22), whereby they address issues of water quantity, water quality, amenity and biodiversity equally.

This chapter will focus on design, maintenance and water quality control aspects, as well as advantages and disadvantages of different PPSs, with the help of recent and relevant case studies. The latest innovations in tackling global climate change (GCC) are discussed, as is current research regarding the combination of PPS, geothermal heating and cooling, water treatment and recycling.

23.2 Rethinking Stormwater Drainage

Most stormwater sewerage networks in large cities in the developed world were first installed during the 19th century. Typically, they comprise an underground, undersized pipe network, which captures storm runoff for distribution to nearby watercourses or sewerage systems. Over the years, some of these systems have become ineffective, inefficient

Water Resources in the Built Environment: Management Issues and Solutions, First Edition.
Edited by Colin A. Booth and Susanne M. Charlesworth.

and a liability to the environment. Furthermore, they are expensive (Schlüter and Jefferies, 2004; Scholz, 2006), OFWAT stating (Worsfold, no date) that upsizing the 309 000 km of overloaded infrastructure in the United Kingdom to the required standards would cost ~ £174 billion, at 2007–2008 prices, and would take centuries to deliver. Instead of simply bolting on 'end-of-pipe' approaches, SUDS challenge the traditional strategy of dealing with the management of surface water by treating the water as a resource and by the development of less-centralised technologies (Balkema *et al.*, 2002).

Various types of PPS are available for utilisation in a variety of residential, commercial and industrial environments, yet they are frequently confined to lightly trafficked areas, even though their capabilities can allow for a much wider range of uses (Scholz and Grabowiecki, 2007). In Scotland, serious consideration has been given to using SUDS devices on highways, including that of 'infiltration pavements' (Pittner and Allerton, 2009). Concern regarding the potential percolation of pollutants through the PPS and into groundwater can be overcome by tanking the system with an impermeable geotextile membrane and discharging the collected effluent into a suitable drainage system (Wilson *et al.*, 2003).

In the past, various barriers have reduced the potential uptake of PPS in England and Wales. However, the pollution control benefits of PPS have provided strong incentives for their utilisation. In the United States, for example, the main priority of the US Environmental Protection Agency (USEPA) is on the control of pollution from stormwater and it therefore requires developers of construction projects greater than 0.4 ha (1 acre) to apply for permits specifying best management practices (BMPs), including PPS, for stormwater runoff management (USEPA, 2007).

23.3 Porous, Permeable and Pervious Pavement Systems

There are three terminologies currently in use to describe a PPS: those that are permeable, porous or pervious. The latter term is generally a catch-all to cover any surface that allows water to pass through it and will not be used in this chapter. 'Permeable' generally refers to PPSs whose surface utilises block pavers (Figure 23.1) where the blocks themselves are impermeable, but slots or notches included in their design allow water to percolate through the gaps between neighbouring blocks and into the underlying substrate. 'Porous', on the other hand, refers to the whole surface allowing water ingress and includes porous asphalt and concrete (Figure 23.2).

A combination of increasing urbanisation (Charlesworth *et al.*, 2011) and global climate change (Gordon-Walker *et al.*, 2007; Schlüter and Jefferies, 2004) have led to the

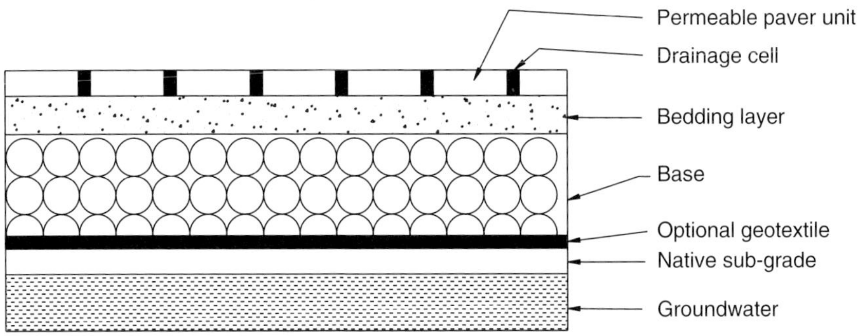

Figure 23.1 Representative layout of a permeable pavement system.

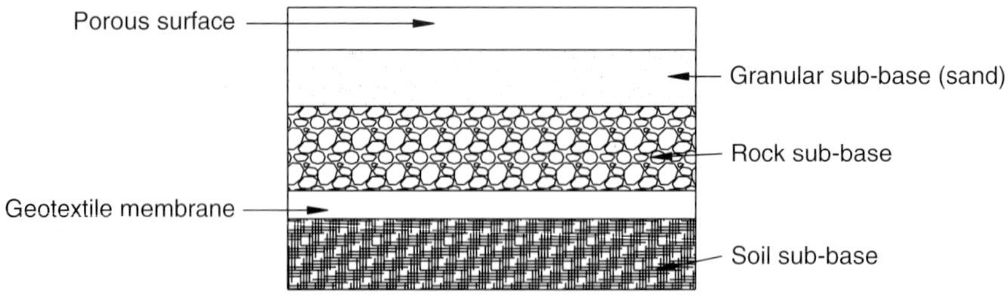

Figure 23.2 Representative layout of a porous pavement system.

recognition that traditional storm sewerage infrastructure has, in some instances, reached its limits. Much has been written about the UK summer floods of 2007, culminating in the Pitt Review (Pitt, 2008) and the fact that 70% of the floodwaters were derived from rain falling on to impermeable surfaces rather than the usual focus for flood resilience measures, which is the river channel. Whilst it is accepted that SUDS alone would not have prevented the flooding, the use of devices, such as PPSs, could have reduced the impact, and certainly reduced the addition of excess surface water (Charlesworth, 2010).

The general principle of a PPS, therefore, is to collect, treat and infiltrate freely, any surface runoff to support groundwater recharge (Woods-Ballard, 2007). In comparison to traditional drainage systems, stormwater retention and infiltration is a more sustainable approach to managing excess surface water (or, in a quote from SUDS Design Manual for Scotland and Northern Ireland (Construction Industry Research and Information Association (CIRIA), 2000): SUDS is 'designed to drain surface water in a more sustainable fashion than some conventional techniques'), is cost-effective and suitable for both new build and retrofit in urban areas (Andersen *et al.*, 1999; Dierkes *et al.*, 2002). Moreover, PPSs have many other potential benefits, such as saving water by recycling, addressing the urban heat island effect and prevention of pollution (Pratt *et al.*, 1999; Charlesworth, 2010).

23.4 Porous Pavements

A variety of both new and traditional materials can be used as porous pavements (Figure 23.2). For example, porous asphalt (or macadam pavement), which looks similar to conventional asphalt, consists of open graded asphalt with a binder layer over an open graded aggregate base located above well-draining soil. In comparison, porous concrete pavement contains coarse stone only, no fine aggregate and frequently a cement binder. The porosity is, therefore, achieved by the omission of fine aggregates. The structural strength of vegetated porous paving can be provided by modular interlocking concrete blocks, which can be pre-cast or cast-in-place, open cell lattices into which soil mixed with grass seeds or porous aggregates are placed (Scholz and Grabowiecki, 2007). Porous concrete can also be made into block paving stones modified by the addition of polymer, as these have better fatigue resistance than those without polymers. However, improvements such as these decrease for low levels of structural stress, sometimes becoming almost negligible in the case of traffic loads on main and highway roads (Pindado *et al.*, 1999). Porous concrete blocks can now be made of recycled materials, such as incinerated sewage sludge or slag of melted sewage sludge and the use of recycled aggregate is now being investigated (Akerojola, 2010; see Section 23.11).

23.5 Cost Implications of Utilising PPS

Engineers often perceive cost to be a major obstacle to adopting PPSs. Although the cost of a PPS surface may be greater than, for example, an asphalt surface, experience in the Northern hemisphere, and recent studies in the United Kingdom, have shown a PPS provides significantly lower initial and whole-of-life project costs than asphalt or cast-in-place concrete pavements (Interpave/Scott Wilson, 2006). This is principally because of the reduction or elimination of subsurface drainage infrastructure. For the SUDS approach generally, Duffy *et al.* (2008) stated that the costings for construction and maintenance of the Dunfermline Eastern Expansion SUDS train (DEX) (see http://www.scotland.gov.uk/Publications/2001/02/awards-ceremony/east-dunfermline-drainage) indicated that, if well designed and maintained, a SUDS approach was more cost-effective than traditional drainage. In a cost–benefit analysis, including specifically PPS for the UK Environment Agency, Gordon-Walker *et al.* (2007) found the benefit-to-cost ratio was 'very positive'. In fact, they found multiple cost benefits for the use of PPS in that, on a full life cycle, PPSs cost less than traditional surfacing, since associated maintenance costs are reduced and outweigh the increased initial capital outlay. Extra excavation may be necessary when installing a PPS, but should any of it require replacing it costs less to replace it in patches, rather than having to dig out a whole asphalt driveway, for instance. Furthermore, there would be no charges from Water Companies for the disposal of surface water runoff from a PPS. Gordon-Walker *et al.* (2007) compute that, were 50% of the current nonroad hard surfacing in the United Kingdom retrofitted with a PPS, ~£1.7bn in savings would be made at their end of life, which would benefit mainly site owners and operators.

23.6 Design Considerations

In general, hard drainage infrastructure is designed for 1 in 30 year storm events to accord with England's Planning Policy Statement 25 (Department for Communities and Local Government (DCLG), 2009) whereby new builds must deal with surface water on site at an average greenfield runoff rate of between 2 and 5 litres per second per hectare. Further details of the calculations necessary can be found in HR Wallingford (2007), Wilson *et al.* (2004) and Gibbs (2004), and these are further reviewed in Balmforth *et al.* (2006).

In designing a PPS, it is fundamentally important to provide and maintain surface infiltration and storage capacity to allow an adequate volume of stormwater to be captured and treated. The structure of a PPS is usually made up of four distinct components (James and von Langsdorf, 2003a): (i) pavers and bedding layer; (ii) unsaturated zone of the base material; (iii) saturated zone of the base material; and (iv) subgrade bottom layer. Typically, subbase depths range from 150 to 350 mm, although this depends on anticipated loadings, designed rainfall events and subsoil characteristics (Gordon-Walker *et al.*, 2007). A PPS incorporating ground source heatpump coils may have a total excavation depth of 650 mm. Aggregates enable the capacity of the reservoir to be about 30% of the volume and, thus, a PPS can be used as part of both retention or detention systems, particularly if the system is tanked. Various recycled aggregates (RAs) can be used to replace primary aggregates, including milestone, slate waste, blast furnace and steel slag, municipal solid waste incinerator ashes, power station ashes and foundry sands (e.g. Anderson *et al.*, 2003). For example, Nishigaki (2000) described specially designed blocks for a PPS using recycled melted slag (more detail on the use of RAs can be found in Section 23.11).

The lifespan of PPSs, in general, depends predominantly on the size of the air voids in their structure. The more contact with air, and hence oxidation, the less durability is achieved

(Choubane *et al.*, 1998). After 6 years of daily parking use, four commercially available PPSs were evaluated for their structural durability, ability to infiltrate precipitation and impacts on effluent water quality (Booth and Leavitt, 1999). None of the PPSs showed any major signs of wear, they infiltrated rainfall effectively and exhibited almost no surface runoff. In comparison with surface runoff from an impermeable asphalt area, infiltrated water passing through the PPSs had significantly lower levels of copper (Cu) and zinc (Zn).

A geotextile is advised by certain PPS manufacturers for use with their products. In general, a fibre area weight of 60 g/m^2 is usually recommended (Omoto *et al.*, 2003). There are certain advantages in adding a geotextile to a PPS substructure, since they can prevent the migration of surface dressing grit and bedding layer gravel down through the structure, leading to blocking or surface rutting, respectively. They also reduce the rutting depth and rate of block breakage in the case of permeable bituminous-stabilised base courses, which enables easy cleaning. Furthermore, most geotextiles can help to retain and degrade oil (see Section 23.10) if there are no potential problems with clogging (e.g. silting) (Newman *et al.*, 2004; Scholz, 2006).

Early tests on PPS hydraulic performance have shown that evaporation, drainage and retention within the permeable structures were mainly influenced by the particle size distribution of the bedding material and by the retention of water in the surface blocks (Andersen *et al.*, 1999; Scholz, 2006). In comparison to conventional surfaces, PPS including porous pavers provide more effective peak flow reductions (up to 42%) and longer discharge times. There is also a significant reduction of evaporation and surface water splashing (Booth and Leavitt, 1999; Pagotto *et al.*, 2000; Abbot and Comino-Mateos, 2003; Scholz, 2006).

A bespoke PPS, constructed in Northern Spain (Santander) (see Figure 23.3) and monitored between 2007 and 2010, revealed storage of rainwater for possible rainwater

Figure 23.3 PPS car park of various designs constructed for the purpose of long-term monitoring, Santander (Spain).

harvesting (see Chapter 12) uses in the PPS was feasible (Gomez-Ullate *et al.*, 2010). Evaporation did not significantly affect the volume of stored water in the PPS in the relatively dry summer, and during periods of heavy winter rain the PPS did not flood. The presence or absence of a geotextile and the composition of these geotextiles in preventing evaporation did not lead to significant differences in stored volumes, indicating that the most important influences on the capacity of the overall storage volume are the proposed usage pattern and the prevailing climate.

23.7 Infiltration through PPSs

There are three possible fates for precipitation initially falling on to the surface of a PPS (James and von Langsdorf, 2003a; Scholz, 2006): (i) infiltration through the surface and percolation to the base material; (ii) evaporation within the structure of the PPS itself or from the surface course; and (iii) runoff (overland flow) should the PPS surface become saturated or suffer clogging. Infiltration through the PPS and bedding layer is usually described using the complex Green-Ampt equations (Mein, 1980), which have physically-based parameters that can be predicted. Infiltration is thus related to the volume of water infiltrated and to the antecedent moisture conditions prevalent in the porous pavers and bedding layer (James and von Langsdorf, 2003a). Green-Ampt provided an approach based on fundamental physics but also give results that match empirical observations of small-scale laboratory experiments. However, it is difficult to upscale to the field or landscape scale (e.g. the suction forces at the wetting front cannot be accurately described) (Scholz and Grabowiecki, 2007).

Water enters the PPS by gravity, vertically downwards from the unsaturated zone (i.e. voids filled with air) to the base layer, or saturated zone (i.e. voids filled with water), and is assumed the sole source of water, unless there is water exchange with the surrounding environment below ground level. Lateral flow is represented by base layer discharge from the saturated zone to the receiving water. Deep percolation represents a lumped sink term for unquantified losses from the base layer. Two primary losses are assumed to be percolation through the confining layer and lateral outflow to somewhere other than the receiving water (James and von Langsdorf, 2003a). Infiltration supports groundwater recharge, decreases groundwater salinity, allows smaller diameters for sewers, if required at all (resulting in cost reduction), and improves the water quality of the receiving environment (Dierkes *et al.*, 2002; Scholz, 2006).

23.8 Maintenance to Enhance Infiltration

Due to their particle retention capacity during filtration, porous asphalt and concrete products can function as pollution sinks; trapped particulates can be removed by cleaning of the pavement utilising various designs and technologies, including vacuum cleaners and power-washers (Dierkes *et al.*, 2002). Earlier porous asphalt and porous concrete pavement systems were prone to void clogging, usually within a few years after installation, due to lack of proper maintenance, leading to lack of porosity. Clogging can be caused by sediment being ground into the porous pavement by traffic before being washed off, as well as waterborne sediment draining on to the pavement surface and shear stress caused by numerous stop–start actions of vehicles at the same spot, resulting in collapse of pores

(Scholz and Grabowiecki, 2007). Suction effects of tyres on fast moving vehicles, however, tend to remove ground-in particulates, thus removing some of the clogging material. If allowed to clog fully, the PPS would have to be removed entirely and replaced; however, studies have shown that routine maintenance can achieve a service life of 15–25 years (Shackel, 2010). For instance, a PPS car park in Massachusetts, United States, constructed in 1977, has never been repaved and continues to effectively infiltrate stormwater (National Asphalt Pavement Association (NAPA), no date). There are also several in the mid-west of the United States that are >20 years old and still working effectively (Adams, 2003). In fact, concerns about expensive long-term maintenance of PPSs to prevent clogging have largely been allayed by tests worldwide (e.g. James and von Langsdorff, 2003b; Borgwardt, 2006).

Efficacy of a maintenance regime on the permeability of concrete grid pavers and permeable interlocking concrete pavers, with in-service ages ranging from 6 months to 20 years, Bean *et al.* (2004) found cleaning the pavers at the end of each experiment improved permeability on 13 out of 14 sites (confidence level = 99.8%); however, those in close proximity to loose fine particles had infiltration rates significantly less than those free of loose fines. Even the minimum infiltration rates were comparable to those of a grassed sandy loam soil. Furthermore, surface infiltration rates of 27 PPS sites tested in North Carolina, Maryland, Virginia and Delaware, whose maintenance regime consisted of removing any residual material on the surface course, revealed post-maintenance infiltration rates improved to 8 cm/hour from pre-maintenance rates of 5 cm/hour. However, pavement locations, their maintenance types and regimes are critical to achieving high surface infiltration rates (Bean *et al.*, 2004).

23.9 Water Quality Improvements

Possible water quality variables of concern have the potential to endanger soil and groundwater resources if they are not sufficiently treated by biodegradation or removal during the infiltration process (Dierkes *et al.*, 2002; Brattebo and Booth, 2003) and include sediment and suspended solids, phosphorus, organic waste with high biochemical oxygen demand, dissolved nutrients and pollutants (including nitrogen, heavy metals, solvents, herbicides and pesticides), oil and grease, and faecal pathogens (D'Arcy *et al.*, 1998; North Carolina Department of Environment and National Resources (NCDENR), 2005; Scholz, 2006). Such pollutants, which are particularly associated with urban areas, have been effectively addressed by PPSs, such that the structure itself has been described as an effective 'in-situ aerobic bioreactor' (Bond *et al.*, 1999; Pindado *et al.*, 1999; Coupe *et al.*, 2003; Scholz, 2006). Hydrocarbons as urban water pollutants have received considerable attention and the effective retention and biodegradation of mineral oils in PPS have been reported (Newman *et al.*, 2002, 2004; Coupe *et al.*, 2003; Puehmeier and Newman, 2008). The research into the relationship between oil contamination and the ability of the PPS to degrade it have concentrated on the microbiological characteristics of biofilms growing on the oil trapped on geotextiles, giving information on the diversity, abundances and ecological interactions between the microorganisms that inhabit the PPS (see Section 23.10).

Reductions in suspended solids, biochemical oxygen demand, chemical oxygen demand and ammonia levels after infiltration through various PPS, in comparison to highway gullies, demonstrate their high treatment efficiency, but also that there is less need for frequent maintenance, unlike with gully pots (Pratt *et al.*, 1999). By installing tanked systems (those bounded by an impermeable geotextile membrane), the PPS can also be used on brownfield sites as water will not pass into the contaminated soil but can either be stored for use elsewhere or conveyed downstream.

Various PPSs have a good track record at removing suspended solids and nitrogen, although those that rely on below-ground infiltration and use of an underdrain system are more successful in removing nitrogen, since the underdrain has been found to be the site of denitrification (NCDENR, 2005).

As well as atmospheric contaminants, sources of harmful pollutants can include roof material and road surfaces; there have been many studies on the polluting potential of gullypot sediments, road and street dusts, whereby metals such as zinc (Zn) and copper (Cu) are now of concern, rather than lead (Pb), whose concentrations have reduced in countries with unleaded petrol (Charlesworth *et al.*, 2011). Street furniture, rainwater gutters and associated pipes often consist of Zn-coated sheets or Cu, and road paint contains cadmium (Cd). Metal roofs usually show high concentrations of heavy metals in the corresponding runoff (if they have not cleaned prior to discharge) (Dierkes *et al.*, 2002; Scholz, 2006). A 3-year weekly study of water quality in a tanked PPS collecting water from the Zn roof of an ecohouse at the Building Research Establishment (BRE), Watford, United Kingdom, found slightly elevated concentrations of Zn, relative to nearby masonry-fired clay tiled roofs. It is noteworthy that the concentrations were within regulatory limits, which shows the ability of a PPS to deal with potential metal pollution.

Whilst oil and diesel fuel contamination are frequently detected on impermeable surfaces, it is not always detected on a PPS (Brattebo and Booth, 2003), and concentrations of Zn and Cu can be significantly lower on a PPS, with Pb being virtually undetectable. In a study of various PPSs, Booth and Leavitt (1999) compared direct surface runoff from an impermeable area and found infiltrated water had significantly lower levels of Cu and Zn. Motor oil was detected in 89% of runoff samples from the impermeable area, but not in any outflow water sample from the PPS. Moreover, diesel fuel was not found in any samples from the PPS.

In order to test how the structure of a PPS can retain pollutants, Wilson *et al.* (2003) set up worst-case combinations of pollution and rainfall and incorporated an oil interceptor into the PPS. Results demonstrated that the PPS retained hydrocarbons, offering improved water quality at the outflow. However, they also found certain detergents present in the PPS were able to cause contamination of the outflow, which may require secondary treatment.

In many studies, a PPS incorporating a geotextile below the bedding layer have retained ~98.7% of applied clean or waste lubricating oils, under a variety of rainfall volumes and intensities (Newman *et al.*, 2006). The experimental programme for most of these analyses has been to assume a minimum 10 times the expected oil application rate on to an urban surface (e.g. an experimental addition rate of 17.8 mg/m^2/week) (Pratt *et al.*, 1999).

There are also many studies that have shown that filtration through a PPS improves water quality by removing suspended solids and, particularly, heavy metals from runoff. For example, Legret *et al.* (1996) showed that suspended solids and Pb can be reduced by permeable systems by up to 64% and 79% respectively, and the Center for Watershed Protection (CWP) (2009) registered up to 80% removal of total suspended solids and metals through a PPS.

Kellems *et al.* (2003) found that enhanced filtration using organic media was an effective alternative to chemical precipitation for the treatment of stormwater, whereby filtration through a specific adsorbent organic medium can remove about 95% of dissolved Cu and Zn. More recent research by Charlesworth *et al.* (2012) showed that biofilms develop on coarse compost, which is as efficient as topsoil in dealing with pollutants and could, therefore, be used in vegetated PPSs. Biofilms are integral to water quality improvements for a PPS. The following section is, therefore, devoted to the biodegradation of pollutants by the microecosystem that develops on geotextile and aggregates, positioned within and on the PPS.

23.10 Biodegradation of Pollutants by Microorganisms

PPSs have been intensively monitored for their biological characteristics, particularly the formation of biofilms in the PPS internal structure that take part in the biodegradation of organic pollution. It has been found that PPS biofilms grow primarily on the geotextile layer and are stable, permanent features of the system. Biofilm maintenance has also been addressed by adding nutrient sources (nitrogen, phosphorus and potassium) to allow the conversion of carbon in oils to cell growth biomass; this process could be maintained for several years (Pratt *et al.*, 1999; Newman *et al.*, 2002).

Coupe *et al.* (2003) showed that a PPS does not require the inoculation of oil degrading microorganisms in order to biodegrade hydrocarbons, because the retention capacity of the geotextile was sufficient to immobilise the oils, grow a biofilm and initiate the treatment process in the presence of bacteria and fungi found naturally on relatively clean aggregates. Subsequent studies have revealed that the added inoculum, although viable at the time and added at a required density, was not detected in molecular biological analysis of PPS effluent. A 'polymerase chain reaction' and 'denaturing gradient gel electrophoresis' were both applied to effluent from proven oil degrading PPS and neither method could provide evidence of inoculum organisms, but detected a robust and biodiverse bacterial population from other sources (Newman *et al.*, 2006). The most probable explanations for this result were protists and metazoa feeding on inoculum organisms and competition from indigenous bacteria and fungi for favourable biofilm sites. The success of autochthonous biodegrader microbes led to an investigation of their origin, which was found to be chiefly the building materials for the PPS, not influenced particularly by taxa imported by the wind or rain. Treatment rates have been estimated at ~40% biodegradation by mass of added oils within 6 months, after which 80% of the remaining undegraded and partially degraded oil was found on the geotextile (Newman *et al.*, 2006).

The bacteria and fungi responsible for biodegradation are, in turn, food for predators, such as protists and metazoa. These organisms are thought to maintain a free draining PPS by penetrating the biofilm and removing blockages using organic materials. A similarity between soil processes and PPS biomass has been suggested, as many microbes are common to both environments and have similar origins (Coupe *et al.*, 2006). The presence of a biodiverse microbial community has also been shown to promote biodegradation, rather than inhibit it, despite the grazing of predators on biodegraders. This is partly due to the regeneration of available sites within the biofilm and the removal of excess growth, but is also due to the secretion of nitrates and other limited nutrients by predators into the oil degrading biofilm (Coupe *et al.*, 2003). A successional series of predators over time, from small abundant species to larger less numerous taxa, has been reported and this has been used as a way of determining organic pollution levels and the state of biodegradation in the PPS. A bioindicator system based on microscopical investigation of the predators in the PPS and other SUDS, sensitive enough to detect changes in pollution composition and concentrations, is a distinct possibility.

Innovative projects in PPS biology have seen that the successful incorporation of phosphorus into geotextiles improves the efficiency of adding nutrients to the biofilm

(Spicer *et al.*, 2006). This innovation was shown to increase the rate of biodegradation in the PPS without adding to downstream eutrophications. Electron microscopy, both scanning and transmission, has been used to observe biofilm growth under different conditions and in response to changing environmental regimes (Coupe *et al.*, 2006).

Assessment of microbiological water quality has been an important process in preventing waterborne diseases, particularly where rainwater harvesting (RwH) has been the desired outcome. The two most common tests carried out are for coliforms and *Escherichia coli*, or faecal coliforms (Barrell *et al.*, 2000). Total coliforms, faecal coliforms, faecal streptococci, heterotrophs, fungi, *Pseudomonas aeruginosa*, *Leptospira*, salmonellae and viruses are often analysed, in an attempt to determine the temporal distribution of bacterial pathogens and viruses in stormwater runoff. However, findings show that it is not possible to accurately predict the time when peak microbial populations, including human pathogens, occur in runoff waters. Chapter 12 in this volume suggests that diversion of the first flush of harvested rainfall can reduce potential contamination.

23.11 Directions of Further Investigation

23.11.1 New Materials

Silica fume and super plasticiser can be added to standard porous concrete ingredients. This usually improves the compressive strength of the porous pavement to allow for higher loads, depending on the application (Yang and Guoliang, 2003; Woods-Ballard, 2007). An additional layer of heat-bonded geotextile was introduced by Newman *et al.* (2004) to a PPS subbase. This liner slowed down the release of minor oil spillages and their subsequent transport through the system. In case of an emergency, however, this solution cannot be used to protect large volumes of released oil, although the oil trap may significantly reduce the amount of released oil.

23.11.2 Recycled Aggregates

The use of recycled aggregates (RA) addresses two problems associated with the use of stone in PPS structures. Firstly, there is increasingly limited availability of landfill space in which to place construction and demolition waste (CDW) and, secondly, a rapidly declining supply of primary aggregates (CIRIA, 2004). The CDW, which could be reused for this purpose, includes recycled concrete (RC), reclaimed asphalt pavement (RAP) and spent railway ballast (SRB). Much research has been carried out on the use of RA in traditional, impermeable road surfaces (e.g. Poon and Chan, 2006), but the PPS is a special case in which surface water passes through the subbase. Thus, there are concerns regarding its use should the RA be contaminated and hence, the potential for groundwater pollution, particularly with hydrocarbons, heavy metals and salts (Rao *et al.*, 2007). In a laboratory-based study of the potential for various RAs to leach contaminant, Akerojola (2010) tested the CDW listed in Table 23.1, and compared them against a virgin aggregate (VA) control obtained from a PPS manufacturer. Results of the study indicated that RA was comparable to VA, in that very minimal contaminants washed through leaching columns to appear in the effluent. However, there have been very few studies carried out specifically for PPS, and more needs to be done in order to provide sufficient data for RA to be used with confidence in the structure of infiltrating pavements.

Table 23.1 Construction and demolition waste (CDW) used to test their polluting potential.

RA type	Source	Type of stone
Spent aggregate	Former steelworks	Granite and limestone
SRB	Railway West Somerset, UK	Granite, limestone
RAP	Recycling facility	Milled asphalt paving
CDW	Recycling facility	Glass, brick, pieces of tile
RC	Building Research Establishment, Watford, UK	Crushed concrete

Source: Akerojola (2010).

23.11.3 Combinations of Technology with PPS

Porous pavements can be combined with water recycling technology. The purpose is to collect treated runoff in a tanked below-ground collection system for subsequent recycling. Applications may include car washing, garden sprinkling and toilet flushing (Scholz and Grabowiecki, 2007). The first PPS to be combined with a ground-source heatpump (GSHP) apparatus was installed in September 2007 at the Building Research Establishment Innovation Park, in Watford, United Kingdom (Coupe *et al.*, 2009). The operation of the system was monitored for 3 years and demonstrated that the combination of water and heating technology could adequately heat a three-bedroom house during two of the coldest winters for 30 years. Tota-Maharaj and Scholz (2010) and Tota-Maharaj *et al.* (2010) reported on the impact on water quality by combining block paver PPS and ground-source heatpumps. In addition to the obvious benefit of either heating or cooling nearby buildings, no significant deterioration in water quality or significant increase in potentially pathogenic organisms due to the change in temperature within the waterlogged subbase was noted (Tota-Maharaj *et al.*, 2010).

23.11.4 PPS and Environmental Impact

PPSs score heavily in both the UK Code for Sustainable Homes (CSH) (DCLG, 2006) and BRE Environmental Assessment Methods (BREEAMs) (see http://www.breeam.org/). Maximum environmental credits can be gained by combining the PPS, RwH and GSHP in one site, as this linking together impacts on the energy, water and surface water elements of both CSH and BREEAM. When applied singly to surface water or RwH, however, PPSs are still effective in meeting mandatory performance levels for sustainable drainage, minimising mains water use and improving the quality of discharged water.

23.11.5 PPS and GCC

Whilst not brought about by global change, but in addition to it, local effects of urbanisation have increased temperatures in cities by the urban heat island effect (UHIE). PPSs can play an extremely useful role in reducing temperatures by promoting evaporation off the surface course and within the internal structure of the pavement, a process called 'evaporative cooling' (Charlesworth, 2010). This effect is promoted in PPSs constructed of water-retaining materials with optimal evaporative properties (Okada *et al.*, 2008), such as slag, bentonite and diatomite. However, since these materials release absorbed water too quickly, 'wet pavements' have been tested in Tokyo (Yamagata *et al.*, 2008) whereby reclaimed wastewater was applied to the pavement surface during the day and results indicated temperatures were reduced both during the day and at night time, by up to 8 °C and 3 °C respectively. PPSs can also be constructed using materials with higher solar reflectivity than conventional paving, so-called 'cool' pavements (USEPA, 2009). Santamouris *et al.* (2012) used a 4500 m^2 cool pavement, installed in a public park in Athens, Greece, to estimate that the peak ambient temperature of a typical summer day across the city could be reduced by up to 1.9 K, and within the park itself by 12 K, while perceptions of comfort were considerably improved. They concluded that reflective pavements efficiently mitigate the UHIE to improve thermal conditions in urban areas. The design and utilisation of a variety of PPSs for urban cooling looks promising but, whilst the efficiency and efficacy of a PPS to reduce storm peaks is well researched and understood, more work is needed to better understand the role PPSs can play in urban cooling (Charlesworth, 2010).

23.11.6 Modelling

With the advent of supportive legislation (e.g. Flood and Water Management Act, 2010), there is a need for the development of computer-based decision support tools or systems (DSSs) for engineers and planners. Recent attempts to incorporate a PPS into a SUDS multicriteria decision support model were led by Ellis *et al.* (2004), who incorporated performance of the system both technically and scientifically, impacts on the environment, its benefits to the community and the economic costs of investing in SUDS infrastructure as a whole. Scholz (2006) and Scholz *et al.* (2006) approached the subject in a more practical sense, applying the decision support matrix to a case study in Glasgow, United Kingdom, and concluding it has the potential to be applied to other cities in their implementation of a SUDS management train (see Chapter 22). Viavattene *et al.* (2008) utilised the geographical information system (GIS) to simplify the ability to communicate such tools to end users, incorporating technical stormwater models into the matrix. More recently, developed tools (e.g. Ellis *et al.*, 2011), have had to integrate perspectives of the user organisation as well as allowing for their personal input in terms of perceived importance of the criteria making up the model. It is admitted, however, that whilst these decision support systems have great value, enabling the selection of a variety of drainage variables, it is still difficult to estimate the sustainability of the system. In a world where environmental improvement has to be quantifiable and the environment is complex, Ellis *et al.* (2011, p. 10) call DSS a 'difficult exercise'.

23.11.7 Sustainability Credentials of PPS

According to Heal *et al.* (2004), there are four main issues when considering the sustainability of SUDS. These are:

1. The fact that they need maintenance, regular inspections and interventions, although traditional drainage also has this requirement.
2. Whilst much is known of the capabilities of the PPS to improve water quality, little is known of the fate of those contaminants, especially toxic metals that do not degrade, unlike some organic compounds.

3. In spite of biodiversity being part of the SUDS triangle, ecological improvement is not associated with PPS unless general enhancement of the downstream environment could be considered, which would encourage enriched biodiversity.
4. There is some field evidence of SUDS failures due to incorrect management or design.

To these issues of concern could be added *social acceptance*, since, if the surface water management approach is unacceptable to those who live in close proximity, it will be unsustainable by default, as land owners or managers will take every opportunity to replace it. Unfortunately, however, minimal research has been carried out on the public perceptions or acceptability of SUDS devices.

23.12 Conclusions

PPSs have become an important and integral part of SUDS systems. In common with most SUDS devices and management trains, PPSs are flexible in approach and provide multiple benefits. Latest research focuses on mitigation of, and adaptation to, GCC, with 'wet' and 'cool' pavements becoming an important part of some cities' management of the urban heat island effect. Multiple benefits can be conferred by the combination of PPS with GSHP and RwH in a tanked system. Concern regarding the quality of water stored in such combinations has particularly led, more recently, to studies of the microbiological water quality of harvested and stored water and also the impacts of certain contaminants (such as oil and herbicide) on biofilm development and maintenance in the PPS. Some of the most important target pollutants, however, remain hydrocarbons, heavy metals, nutrients (such as nitrogen and phosphorus), as well as microbiological concerns.

As stated by Ellis *et al.* (2011), infrastructure associated with surface water management in urban areas is 'wicked'. In other words, it is complicated, has no obvious solution and has a collection of stakeholders whose interests are diverse. Recent legislation such as the Flood and Water Management Act, 2010, has encouraged the development of decision support tools or systems to enable better design of SUDS in urban areas. These tools will be necessary for those who design and implement sustainable drainage and to ensure any mistakes made in the past are not repeated. However, without doubt, more research is needed to optimise their use since they are limited by the present state of knowledge.

References

Abbot, C.L. and Comino-Mateos L. (2003) In-situ hydraulic performance of a permeable pavement sustainable urban drainage system. *Journal of the Chartered Institution of Water and Environmental Management*, **17** (3), 187–190.

Adams, M. (2003) Porous asphalt pavement with recharge beds: 20 years and still working. *Stormwater*. Available at: http://www.stormh2o.com/SW/Articles/Porous_Asphalt_Pavement_With_Recharge_Beds_20_Year_228.aspx.

Akerojola, T.T. (2010) *The Contaminant Potential of Recycled Aggregate When Used in Porous Pavement*. Unpublished MSc thesis, Coventry University.

Andersen, C.T., Foster, I.D.L. and Pratt, C.J. (1999) The role of urban surfaces (permeable pavements) in regulating drainage and evaporation: development of a laboratory simulation experiment. *Hydrological Processes*, **13** (4), 597–609.

Anderson, P., Cunningham, C.J., Hearnden, R.A., Barry, D.A. and Philp, J.C. (2003) Optimisation and assessment of different railway ballast cleaning systems. *Land Contamination and Reclamation*, **11** (4) 397–404.

Balkema, A.J., Preisig, H.A., Otterpohl, R. and Lambert, F.J.D. (2002) Indicators for the sustainability assessment of wastewater treatment systems. *Urban Water*, **4** (2), 153–161.

Balmforth, D., Digman, C., Kellagher, R. and Butler, D. (2006) *Designing for Exceedance in Urban Drainage – Good Practice (C635)*. Construction Industry Research and Information Association, London.

Barrell, R.A.E., Hunter, P.R. and Nichols, G. (2000) Microbiological standards for water and their relationship to health risk. *Communication and Discussion in Public Health*, **3** (1), 8–13.

Bean, E.Z., Hunt, W.F. and Bidelspach, D.A. (2004) Study on the surface infiltration rate of permeable pavements. In: Sehlke, G., Hayes, D.F. and Stevens, D.K. (eds), *Proceedings of the American Society of Civil Engineers and EWRI 2004 World Water and Environmental Resources Congress*, Salt Lake City, Utah.

Bond, P.C., Pratt, C.J. and Newman, A.P. (1999) A review of stormwater quantity and quality performance of permeable pavements in the UK. In: *Proceedings of 8th International Conference on Urban Storm Drainage*, Sydney, Australia, pp. 248–255.

Booth, D.B. and Leavitt J. (1999) Field evaluation of permeable pavement systems for improved storm water management. *American Planning Association Journal*, **65** (3), 314–325.

Borgwardt, S. (2006) Long-term in-situ infiltration performance of permeable concrete block Pavement. In: *Proceedings of 8th International Conference on Concrete Block Paving*, San Francisco, California.

Brattebo, B.O. and Booth, D.B. (2003) Long-term storm water quantity and quality performance of permeable pavement systems. *Water Research*, **37** (26), 4369–4376.

Center for Watershed Protection (CWP) (2007) *National Pollutant Removal Performance Database*, Version 3. Center for Watershed Protection. Ellicott City, Maryland.

Center for Watershed Protection (CWP) (2009) *Coastal Stormwater Supplement to the Georgia Stormwater Management Manual*, First Edition. Chatham County – Savannah Metropolitan Planning Commission, Savannah, Georgia.

Charlesworth, S.M. (2010) A review of the adaptation and mitigation of global climate change using sustainable drainage in cities. *Journal of Water and Climate Change*, **1**, 165–180.

Charlesworth, S.M. and Warwick, F. (2012) Adapting and mitigating floods using sustainable drainage (SUDS). In: Lamond, J.E., Booth, C.A., Hammond, F.N. and Proverbs, D.G. (eds), *Flood Hazards, Impacts and Responses for the Built Environment*. Chapter **15**. Taylor and Francis CRC Press, pp. 207– 234.

Charlesworth, S.M., De Miguel, E. and Ordóñez, A. (2011) A review of the distribution of particulate trace elements in urban terrestrial environments and its application to considerations of risk. *Environmental Geochemistry and Health*, **33**, 103–123.

Charlesworth, S.M., Nnadi, E., Oyelola, O., Bennett, J., Warwick, F., Jackson, R. and Lawson, D. (2012) Laboratory based experiments to assess the use of green and food based compost to improve water quality in a sustainable drainage (SUDS) device such as a swale. *Science of the Total Environment*, **424**, 337–343.

Choubane, B., Page, G.C. and Musselman, J.A. (1998) Investigation of water permeability of coarse graded superpave pavements. *Journal of the Association of Asphalt Paving Technology*, **67** (2), 254–276.

Construction Industry Research and Information Association (CIRIA) (2000) *Sustainable Urban Drainage Systems Design Manual for Scotland and Northern Ireland*, C521. CIRIA, London.

Construction Industry Research and Information Association (CIRIA) (2004) *Potential Use of Alternatives to Primary Aggregates in Coastal and River Engineering*. CIRIA, London.

Coupe, S.J., Smith, H.G., Newman, A.P. and Puehmeier, T. (2003) Biodegradation and microbial diversity within permeable pavements. *European Journal of Protistology*, **39** (4), 495–498.

Coupe, S.J., Newman, A.P. and Robinson, K. (2006) Analysis of permeable pavement microbial biofilms by electron microscopy. In: *8th International Conference on Concrete Block Paving*, San Francisco, California.

Coupe, S.J., Faraj, A.S. and Charlesworth, S.M. (2009) Combining permeable paving with renewable energy devices: installation, performance and future prospects. In: *9th International Conference on Concrete Block Paving*, Buenos Aires, Argentina.

D'Arcy, B.J., Usman, F., Griffiths, D. and Chatfield, P. (1998) Initiatives to tackle diffuse pollution in the UK. *Water Science and Technology*, **38** (10), 131–138.

Department for Communities and Local Government (DCLG) (2006) *Code for Sustainable Homes: A Step-Change in Sustainable Home Building Practice*, 31 pp.

Department for Communities and Local Government (DCLG) (2009) *Planning Policy Statement 25: Development and Flood Risk Practice Guide.*

Dierkes, C., Kuhlman, L., Kandasamy, J. and Angelis G. (2002) Pollution retention capability and maintenance of permeable pavements. In: Strecker, E.W. (ed.), *Proceedings of the 9th International Conference on Urban Drainage*, Portland, Oregon.

Duffy, A., Jefferies, C., Waddell, G., Shanks, G., Blackwood, D. and Watkins A. (2008) A cost comparison of traditional drainage and SUDS in Scotland. *Water Science and Technology*, **57** (9), 1451–1459.

Ellis, J.B., Deutsch, J.-C., Mouchel, J.-M., Scholes, L. and Revitt, M.D. (2004) Multicriteria decision approaches to support sustainable drainage options for the treatment of highway and urban runoff. *Science of the Total Environment*, **335**, 251–260.

Ellis, J.B., Lundy, L. and Revitt, M. (2011) An integrated decision support approach to the selection of sustainable urban drainage systems (SUDS). In: *SWITCH Conference: The Future of Urban Water; Solutions for Liveable and Resilient Cities*. UNESCO, Paris.

Gibbs, G. (2004) *Preliminary Rainfall Runoff Management for Developments (EA/DEFRA W5–074/A), Summary Information*. Available at: http://www.ciria.org.uk/suds/pdf/preliminary_rainfall_runoff_mgt_for_development.pdf.

Gomez-Ullate, E., Bayon, J.R., Coupe, S.J. and Castro-Fresno, D. (2010) Performance of pervious pavement parking bays storing rainwater in the north of Spain. *Water Science and Technology*, **62**, 615–621.

Gordon-Walker, S., Harle, T. and Naismith, I. (2007) *Cost–Benefit of SUDS Retrofit in Urban Areas*. Science Report – SC060024. Environment Agency.

Heal, K., McLean, N. and D'Arcy, B. (2004) SUDS and sustainability. In: *26th Meeting of the Standing Conference on Stormwater Source Control*, Dunfermline, Scotland.

HR Wallingford (2007) *Preliminary Rainfall Runoff Management for Developments*. EA/DEFRA Technical Report W5–074/A.

Interpave/Scott Wilson (2006) *The Costs of Paving – Comparative Initial and Whole of Life Costs for Pavements*. Interpave, London.

James, W. and von Langsdorf, H. (2003a) Computer aided design of permeable concrete block pavement for reducing stressors and contaminants in an urban environment. In: *Proceedings of the 7th International Conference on Concrete Block Paving*, Sun City, South Africa.

James, W. and von Langsdorf, H. (2003b) The use of permeable concrete block pavement in controlling environmental stressors in urban areas. *Proc. 7th Int. Conference on Concrete Block Paving*, Sun City, South Africa.

Kellems, B.L., Johnson, P.E., Sanchez, F. and Crowser, H. (2003) Design of emerging technologies for control and removal of storm water pollutants. In: Bizier, P. and DeBarry P. (eds), *Proceedings of the Water World and Environmental Resources Congress*, Philadelphia, Pennsylvania.

Legret, M., Colandini, V. and LeMarc, C. (1996) Effects of a porous pavement with reservoir structure on the quality of runoff water and soil. *Science of the Total Environment*, **190**, 335–340.

Mein, R.G. (1980) Recent developments in modeling of infiltration: extension of the Green-Ampt Model. In: *Hydrology ad Water Resources Symposium*, Adelaide, Australia.

National Asphalt Pavement Association (NAPA) (nd) *Porous Asphalt*. Available at: http://www.asphaltpavement.org/index.php?option=com_content&task=view&id=359&Itemid=863.

Newman, A.P., Coupe, S.J., Puehmeier, T., Morgan, J.A., Henderson, J. and Pratt, C.J. (2002) Microbial ecology of oil degrading porous pavement structures. Global Solutions for Urban Drainage. In: Strecker, E.W. (ed.), *Proceedings of the 9th International Conference on Urban Drainage*, pp. 1–12, Portland, Oregon.

Newman, A.P., Puehmeier, T., Kwok, V., Lam, M., Coupe, S.J., Shuttleworth, A. and Pratt, C.J. (2004) Protecting groundwater with oil retaining pervious pavements: historical perspectives, limitations and recent developments. *Quarterly Journal of Engineering Geology and Hydrogeology*, **37** (4), 283–291.

Newman, A.P., Coupe, S.J., Smith, H.G., Puehmeier, T. and Bond, P. (2006) The Microbiology of Permeable Pavements. In *8th International Conference on Concrete Block Paving*, California, US.

Nishigaki, M. (2000) Producing permeable blocks and pavement bricks from molten slag. *Waste Management*, **20** (2–3), 185–192.

North Carolina Department of Environment and Natural Resources (NCDENR) (2005) *Updated Draft Manual of Storm Water Best Management Practices*. Public Consultation Document DOC-7-1. North Carolina Department of Environment and Natural Resources (NCDENR), Division of Water Quality, North Carolina.

Okada, K., Matsui, S., Isobe, T., Kameshima, Y. and Nakajima, A. (2008) Water-retention properties of porous ceramics prepared from mixtures of allophone and vermiculite for materials to counteract heat island effects. *Ceramics International*, **34**, 345–350.

Omoto, S., Yoshida, T. and Hata, S. (2003) Full-scale durability evaluation testing of interlocking block pavement with geotextile. In: *Proceedings of the 7th International Conference on Concrete Block Paving*, Sun City, South Africa.

Pagotto, C., Legret, M. and Le Cloirec, P. (2000) Comparison of the hydraulic behaviour and the quality of highway runoff water according to the type of pavement. *Water Research*, **34** (18), 4446–4454.

Pindado, M.A., Aguado, A. and Josa, A. (1999) Fatigue behavior of polymer-modified porous concretes. *Cement and Concrete Research*, **29** (7), 1077–1083.

Pitt, M. (2008) *Learning the Lessons from the 2007 Floods*, 462 pp. Available at: http://webarchive.nationalarchives.gov.uk/20100807034701/http://archive.cabinetoffice.gov.uk/pittreview/_/media/assets/www.cabinetoffice.gov.uk/flooding_review/pitt_review_full%20pdf.pdf.

Pittner, C. and Allerton, G. (2009) *SUDS for Roads*. SUDS Working Party. Available at: http://scots.sharepoint.apptix.net/roads/General%20Publications/SuDS%20for%20Roads/20100805%20SUDS%20for%20Roads%20-%20FINAL%20Version.pdf.

Poon, C.S. and Chan, D. (2006) Feasible use of recycled concrete aggregates and crushed clay brick as unbound roadsub-base. *Construction and Building Materials*, **20**, 578–585.

Pratt, C.J., Newman, A.P. and Bond P.C. (1999) Mineral oil biodegradation within a permeable pavement: long term observations. *Water Science and Technology*, **39** (2), 130–109.

Puehmeier, T. and Newman, A.P. (2008) Oil retaining and treating geotextile for pavement applications. In: *11th International Conference on Urban Drainage*, Edinburgh, Scotland.

Rao, A., Jha, K.N. and Misra, S. (2007) Use of aggregates from recycled construction and demolition waste in concrete. *Resources, Conservation and Recycling*, **50**, 71–81.

Santamouris, M., Gaitani, N., Spanou, A., Saliari, M., Giannopoulou, K., Vasilakopoulou, K. and Kardomateas, T. (2012) Using cool paving materials to improve microclimate of urban areas – design realization and results of the Flisvos project. *Building and Environment*, **53**, 128–136.

Schlüter, W. and Jefferies, C. (2004) Modelling the outflow from a porous pavement. *Urban Water*, **4** (3), 245–253.

Scholz, M. (2006) Practical sustainable urban drainage system decision support tools. *Proceedings of the Institution of Civil Engineers – Engineering Sustainability*, **159** (3), 117–125.

Scholz, M. and Grabowiecki, P. (2007) Review of permeable pavement systems. *Building and Environment*, **42** (11), 3830–3836.

Scholz, M., Corrigan, N.L. and Kazemi-Yazdi, S. (2006) The Glasgow SUDS Management Project: case studies (Belvidere Hospital and Celtic FC Stadium areas). *Environmental Engineering Science*, **23** (6), 908–922.

Shackel, B. (2010) The design, construction and evaluation of permeable pavements in Australia. In: *24th ARRB Conference*, Melbourne, Australia.

Spicer, G.E., Lynch, D.E. and Coupe, S.J. (2006) The development of geotextiles incorporating slow-release phosphate beads for the maintenance of oil degrading bacteria in permeable pavements. *Water Science and Technology*, **54** (6–7) 273–280.

Tota-Maharaj, K. and Scholz, M. (2010) Efficiency of permeable pavement systems for the removal of urban runoff pollutants under varying environmental conditions. *Environmental Progress and Sustainable Energy*, **29** (3), 358–369.

Tota-Maharaj, K., Scholz, M., Ahmed, T., French, C. and Pagaling, E. (2010) The synergy of permeable pavements and geothermal heat pumps for stormwater treatment and reuse. *Environmental Technology*, **31** (14), 1517–1531.

United States Environmental Protection Agency (USEPA) (2009) *Reducing Urban Heat Islands: Compendium of Strategies, Cool Pavements Chapter*. USEPA, 39 pp.

United States Environmental Protection Agency (USEPA) (2007) *Developing your Stormwater Prevention Plan: A Guide for Construction Sites*. EPA–833–R–06–004, 50 pp.

Viavattene, C., Scholes, L., Revitt, D.M. and Ellis, J.B. (2008) A GIS based decision support system for the implementation of stormwater best management practices. In: *11th International Conference on Urban Drainage*, Edinburgh, Scotland.

Wilson, S., Newman, A.P., Puehmeier, T. and Shuttleworth, A. (2003) Performance of an oil interceptor incorporated into a pervious pavement. *Proceedings of the Institution of Civil Engineers – Engineering Sustainability*, **156** (1), 51–58.

Wilson, S., Bray, R. and Cooper, P. (2004) *Sustainable Drainage Systems: Hydraulic, Structural and Water Quality Advice, C609*. Construction Industry Research and Information Association, London.

Woods-Ballard, B., Kellagher, R., Martin, P., Jefferies, C., Bray, R. and Shaffer, P. (2007) *The SUDS Manual, C697*. Construction Industry Research and Information Association, London.

Worsfold, M. (no date) *Sustainable Drainage*. OFWAT. Available at: http://www.ofwat.gov.uk/media-centre/speeches/prs_spe_100303drainage.pdf.

Yamagata, H., Nasu, M., Yoshizawa, M., Miyamoto, A. and Minamiyama, M. (2008) Heat island mitigation using water retentive pavement sprinkled with reclaimed wastewater. *Water Science and Technology*, **57** (5), 763–772.

Yang, J. and Guoliang, J. (2003) Experimental study of pervious concrete pavement materials. *Cement and Concrete Research*, **33** (1), 381–386.

24

Multiple Benefits of Green Infrastructure

Rebecca Wade and Neil McLean

24.1 Introduction

This chapter will consider many aspects of green infrastructure (GI), not only water management. This is essential in considering water resource issues in the built environment as urban settings are complex and often space-limited with multiple users and stakeholders; thus, whilst GI can address aspects of water management, it must also deliver multiple benefits.

Urban areas can contain rich flora that contribute significantly to biodiversity, and thus also provide benefits to humans, but loss and isolation of habitats due to urban sprawl threatens biodiversity (Kong *et al.*, 2010). In the United Kingdom, >80% of the population lives in urban areas (United Kingdom National Ecosystem Assessment (UKNEA), 2011), which may only occupy a few percent of the total land area but has a large impact on catchment hydrology, habitat quality (Walsh *et al.*, 2005) and their populations have an ecological footprint that extends far beyond the urban zone. Associated with urban expansion in recent decades is a decline in biodiversity in urban catchments. Housing density on previously developed land has doubled in the past decade from 22 to 44 dwellings per hectare (in northern England) and increased almost threefold in London, from 47 dwellings per hectare to 122 (UKNEA, 2011). High-density development reduces accessible green space in many central areas, thus compromising ecosystem goods and services (see Box 24.1). Yet the connectivity provided between urban green spaces offers habitats and corridors that help conserve biodiversity (Kong *et al.*, 2010). If GI is proactively planned, developed and maintained it has the potential to guide urban development by providing a framework for economic growth and nature conservation (Walmsley, 2006; Schrijnen, 2000; Tzoulas *et al.*, 2007). Indeed, an Institution of Civil Engineers (ICE) publication (Mell, 2009) suggested that GI can play a pivotal role in urban renaissance by providing a complementary green matrix of spaces that offer multilevel benefits for human populations.

Water Resources in the Built Environment: Management Issues and Solutions, First Edition.
Edited by Colin A. Booth and Susanne M. Charlesworth.

Box 24.1 Ecosystem Services

The term 'Ecosystem Service' refers to the delivery, provision, protection or maintenance of goods and benefits that humans obtain from ecosystem functions (Millennium Assessment, 2003; Bolund and Hunhammar, 1999). Ecosystem Services are defined by the type of benefits (goods and services) provided to humans by the environment – these are categorised as: Supporting, Provisioning, Regulating and Cultural services.

In the United Kingdom, the National Ecosystem Assessment (UKNEA, 2011) has established the current state of understanding regarding these benefits and how they relate to broad habitats in the United Kingdom. The UKNEA includes a chapter on Urban Habitats (Chapter 10 of UKNEA, 2011), which has a strong focus on urban blue and green spaces, the current state of provision, how these spaces are organised and how they contribute to human health and well-being, biodiversity and urban development. UKNEA is the first comprehensive attempt in the United Kingdom to value a wide range of ecosystem services and presents many important key findings and identifies many gaps in our knowledge associated with the provision of ecosystem services.

Some UKNEA (Urban Chapter) Key Findings

Access to Urban greenspace is essential for good mental and physical health, childhood development, social cohesion and other important cultural services. More than 6.8% of the UK's land area is now classified as 'urban', with more than 10% of England, 1.9% of Scotland, 3.6% of Northern Ireland and 4.1% of Wales contributing to this habitat. About 80% of the population reside in these areas, where the amount of mean accessible greenspace is 2 hectares (ha) per 1000 people in England and 16 ha per 1000 people in Scotland. Deprived areas systematically fare worse in terms of quantity and quality of greenspace.

Urban ecosystem services could be significantly enhanced to improve climate mitigation and adaptation. Temperatures in cities are higher than in rural areas with consequences for human well-being and the environment. London's maximum daytime and nocturnal Urban Heat Intensity can reach 8.0°C and 7.0°C respectively. The process of urbanisation and development alters the natural energy balance, mainly due to the loss of cooling from vegetated surfaces when they are replaced by impervious materials used in the construction of buildings and roads.

Therefore, it is vitally important to look at the ways in which urban spaces are managed and to identify how urban areas contribute to ecosystem service provision and how GI can help achieve this.

24.2 What is Green Infrastructure?

GI can be considered to include various assets or elements within green and blue spaces in our urban and peri-urban environments. These might include parks, gardens, allotments, public plazas, ponds, wetlands, wildlife sites, living roofs and walls, rain gardens, street trees, urban farms and woodlands, and others. GI is described as an approach that encompasses many disciplines and practices (including planners, developers, architects and landscape architects, utilities providers and regulators, engineers, ecologists, housing providers, transport and green space managers, environmental services, community groups and policy-makers) and that delivers multiple functions, such as storm and surface water management, improvement of air and water quality, energy demand reduction, recreation, sustainable transport, health and amenities – as well as supporting native biodiversity and biodiverse habitats.

There are now many pieces of guidance and policy regarding GI, but because it is such a multidisciplinary approach it is important to first of all establish how GI is defined.

24.2.1 Definitions of Green Infrastructure

(a) European Environment Agency (2011)

'GI is a concept addressing the connectivity of ecosystems, their protection and the provision of ecosystem services, while also addressing mitigation and adaptation to climate change.... Its ultimate aim is contributing to the development of a greener and more sustainable economy by investing in ecosystem-based approaches delivering multiple benefits in addition to technical solutions, and mitigating adverse effects of transport and energy infrastructure.'

(b) Natural England Definition of GI (2009)

'GI is a strategically planned and delivered network comprising the broadest range of high quality green spaces and other environmental features.... GI includes established green spaces and new sites and should thread through and surround the built environment and connect the urban area to its wider rural hinterland.'

(c) Construction Industry Research and Information Association (CIRIA) (2011)

'GI is a strategically planned and delivered network of natural and man-made green (land) and blue (water) spaces that sustain natural processes. It is designed and managed as a multifunctional resource capable of delivering a wide range of environmental and quality of life benefits for society.'

(d) Scottish Government (2011)

'GI is not just about green spaces like parks and open spaces, it also incorporates blue infrastructure including SUDS, wetlands, rivers, canals and their banks.'

(e) Sandström (2002)

'The concept of GI has been introduced to upgrade urban green space systems as a coherent planning entity.'

(f) *Tzoulas et al. (2007)*

'GI can be considered to comprise all natural, semi-natural and artificial networks of multi-functional ecological systems within, around and between urban areas, at all spatial scales.'

(g) *Benedict and McMahon (2002)*

'GI is an interconnected network of green space that conserves natural ecosystem values and functions and provides associated benefits to human populations.'

(h) *Wise (2008)*

'GI is the interconnected network of open spaces and natural areas – such as greenways, wetlands, parks, forest reserves, and native plant vegetation – that naturally manages stormwater, reduces the risk of floods, captures pollution, and improves water quality.'

In summary GI should connect all types of green spaces including grassland, wetland and woodland to other green spaces in a way that will deliver a quality habitat and amenity. This is within, between and beyond the urban landscape and should provide multiple benefits.

In order to understand the scales at which GI is considered and also how those benefits can be planned and managed it is helpful to break down the major components. A hierarchy of green spaces can be defined:

- Green (and blue) spaces – individual areas of vegetated land (and/or water).
- Green networks – a system of connected green spaces in a more strategic and wider context, often including watercourses, wetlands and managed water.
- Green infrastructure – which includes the connections between the blue and green spaces, the green and blue spaces themselves, the quality of these elements and how they are used.

The term Integrated Green Infrastructure (IGI) is now starting to be used to indicate a 'joined-up' approach to understanding, defining and planning urban green and blue spaces, which are multi-functional by design and which provide multiple benefits.

24.3 Benefits of Green Infrastructure

There is general agreement that the ecosystem functions of GI can provide a range of benefits that will help to maintain more sustainable conditions in the urban environment. These functions include water management through the provision of sustainable urban drainage systems (SUDS), which manage water quality, quantity and provide amenity; enhanced flood alleviation; local climatic amelioration; improved air quality; providing conditions for urban biodiversity; provision of green space for public uses; and associated increased health and well-being benefits.

Table 24.1 The use of GI in urban design and planning (adapted from Town and Country Planning Association (TCPA), 2004).

GI already in existence	The built environment and GI	GI in new-build urban developments
• Large areas of parkland in a regional setting, green grid networks and community forests • Greenway corridors, e.g. woodlands and wetlands • Natural green spaces • Formal, smaller parks	• Existing and new street trees • Community green space • Neighbourhood green space • Green roofs associated with the built environment	• GI newly created in new urban developments • Greenway corridors and SUDS

GI is increasingly being incorporated into construction projects in the United Kingdom and in many other countries. This reflects the increased awareness of the need to protect biodiversity, to meet National and International sustainability targets and gain associated environmental, social and economic benefits derived from ecosystem services (CIRIA, 2011).

There are a number of levels at which GI can be used in urban design and planning. These are summarised well by the Town and Country Planning Association in its design guide for biodiversity (Town and Country Planning Association (TCPA), 2004) and are summarised in Table 24.1.

GI can be viewed as simultaneously providing natural resource sinks to assist urban climate control, water management and provide important green networks in an increasingly urbanised landscape (Mell, 2009). Due to the potential of GI to be 'retrofitted' into most environments, Mell (2009) argues that GI can be delivered across diverse urban environments in the United Kingdom to promote sustainable communities and landscape management.

To be successful GI should deliver multiple functions. From a developer's viewpoint and on simple economic terms it makes sense to make efficient use of land. One component of planning-in multiple function delivery that is increasingly applied is the use of green spaces to manage surface water.

24.3.1 Surface Water Management

In many countries, there is an obligation to meet surface water management legislation sustainably, from both a water quantity (flooding and drought) aspect and a water quality (pollution) perspective. In Europe, the EU Water Framework Directive has been in place for several years and provides an overarching piece of legislation that aims to harmonise existing European water policy and to improve water quality in all of Europe's aquatic environments (Kaika and Page, 2003). The Water Framework Directive (European Commission, 2000), affecting 27 countries, marks an important trend towards an ecosystem-based approach for water policy and water resource management (Kallis and Butler, 2001).

In addition to the need to address current surface water management challenges, there is a need to develop more resilient systems that are capable of responding to changing climatic conditions. With the advent of climate change, variations in rainfall seasons are to be expected as well as changing intensities, increasing the likelihood of flooding (Anderson and Bausch, 2006). Using GI to offset this makes sense (Chartered Society of Designers

Figure 24.1 SUDS detention basin and amenity area, Dunfermline (Scotland).

(CSD), 2011) and in order to gain maximum benefit and flexibility of response to changes, the inclusion of SUDS should be incorporated into GI planning.

Indeed, it is the driver of water management that can actually enable a more successful GI. Whether a development is new, or regeneration, there will be a requirement to address surface water. Taking this a step further, drainage assets can be installed that will attenuate and treat surface water, but during times of rainfall, including significant potential flooding events, the GI can also deliver a means of flood routing using, for example, swales and linear wetlands. In turn these flood routing components and surface water conveyance routes may provide an excellent habitat for different types of biodiversity and also allow connection between various areas of green space that might otherwise be remote and disconnected.

The management of surface water through blue and green networks provides an ideal opportunity for delivery of GIs. Many studies have now been published that indicate some of the surface water management benefits that enhanced green space and GI provision could be expected to deliver. For example, a study by the University of Manchester revealed that by increasing the green space in the residential areas studied in Greater Manchester by 10%, runoff would be reduced by 4.9% for the highest future rainfall scenarios considered by the study. In addition, placing tree cover in this green space would reduce runoff by 5.7% (Gill *et al.*, 2007).

It is becoming common practice now, especially in new developments, to install measures that will provide attenuation of surface water runoff and/or water quality treatment. The requirement in many countries is to deliver this through SUDS, taking account of the associated philosophy, to mimic natural drainage. This driver can then begin to incorporate the delivery of GI (Figure 24.1).

Box 24.2 The 'Pipe-Free' Network

An approach to mimic natural drainage would be to assume a drainage network without pipes, or the 'pipe-free' network. This mindset encourages designers to avoid routing surface water underground. This has several benefits:

- Construction costs are lower.
- Conveyance routes will have a higher capacity for flood storage.
- Inappropriate and illegal connections are easily identified.
- Hydraulic failure, either infiltration or exfiltration is readily identified.
- Additional water quality treatment occurs through natural treatment processes.
- A useful habitat is created.

Figure 24.1 shows a basin in Dunfermline which can store surface water generated by the surrounding residential development during wet periods and can be used as a play area for local children during dry periods. A variety of appropriate plantings have been added to the edges of the detention basin to increase the vegetation structure and biodiversity potential of the site. The detention basin is part of a regional SUDS network that links the blue and green features and functions of this site with other parts of the development and surrounding area.

Delivering surface water management for urban areas in a way that mimics nature has been described as a 'pipe-free' network approach. This concept is further explored in Box 24.2.

The multiuse of land, driven by the requirement to deliver surface water management is therefore a driver for GI. Indeed, many programmes are now in place to remove culverted watercourses (watercourses that have been covered over). This process is sometimes referred to as 'daylighting'. The Chartered Institution of Water and Environmental Management have a policy position statement of promoting de-culverting (Chartered Institution of Water and Environmental Management (CIWEM), 2007); regulators also discourage the construction of new culverts and generally have a presumption against culverting (Environment Agency (EA), 2010; Scottish Environment Protection Agency (SEPA), 2006). Where redevelopment takes place and where culverts are already in place daylighting is often sought for the new development. The following section gives a case study example of a pipe-free development in Central Scotland.

24.3.2 Access Networks

Where such water conveyance routes that form part of GI exist, they can provide access routes for people as well as wildlife. Green transport routes, such as cycleways and walking routes, can run alongside linear landscape elements such as streams, rivers, canals and swales and through green spaces. These routes provide both opportunities for sustainable transport (e.g. cycle lanes) and often distance the users from some of the negative aspects of road traffic (noise, exhaust fumes, stop–start progress). However, the use of GI elements in the conventional transport network is also very important, indeed integration with transportation plans is a common element of many GI programmes. By designing vegetated

Case Study: J4M8

An excellent example of the pipe-free network is at the 'J4M8' Distribution Park in Central Scotland. Its title comes from the adjacent junction (number 4) of the M8 motorway midway between Glasgow and Edinburgh, in Scotland.

This site is in development and comprises several large distribution depots with additional plots for more as opportunities arise. The site was planned prior to any legislation requiring SUDS; however, the planning process required water quality measures in the form of SUDS to be installed. Strident steps were taken as the receiving watercourse is the River Almond in West Lothian, which until fairly recently had the dubious record of being the most polluted river in Scotland. Progress had been made to remove and improve poorly treated sewage discharges into the river with significant recovery taking place.

Under the Water Framework Directive (European Commission (EC), 2000) improvements must be made to the water environment, but importantly there must be 'no deterioration' of existing watercourses. The construction of a large distribution park with many heavy vehicle movements and the risk of spillage of highly polluting material, and thereby deterioration, should not be allowed to go ahead without preventative measures in place.

Agreement was reached between the developer and regulator to see that SUDS were installed. Each plot and therefore each occupier presently has in-curtilage water management assets; the first level of treatment (ponds, permeable paving systems and basins) eventually runs to one of two regional ponds serving the whole park – the third level of treatment.

The connection between the local and regional facilities is achieved with a network of swales and linear wetlands alongside the service roads and paths in the site (see Figure 24.2). The only pipes in the system are at road crossings at three locations, leaving the rest of the two

Figure 24.2 Case Study J4M8 (junction four (J4) of the M8 motorway) (UK).

kilometres of open conveyance routes to serve the distribution park – the second level of treatment.

The linear wetland provides water quality treatment and acts as a green corridor connecting otherwise remote elements of green network. It receives flows from a first-level pond in the distance (not seen) plus runoff directly from the parking lot and road. Flows are conveyed for all rainfall events and significant additional storage is provided during larger events. Good quality biodiversity has been identified along its length.

What were shown as swales 'on paper' became highly appealing linear wetlands when constructed. In time these wetlands have become well established. The fear of high maintenance has never been realised and although grass cutting and landscape maintenance does take place the wet areas and floor of the conveyance system receive no attention, with only the easier to reach 'shoulders' of the system receiving grass cutting, which would have to be performed in any case.

This is a highly successful, appealing and excellent quality GI element, which provides several functions and is a robust habitat for different species of flora and fauna. The linear configuration of the drainage network limits certain aspects of biodiversity, but provides resilient connections between the larger elements of ponds and basins at each end of the network.

There is good evidence of water voles in the area and the character of this threatened species means that its instinct to move beyond its current habitat location to adjacent ones is ideally suited to this system by using the cover and habitat of the linear wetlands.

drainage and porous materials into streets, alleys, rights of way and parking lots, cities can increase on-site neighbourhood stormwater capacity (Wise, 2008) whilst also delivering access routes and networks.

In Scotland the Glasgow and Clyde Valley (GCV) Green Network Partnership use an Integrated Green Infrastructure (IGI) approach (GCV Green Network Partnership, 2012) to introduce active travel routes for walking and cycling in a way that is designed to encourage a more active lifestyle for the community. If a development can be configured such that it is easier to walk five minutes to the shops or to school rather than longer by car then it becomes more likely that the residents will become more active and, thereby, healthier. Regional and national governments are promoting this as a means of encouraging increasing activity, this being particularly important in an ageing population and to the growing numbers of people living a more passive life.

By installing active travel routes alongside these green conveyance connections it adds another opportunity for multiuse of land. These active travel routes may have hard-standing surfaces that will need to be drained, logically using the swale or linear wetland, or indeed maybe a softer surface perhaps as the 'shoulder' of the swale or linear wetland.

Various spatial tools are available that can inform and support the making of GI policy. These can be used to help define access networks and opportunities for gaining multiple benefits. One such tool from the GCV Green Network Partnership, called Opportunities Mapping, overlies various spatial GIS datasets to identify four key scenarios:

- the opportunities to improve the existing Green Network resource;
- the priority areas to expand the Green Network in terms of access.;
- the priority areas to expand the Green Network in terms of biodiversity; and
- the major areas of land use change and social need.

By overlying each of these datasets, 'hotspot' areas have been revealed that would provide the greatest benefit to areas that needed GI most. Reports for each of the eight local authorities in the Glasgow and Clyde Valley area provide greater detail and are being used by planners to place GI where it is best needed and can be provided (GCV Green Network Partnership, 2012).

24.3.3 Habitat Networks and Ecosystem Health

Managing urban blue and green space for habitat networks has had mixed success and there are many varied approaches. Sandström *et al.* (2006) conducted in-depth interviews with 18 urban planners in Sweden, with results indicating that legislation was an important driver for green space planning, that they paid attention to new knowledge concerning recreation values and public health, but that biodiversity maintenance was not a high priority. There was general agreement that local governments lacked the necessary resources to plan for biodiversity. However, at the same time there was widespread acknowledgement that the maintenance of biodiversity was of primary concern for the ecological dimension of sustainable development (Heywood, 1995).

In understanding ecological and habitat health, biodiversity can be defined as being composed of compositional, structural and functional elements of an ecosystem (Larsson *et al.*, 2001). These three elements of biodiversity are interlinked. For example, components such as species are linked with structures such as habitat diversity, which in turn is often dependent on the function of natural processes for its renewal. It is the loss of these vegetation structures, and therefore the quality of habitat that they can provide, that poses a limitation to species diversity in urban settings (Fahrig, 2002).

To reduce the isolation of habitat fragments, ecologists and conservation biologists recommend maintaining habitat connectivity by preserving corridors that permit movement of species between remaining habitats and by developing urban green space networks (e.g. Parker et al., 2008; Esbah et al., 2009). Development of these networks is increasingly considered a suitable approach to improve the ecological value of urban green space. Landscape-level habitat connectivity plays an important role in population viability by maintaining gene flow and facilitating migration, dispersal and recolonisation (Kong *et al.*, 2010).

In terms of ecosystem health, elements of GI can preserve and enhance diversity in ecosystems with respect to habitats, species and genes. Diversity is one of the most important indicators of ecosystem health (Rapport *et al.*, 1995). Species-rich heterogeneous habitats are considered to be more resilient than homogeneous habitats (Bengtsson *et al.*, 2005). GI can influence urban and peri–urban ecosystem health by contributing to ecosystem resilience, organisation and vigour (Tzoulas *et al.*, 2007).

At Coventry University (United Kingdom), traditional ornamental flower beds have been replaced with edible fruits and vegetable plants and pollinating plants (Figure 24.3). Students and staff can not only enjoy the aesthetics of the flowering plants in the green space but can also share in the produce. The campus managers have altered their estate management approach to make spaces more directly productive and multifunctional.

24.3.4 Human Health

There is considerable published evidence that green space and GI provision can be linked to human health. Greenspace Scotland (2004) has made strong links between the value of green space for health and the wider issues of environmental planning. Commission for

Figure 24.3 The 'edible campus' at the University of Coventry (UK).

Architecture and the Built Environment (CABE) Space (2004) similarly emphasises the health value of green space in its analysis of how high-quality parks and public spaces create economic, social and environmental value (Goode, 2006). In 2010, The Chartered Society of Designers (2011) undertook an assessment of the contribution parks and green spaces can make to people's lives and suggested that the benefits of green space are very cost-effective in promoting health and well-being, as well as providing a mechanism for increasing community and citizen involvement in volunteering. They went on to suggest that 'imaginative use of green space also provides in many cases an excellent Social Return on Investment whilst increasing individual and community engagement' (CSD, 2011).

The mechanisms by which this link has been established are worth exploring. Douglas (2005) reviewed scientific evidence (from controlled experiments, tests using slides and videos, and attitudinal surveys) regarding the impact of natural areas on mental health and well-being. The results of these were presented as clear evidence that among many sectors of society there are positive benefits to be gained from both active and passive involvement with natural areas in towns and cities. Regular access to restorative, natural environments can halt or slow processes that negatively affect mental and physical health.

There are social implications to these linkages as well as health implications. Populations exposed to greener environments also enjoy lower levels of income deprivation related health inequality. Physical environments that promote good health may be important in the fight to reduce socioeconomic health inequalities (Mitchell and Popham, 2008). It would therefore seem that proximity and exposure to green elements of the landscape can be demonstrated to promote health as well as cultural, psychological and other nonmaterial benefits. Humans obtain these benefits from contact with these ecosystems, which are of particular importance in urban settings (Butler and Olouch-Kosura, 2006). The link to benefits that are not related to physical activity and health is important as some authors have pointed out that it can be difficult to make direct links between the amount of green space in the living environment and levels of physical activity. Furthermore, the amount of physical activity undertaken in greener living environments does not directly explain the relationship between green space and health (Mass *et al.*, 2008).

In the United Kingdom, urban green space is seen as having significant health benefits. Parks and green spaces may foster active lifestyles that combat obesity-related diseases and premature death. However, such uses and benefits are available and availed upon unequally. From research in the United Kingdom (urban parks) and in the United States (urban national parks) provision of these spaces was designed to bring nature and recreational opportunities to socioeconomically disadvantaged communities, but the extent to which this has been achieved has been questioned recently by Byrne and Wolch (2009).

A major challenge for health equity is to work with communities not only to find out what they want and then provide it but to enable them to take control and provide their own solutions. Communities need to be involved in the delivery of services, behaviour change initiatives and solutions, as well as in their design. This enablement concept is the ethos of co-production (New Economics Foundation, 2006). Therefore, how is it possible to ensure that the design, location and management of open space in urban developments can facilitate behavioural change, and hence enhance health and well-being, particularly in disadvantaged communities? Communication strategies are required to ensure that the design and use of open space go beyond merely raising awareness of the positive interaction between use of green space and health benefits but result in a shared understanding of the key issues and measurable behaviour change as evidenced by the use of the space.

Tzoulas *et al.* (2007) have attempted to formulate a conceptual framework of associations between urban green space and ecosystem and human health. They examined possible contributions of GI to both ecosystem and human health, finding many dynamic factors and complex interactions affecting ecosystem health and human health in urban areas (Tzoulas *et al.*, 2007). As with many urban space/human health interactions, the links are not always clear and direct correlations can be difficult to make, but there is an increasing weight of evidence that indicates the important role that GI plays in helping to deliver health and well-being benefits.

24.4 Barriers to Implementation of GI

One of the key barriers for GI provision is that the delivery of multifunctional spaces and places falls within the remit of multiple organisations, authorities, regulators and land owners. Climate, water management, biodiversity and health each has a part to play in terms of ecosystem service provision. Often these services are not dealt with adequately in the planning process, because responsibilities for management or provision of these lie with multiple parties, and where they are, the linkages are frequently not well defined or recognised. More examples of integrated guidance are now becoming available. In a report to the Royal Commission on Environmental Pollution, Goode (2006) concluded that cross–cutting guidance is needed, based firmly on the multifunctional green space network and ensuring that it provides the necessary advice for sustainability. Similarly in the case of sustainable surface water management and SUDS, guidance needs to include all aspects of these systems and the potential benefits they can provide. At other levels there are clear links between green roofs, SUDS and biodiversity, which demonstrate the need for integrated guidance (Goode, 2006).

Often convention is maintained in the approaches taken and there can be resistance to change in the way the environment is managed. To move convention and apply change generally needs one of two things: an incentive or benefit, which may need to be significant to overcome any risk, and/or a requirement (obligation), where regulation and enforcement need to be introduced.

Several key areas can be defined that can contribute to GI delivery. These are planning, shared guidance, economics and procurement.

24.4.1 Planning

Planning legislation and approaches are often cited as the problem or the solution to the introduction of new means of environmental management. An open-minded and willing planner will still need convincing that GI is something that is essential to the success of a place. Planners cannot be experts in all aspects of their sector and there is a need to share knowledge with other technical departments in any planning authority (Geldof and Stahre, 2006). GI is a tool that spans several different disciplines and requires multiple stakeholder participation for successful delivery.

24.4.2 Shared Guidance

Proposals can often fail because separate parts of the planning and approval system have different or conflicting guidance or policies. Often proposals have been passed by a development control officer only to fail when it reaches the roads engineer or vice versa. Shared guidance across departments could reduce this problem and ameliorate, to some extent, this barrier. One such example is Glasgow City Council's *Guidance for New Residential Areas* (Glasgow City Council, 2012) which was co-written by planning officers and roads engineers, and will be of great benefit to developers in the area to make proposals that are acceptable to both the engineer and the planner.

24.4.3 Economics

The cost of GI needs to be understood. All too often what starts off as a comprehensive green network in a development becomes compromised and constrained due to the perception of higher cost. This is both capital cost and maintenance costs, but the economic benefit that GI can bring are still not always appreciated. With current increasing emphasis on the need to recognise, value and provide ecosystem services in all habitats, including urban settings, it is possible that GI that can deliver multiple ecosystem goods and benefits will receive greater recognition and importance.

24.4.4 Procurement

Through the tendering process all too often a brief will be too constrained for any attempt by designers to consider GI. It is therefore essential that GI is fully considered during the procurement process. For example, the tendering for a master plan should state clearly that GI must become part of the other essential infrastructure components such as transport, waste, energy and water.

24.5 The Value of GI

In order to facilitate the delivery of GI, and also to align the assessment of environmental value with ecosystem service value, several studies have investigated mechanisms for determining the perceived or functional value of green spaces and GI. Rouwendal and Van Der Straaten (2008) investigated the costs and benefits of providing open space in cities by examining house prices in relation to closeness to public green space. The study found that willingess to pay goes up with income but not as steeply as it does for floor space; the quality of the green space did not seem to be a factor in determining value. The study was carried out in the Netherlands where there is a scarcity of residential land and as such the price for such land is high and local government puts restrictions on lot sizes. Luttik (2000) also investigated the impact that a view of open space had on house prices. The study was of a limited number of relatively small areas and found that a view on to open space increased the value of a house by between 6 and 12%. The proximity to open space can, therefore, be important in terms of house value, and, in a study based in Cardiff, United Kingdom, Orford (1999) found the effect of living close to a public park increased house values for properties that looked onto the park but reduced prices by up to 50%, even just a short distance from the park.

Sander and Polasky (2009) also reported increased house prices with proximity to green space, and with views of green space. The larger the area of green space views, the higher the price. An interesting point presented by this work is that a preference was found for views with fewer land cover types, with those including greater areas of water or grassy land cover preferred. Whilst this preference is helpful in terms of encouraging more space for water management in urban settings, it also indicates that diversity of landcover types, which is good for ecological health and climate change adaptation, is not necessarily seen as desirable.

Many other studies have been conducted on related topics such as willingness to pay for urban conservation in Hong Kong (Lo and Jim, 2010). Reasons for willingness to pay ranked from (1) air purification, (2) amenity and (3) recreation/social function, with the lowest ranking being to benefit members of the community. In another study, Chen, Bao and Zhu (2006) assessed the willingness of the public to pay to conserve urban green space (this time in Hangzhou City, China). The results showed that willingness to pay is directly related to gender, income level and home ownership status, but not to age or educational level. The perceived benefits ranked most highly were 'aesthetic/visual' and 'gives shade, reducing glare and energy consumption', with other benefits of 'provides recreation places', 'attracts birds and other wildlife', 'enhances city/urban climate', 'increases privacy' and 'increases property value' also being mentioned. It was found that 75% of all respondents in their study were willing to pay to conserve urban green space.

24.6 Conclusions

This chapter has illustrated that the real strength of GI lies in the multiple benefits it can provide. Well-designed and planned GI can deliver not only surface water management objectives but also other environmental, social, economic and cultural benefits. Current policy and legislative drivers support the implementation of GI in new developments and in the redevelopment of already developed areas. GI provision can deliver a 'joined-up' network of measures incorporating green and blue spaces that can contribute to climate adaptation, to transitioning of cities to more water-sensitive places and to the economic development and health and well-being of urban places.

References

Anderson, J. and Bausch, C. (2006) *Climate Change and Natural Disasters: Scientific Evidence of a Possible Relation between Recent Natural Disasters and Climate Change*. Policy Brief for the EP Environment Committee, IP/A/ENVI/FWC/2005–35, Brief Number 02a/2006.

Benedict, M.A. and McMahon, E.T. (2002) GI: smart conservation for the 21st century. *Renewable Resources Journal*, **20**, 12–17.

Bengtsson, J., Ahnström, J. and Weibull, A.-C. (2005) The effects of organic agriculture on biodiversity and abundance: a meta-analysis. *Journal of Applied Ecology*, **42**, 261–269.

Bolund, P. and Hunhammar, S. (1999) Ecosystem services in urban areas. *Ecological Economics*, **29**, 293–301.

Butler, C. and Olouch-Kosura, W. (2006) Linking future ecosystem services and future human well-being. *Ecology and Society*, **11**, 30–46.

Byrne, J. and Wolch, J. (2009) Nature, race and parks: past research and future directions for geographic research. *Progress in Human Geography*, **33**, 743–765.

Chartered Institution of Water and Environmental Management (CIWEM) (2007) *De-culverting of Watercourses*. Policy Position Statement. Available at: http://www.ciwem.org/policy-and-international/policy-position-statements/de-culverting-of-watercourses.aspx.

Chartered Society of Designers (CSD) (2011) *Understanding the Contribution Parks and Green Spaces Can Make to People's Lives*. Chartered Society of Designers. Available at: http://csd.org.uk/uploadedfiles/files/value_of_green_space_report.pdf.

Chen, B., Bao, Z. and Zhu, Z. (2006) Assessing the willingness of the public to pay to conserve urban green space: the Hangzhou City, China, case. *Journal of Environmental Health*, **69**, 26–30.

Commission for Architecture and the Built Environment (CABE) Space (2004) *The Value of Public Space: How High Quality Parks and Public Spaces Create Economic, Social and Environmental Value*. Commission for Architecture and the Built Environment, London.

Construction Industry Research and Information Association (CIRIA) (2011) *Delivering Biodiversity Benefits through Green Infrastructure* (Authors Dale, K., Thomson, C., Kelly, J., Hay, D. and MacDougall, K. CIRIA C711 | 978-0-86017-713-5 | PB | 2011.

Douglas, I. (2005) *Urban Greenspace and Mental Health*. Discussion Paper of the UKMAB Urban Forum.

Environment Agency (EA) (2010). *Fluvial Design Guide*, Chapter **8**, Section 8.6.1. Environment Agency, London.

Esbah, H., Cook, E. and Ewan, J. (2009) Effects of increasing urbanization on the ecological integrity of open space preserves. *Environmental Management*, **43**, 846–862.

European Commission (2000) *Directive 2000/60/EC of the European Parliament and of the Council of 23rd October 2000: Establishing a Framework for Community Action in the Field of Water Policy*. Official Journal 22 December L 327/1. European Commission, Brussels.

European Environment Agency (EEA) (2011) *GI and Territorial Cohesion; The Concept of GI and Its Integration into Policies Using Monitoring Systems*. EEA Technical Report No. 18/2011. European Environment Agency, Copenhagen/Luxembourg.

Fahrig, L. (2002) Effect of habitat fragmentation on the extinction threshold: a synthesis. *Ecological Applications*, **12**, 346–353.

GCV Green Network Partnership (2012) *Integrated GI*. Available at: http://www.gcvgreennetwork.gov.uk/Integrating-Green-Infrastructure.html.

Geldof, G.D. and Stahre, P. (2006) *On the road to a new stormwater planning approach: from Model A to Model B*. Water Practice & Technology, IWA Publishing.

Gill, S.E., Handley, J.F., Ennos, A.R. and Pauleit, S. (2007). Adapting cities for climate change: the role of GI. *Built Environment*, **33**, 115–133.

Glasgow City Council (2012) *Glasgow City Council Design Guide: New Residential Areas*. Available at: http://www.glasgowconsult.co.uk/glasgow/KMS/dmart.aspx?strTab=PublicDMartCompleted&PageContext=PublicDMart&PageType=item&DMartId=666&breadcrumb_pc=PublicDMartCompleted&breadcrumb_pg=search&breadcrumb_pn=dmart.aspx&filter_Status=2.

Goode, D. (2006) *GI*. Report to the Royal Commission on Environmental Pollution, March 2006.

Greenspace Scotland (2004) *Making the Links – Greenspace and the Partnership Agreement*, October 2004.

Heywood, V.H. (1995) *Global Biodiversity Assessment*. CIHEAM – Options Mediterraneennes. Available at: http://ressources.ciheam.org/om/pdf/c23/CI011056.pdf.

Kaika, M. and Page, B. (2003) The EU Water Framework Directive: Part 1. European policy-making and the changing topography of lobbying. *European Environment*, **13**, 314–327.

Kallis, G. and Butler, D. (2001) The EU Water Framework Directive: measures and implications. *Water Policy*, **3**, 125–142.

Kong, F., Yin, H., Nakagoshi, N. and Zong, Y. (2010) Urban green space network development for biodiversity conservation: identification based on graph theory and gravity modeling. *Landscape and Urban Planning*, **95**, 16–27.

Larsson, T.-B., Angelstam, P., Balent, G. *et al.* (2001) Biodiversity evaluation tools for European Forests. *Ecological Bulletins*, **50**, 171–221.

Lo, A.Y. and Jim, C.Y. (2010) Willingness of residents to pay and motives for conservation of urban green spaces in the compact city of Hong Kong. *Urban Forestry and Urban Greening*, **9**, 113–120.

Luttik, J. (2000) The value of trees and open space as reflected by house prices in the Netherlands. *Landscape and Urban Planning*, **48**, 161–167.

Maas, J., Verheij, R.A., Spreeuwenberg, P. and Groenewegen, P.P. (2008) Physical activity as a possible mechanism behind the relationship between green space and health: a multilevel analysis. *BMC Public Health*, **8** (206), 1–13.

Mell, I.C. (2009) Can GI promote urban sustainability? *Proceedings of the ICE – Engineering Sustainability*, **162**, 23–34.

Millennium Assessment (2003). *Ecosystems and Human Well-Being: A Framework for Assessment*. Millennium Ecosystem Assessment Series. Island Press, Washington, DC.

Mitchell, R. and Popham, F. (2008) Effect of exposure to natural environment on health inequalities: an observational population study. *The Lancet*, **372**, 1655–1660.

Natural England (2009) *GI Guidance*. Natural England, Sheffield.

New Economics Foundation (2006) Available at: http://www.neweconomics.org/.

Orford, S. (1999) *Valuing the Built Environment, GIS and House Price Analysis*. Ashgate.

Parker, K., Heada, L., Chisholm, L.A. and Feneley, N. (2008) Conceptual model of ecological connectivity in the Shellharbour Local Government Area, New South Wales, Australia. *Landscape Urban Planning*, **86**, 47–59.

Rapport, D.J., Böhm, G., Buckingham, D., Cairns Jr, J., Costanza, R., Karr, J.R., De Kruijf, H.A.M., Levins, R., McMichael, A.J., Nielsen, N.O. and Whitford, W.J. (1999) Ecosystem health: the concept, the ISEH, and the important tasks ahead. *Ecosystem Health*, **5**, 82–90.

Rouwendal, J. and Van Der Straaten, W. (2008) *The Costs and Benefits of Providing Open Space in Cities*. Tinbergen Institute Discussion Paper No. 2008–001/3, January 2008. Available at: SSRN: http://ssrn.com/abstract=1085202.

Sander, H.A. and Polasky, S. (2009) The value of views and open space: estimates from a hedonic pricing model for Ramsey County, Minnesota, USA. *Land Use Policy*, **23**, 837–845.

Sandström, U.G., Angelstam, P. and Khakee, A. (2006) Urban comprehensive planning – identifying barriers for the maintenance of functional habitat networks. *Landscape and Urban Planning*, **75**, 43–57.

Schrijnen, P.M. (2000) Infrastructure networks and red–green patterns in city regions. *Landscape Urban Planning*, **48**, 191–204.

Scottish Government (2011) *GI; Design and Placemaking*. Scottish Government, Edinburgh.

Scottish Environment Protection Agency (SEPA) (2006) *SEPA Position Statement to Support the Implementation of the Water Environment (Controlled Activities) (Scotland) Regulations 2005: Culverting of Watercourses*.

Town and Country Planning Association (TCPA) (2004) *Biodiversity by Design*. A Guide for Sustainable Communities. Town and Country Planning Association.

Tzoulas, K., Korpela, K., Venn, S., Ylipelkonen, V., Kazmierczak, A., Niemela, J. and James, P. (2007) Promoting ecosystem and human health in urban areas using GI: A Literature Review. *Landscape and Urban Planning*, **81**, 167–178.

UK National Ecosystem Assessment (UKNEA) (2011) *The UK National Ecosystem Assessment Technical Report*. UNEP–WCMC, Cambridge.

Walmsley, A. (2006) Greenways: multiplying and diversifying in the 21st century. *Landscape Urban Planning*, **76**, 252–290.

Walsh, C.J., Fletcher, T.D. and Ladson, A.R. (2005) Stream restoration in urban catchments through redesigning stormwater systems: looking to the catchment to save the stream. *Journal of the North American Benthological Society*, **24**, 690–705.

Wise, S. (2008) *GI Rising – Best Practices in Stormwater Management*. American Association of Planners, August/September 2008.

25

Constructed Wetlands for Wastewater Management

Kate V. Heal

25.1 Introduction

A constructed wetland is 'an artificial wetland engineered to achieve biological and physicochemical improvement in the environment' (Nuttall *et al.*, 1997). Although constructed wetlands are sometimes referred to as 'reedbeds', the term reedbed refers specifically to constructed wetlands in which the dominant plants are reeds. Throughout this chapter the term 'constructed wetland' is used to describe all types of constructed wetlands, including reedbeds. Constructed wetlands for the management of different types of wastewater can take various forms but all comprise water, vegetation (emergent and/or floating) and substrates that are saturated for much of the time.

Whilst there is a long history of wastewater improvement by passage through natural wetlands (see Kadlec and Wallace, 2009, for examples and references), today only constructed wetlands should be used for wastewater management in order to protect existing ecosystems. Constructed wetlands have been used increasingly from the 1970s onwards for managing wastewater in the built environment, initially in Europe, North America, Australasia and north-east Asia, but now also worldwide. An example of this exponential increase is provided by the database of the Constructed Wetlands Association, which charted an increase in the number of these wetlands in the United Kingdom from 154 in 1996 to 1012 in 2007 (Cooper, 2009). The types of wastewaters managed in constructed wetlands range from point source pollution (e.g. industrial and domestic sewage effluent) to nonpoint source pollution (e.g. stormwater runoff from relatively impermeable urban surfaces). The sizes of constructed wetlands used in the built environment also vary, from ~10 m^2 for treatment of domestic sewage from a single dwelling (e.g. a compact vertical flow reedbed treating sewage from two households comprising 8 residents is described by Weedon, 2003) to >1000 m^2 wetlands for tertiary treatment at wastewater treatment works (WWTW) for hundreds to thousands of homes (e.g. constructed reed beds treating

Water Resources in the Built Environment: Management Issues and Solutions, First Edition.
Edited by Colin A. Booth and Susanne M. Charlesworth.

up to 2000 population equivalents of sewage per day are described by Green and Upton, 1994).

The increasing use of constructed wetlands for managing wastewater in the built environment can be attributed to many drivers (Rousseau *et al.*, 2008), which include:

- Lower construction and maintenance costs compared to more highly engineered wastewater infrastructure (see Chapter 26).
- Smaller carbon and nitrogen footprints (Chen *et al.*, 2011; Fuchs *et al.*, 2011) compared to more highly engineered wastewater infrastructure.
- Greater sustainability and resilience of constructed wetlands compared to more highly engineered wastewater infrastructure.
- Provision by constructed wetlands of high-quality water required by increasingly stringent standards and legislation (see Chapter 2).
- In some situations constructed wetlands provide other ecosystem services, such as biodiversity and amenity values (Yang *et al.*, 2008), reduction in flood risk (see Chapter 17), and high-quality water for recycling, thus reducing water demand (e.g. Radcliffe, 2010).
- Existence of detailed design guidance for constructed wetlands (e.g. Kadlec *et al.*, 2000; Woods-Ballard *et al.*, 2007), based on 30 years of experience in researching, designing and managing constructed wetlands, so that constructed wetlands are now regarded as 'conventional' treatment.

This chapter presents an overview of the use of constructed wetlands for wastewater management in the built and urban environments. The term 'wastewater' is used broadly to include not only domestic and municipal wastewaters (sewage) but also contaminated waters generated by urban stormwater runoff and other diffuse pollution sources. Because of the large volume of detailed literature already published on this topic (e.g. Kadlec and Wallace, 2009), the chapter does not aim to be fully comprehensive. References to other publications will enable topics to be explored in more detail if desired. In the first section of the chapter, the treatment processes occurring in constructed wetlands are outlined, because it is important to understand these in order to select the appropriate constructed wetland type and optimise its design, construction and maintenance, which are then described in the following sections. The final section of the chapter discusses the costs of constructed wetlands for wastewater management, before commenting on the future use of constructed wetlands in the built environment.

25.2 Treatment Processes in Constructed Wetlands

Constructed wetlands are used to treat a wide range of contaminants that may occur in runoff and aquatic discharges from built environments, including: total suspended solids, biochemical oxygen demand (BOD) and chemical oxygen demand (COD), hydrocarbons (e.g. oils and polycyclic aromatic hydrocarbons (PAHs)), potentially toxic metals, nutrients (nitrate, ammonium, phosphate), pesticides, pathogens and endocrine disruptors (e.g. pharmaceuticals and personal care products; Onesios *et al.*, 2009). Constructed wetlands are also being increasingly deployed in the management of sludge from wastewater treatment (Uggetti *et al.*, 2010).

Knowledge of the treatment processes in constructed wetlands is necessary in order to design wetlands that will provide the appropriate level of treatment for the contaminants of concern. Treatment of contaminants from wastewater in constructed wetlands occurs

Table 25.1 The predominant removal processes for different contaminants in a constructed wetland (adapted from Nuttall *et al.*, 1997, and Woods-Ballard *et al.*, 2007).

Contaminant	Removal processes
BOD, COD	Oxidation, plant uptake, filtration, sedimentation, biodegradation
Suspended solids	Filtration, sedimentation
Nitrogen	Sorption, assimilation/uptake by plants, nitrification, denitrification, anaerobic ammonia oxidation (Anammox)
Phosphorus	Sorption, plant/microbial uptake, precipitation
Hydrocarbons, PAHs	Sorption, biodegradation, photolysis, filtration, volatilisation
Pesticides	Biodegradation, sorption, volatilisation
Potentially toxic metals (e.g. chromium, lead, zinc)	Sorption, cation exchange, bioaccumulation/uptake by plants, oxidation, precipitation, sedimentation
Sulphur	Microbial oxidation or reduction
Endocrine disruptors	Sorption, biodegradation
Bacteria and other pathogens	Sorption, predation, die-off, sedimentation, inactivation by UV

through a combination of physical, chemical and biological processes, summarised in Table 25.1.

Constructed wetlands are complex systems with the nature and rate of removal processes occurring within a specific constructed wetland dependent upon the wetland design, the composition and flow rate of the water entering the wetland, and environmental conditions such as temperature, pH, redox potential and water depth. The majority of contaminants are removed in the water column or substrate of constructed wetlands, rather than by plant uptake. Thus, the role of plants in constructed wetlands is primarily to stabilise the substrate and to create the environment that facilitates contaminant removal processes (e.g. provision of surfaces and organic matter for microbial activity, release of oxygen from plant roots to the substrate and water column, and insulation of the wetland surface in cold conditions).

25.3 Different Types and Deployment of Constructed Wetlands and Reedbeds

This chapter focuses on 'offline' constructed wetlands for wastewater management; it does not discuss online, in-stream systems, which have been found to be less effective at treating diffuse water pollution (D'Arcy *et al.*, 2007). The two main types of constructed wetlands for wastewater management are surface flow and subsurface flow wetlands. Typical components of these different systems are illustrated schematically in Figure 25.1, while Figure 25.2 contains images of constructed wetlands applied to manage different types of wastewater.

25.3.1 Surface Flow Constructed Wetlands

Surface flow (or free water surface) wetlands (Figure 25.1a) are characterised by areas of open water and are most similar to natural wetlands, sometimes containing floating and

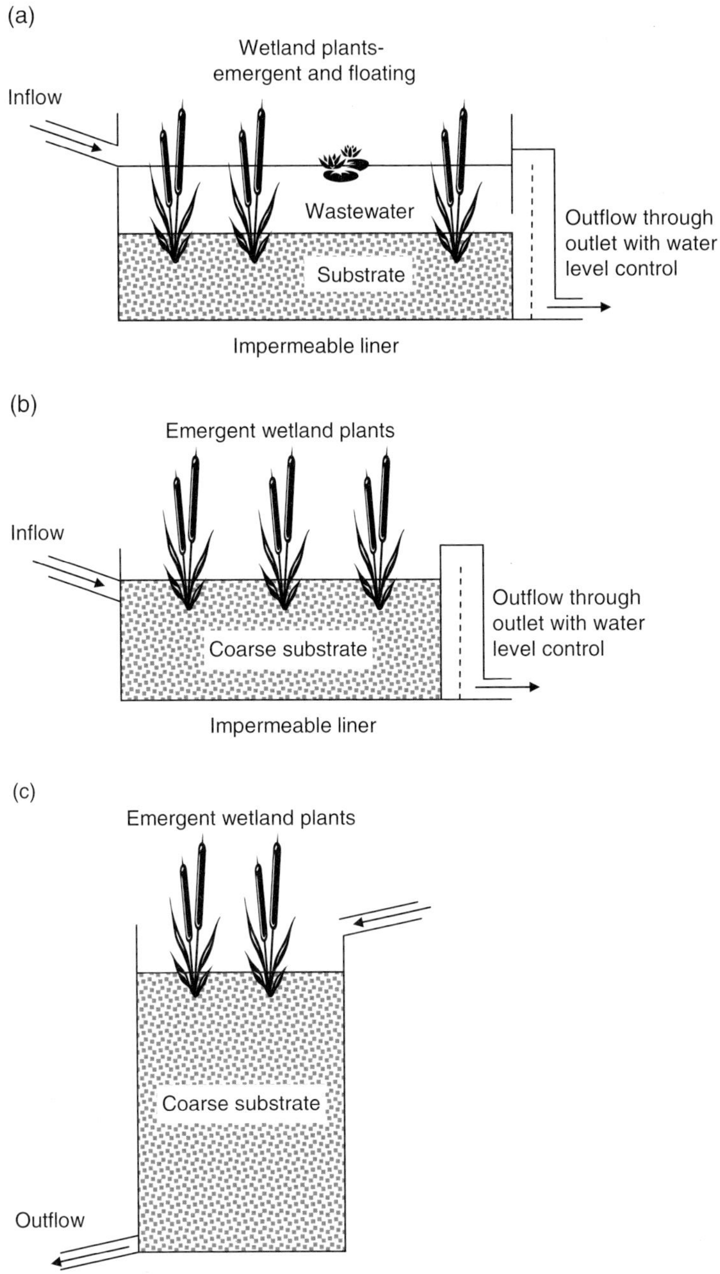

Figure 25.1 Basic elements of different constructed wetland types: (a) a surface flow wetland, (b) a horizontal subsurface flow wetland and (c) a vertical subsurface flow wetland (adapted from Nuttall *et al.*, 1997).

Figure 25.2 Examples of constructed wetlands for wastewater management: (a) combined pond and wetland system for treatment of urban stormwater, Edmonton (Canada); (b) horizontal subsurface flow wetland with two parallel treatment lines treating gully pot effluent, Fife (Scotland); (c) surface flow wetland planted with sedge grass (*Cyperus tegetiformis*) treating wastewater from a brewery near Khon Kaen (Thailand). The treated water is recycled for irrigation of a golf course (photo reproduced by permission of Dr Netnapid Tantemsapya, Khon Kaen University, Thailand); (d) mature constructed wetland, an integral part of an ecological treatment system for managing distillery wastewater, Scotland, and designed by Living Water (photo reproduced by permission of Jane Shields, Living Water Ecosystems Ltd).

submerged plants in addition to emergent plants. They are used to treat a wide range of wastewaters, particularly urban and industrial stormwater, since they can accommodate variable inflows. Surface flow wetlands are also often used to provide tertiary treatment, or 'polishing' of wastewaters after primary and secondary treatment (see Chapter 26). Compared to subsurface flow wetlands, surface flow wetlands typically have lower capital and maintenance costs and more associated benefits, such as biodiversity and amenity values, although the potential for contact between humans, wildlife and wastewater is increased (Kadlec, 2009). The ancillary benefits of surface flow wetlands can be enhanced by thoughtful design, as discussed in Kadlec and Wallace (2009).

25.3.2 Subsurface Flow Constructed Wetlands

In contrast, subsurface flow wetlands are more controlled systems in which flow occurs predominantly through the substrate, either as horizontal subsurface flow or vertical flow (HSSF and VF wetlands respectively). HSSF wetlands are designed so that flow occurs horizontally below the surface of the substrate, which is typically gravel or sand (Figure 25.1b). In VF wetlands, wastewater applied either intermittently or continuously to the surface percolates vertically downwards and drains from the base of the wetland (Figure 25.1c). Subsurface flow wetlands are often used to treat smaller wastewater flows compared to surface flow systems, due to their higher construction and maintenance costs. They are typically used for secondary municipal wastewater treatment for small communities or for specialised wastewaters from industrial processes. For example, due to the higher rate of oxygen transfer in VF wetlands they can be used to treat very concentrated wastewaters, such as landfill leachate and food processing wastewaters with very high ammonium concentrations. Because standing water is minimised in subsurface flow wetlands they are more effective in freezing conditions than surface flow systems (Kadlec, 2009).

25.3.3 Use of Constructed Wetlands in a Treatment Train

Constructed wetlands are not stand-alone treatment plants but should be deployed as part of a treatment train for wastewater management. Normally, pre-treatment of wastewater is necessary before passage through a constructed wetland, particularly for surface flow systems. For example, wastewater is often passed through a sedimentation tank and a lagoon prior to tertiary treatment in a surface flow wetland, whilst an HSSF wetland may be deployed to treat domestic wastewater from a single dwelling after passage through a septic tank. Constructed wetlands can also be deployed in combination with other low- and high-tech methods for wastewater treatment to utilise the specific advantages of each. For example, a VF wetland followed by an HSSF system may be deployed to provide nitrification and denitrification respectively, for removal of different forms of nitrogen from wastewaters.

25.4 Performance of Constructed Wetlands for Wastewater Management

The performance of constructed wetlands for wastewater management is reported from numerous laboratory and field studies. Tables summarising the results from these studies are not included here because of the large volume of literature on this topic. For example, performance data are available in Kadlec and Wallace (2009) for constructed wetlands managing many types of contaminants, mainly from US studies, in the Constructed Wetlands Association's database of constructed wetlands in the United Kingdom, which primarily contains information on wetlands for sewage treatment (Cooper, 2008) and in Onesios *et al.* (2009) regarding removal of pharmaceuticals and personal care products in constructed wetlands.

High performances of constructed wetlands for wastewater treatment have been reported for a wide range of contaminants, in which >50% and often >80% of a contaminant is removed between inlet and outlet, and the outflow meets the required water quality standard (see performance tables in Kadlec and Wallace (2009), Kadlec *et al.* (2000) and Liu *et al.* (2009) for examples). However, in order to interpret the performance data reported for constructed wetlands it is important to understand how performance is calculated.

25.4.1 Percentage Concentration Reduction

The most widely used metric of constructed wetland performance is the concentration reduction, the percentage reduction of the contaminant concentration between the inlet and outlet of the wetland:

$$\text{Percentage concentration reduction} = 100 \times \left(\frac{C_i - C_o}{C_i} \right) \quad (25.1)$$

where C is the contaminant concentration and the subscripts 'i' and 'o' refer to the wetland inlet and outlet respectively. Although percentage contaminant concentration reduction is the most straightforward measure of constructed wetland performance to determine, it does not take account of the flows of water containing the contaminant in and out of the wetland. A further inaccuracy is that contaminant concentrations are often measured in water samples taken from the inlet and outlet almost contemporaneously, so no account is taken of the hydraulic residence time in the wetland. Nevertheless, this procedure is the norm for practical reasons.

25.4.2 Mass Removal of Contaminants

A more informative metric of constructed wetland performance is the mass removal of a contaminant, which takes account of the flows at the wetland inlet and outlet, as well as the contaminant concentration:

$$\text{Mass removal} = Q_i C_i - Q_o C_o \quad (25.2)$$

where Q is the flow at the wetland inlet or outlet. The mass removal is often expressed as a percentage mass removal of the inlet mass and as a rate per wetland area (e.g. mass removal per m^2 wetland area per day). The latter metric is often used in the design of constructed wetlands (see Section 25.5).

25.4.3 Interpreting Performance Data

It is also important to consider the time period over which performance of a constructed wetland has been assessed, since it can vary for a number of reasons. In cold or temperate climates, removal of nitrogen is normally greater in summer than winter because of the effect of temperature on microbial activity. Performance of a constructed wetland is often lower 2–3 years after start-up as the plant and microbial communities become established. Conversely, removal of phosphorus in a constructed wetland is often high initially and then declines several years after start-up as surfaces for phosphorus adsorption (the main removal mechanism) are exhausted (Arias and Brix, 2005).

When assessing constructed wetland performance it is always valuable to compare the contaminant concentration measured at the wetland outlet with the appropriate water quality target, to ensure that it is being met, since protection of receiving waterbodies is the main aim of constructed wetlands for wastewater management. Very high percentage concentration or mass removal reduction figures do not always mean the constructed wetland is achieving its purpose. These figures show only that the wetland itself is performing very well, but do not yield information on whether the water quality target is being achieved.

25.5 Design Considerations for Constructed Wetlands

This section gives an overview of the design process for constructed wetlands for wastewater management. The design process is identical for constructed wetlands in new developments and retrofitted wetlands. Detailed guidance on the design of constructed wetlands for different purposes (e.g. Kadlec *et al.*, 2000; Woods-Ballard *et al.*, 2007) should be used to inform the design of specific constructed wetlands. Whilst process-based models of constructed wetlands exist (e.g. Wynn and Liehr, 2001; Langergraber *et al.*, 2009) they are not normally used for constructed wetland design. The design of constructed wetlands can be divided into two stages: sizing calculations and physical specifications.

25.5.1 Wetland Sizing Calculations

The basic information required to calculate the size of a constructed wetland for treatment of a specific contaminant is:

- The contaminant concentration at the wetland inlet.
- The flow rate at the wetland inlet.
- Other water inputs or outputs from the wetland (normally precipitation input and evapotranspiration and seepage outputs, though seepage will be minimal if synthetic liners are used).
- The water quality target for the wetland outlet (normally contaminant concentration or mass flux rate).

This information is used to calculate constructed wetland size using either a first-order modelling approach or a loading approach. These approaches are outlined below, but a more detailed discussion can be found in Chapters 15, 17 and 20 of Kadlec and Wallace (2009). Since primary treatment of wastewater normally occurs prior to the wetland, estimated wetland inputs should be based upon the pre-treated wastewater. In the first-order modelling approach it is assumed that there is plug flow within the wetland, with no back mixing, and also that there is an exponential decline in contaminant concentration through the wetland to a background concentration. Thus, wetland area can be calculated, taking account of adjustments for units, as

$$A = \frac{Q_i}{k} \ln\left(\frac{C_o - C^*}{C_i - C^*}\right) \tag{25.3}$$

where C_i and Q_i are as defined before, A is the wetland area, k is the rate constant for contaminant removal, C_o is the target contaminant concentration in the wetland outflow and C^* is the irreducible background contaminant concentration in wetland water.

Appropriate values of k and C^* are selected from the literature, depending on the type of wetland that is planned, that is surface or subsurface flow; k values may be adjusted for temperature, especially in cold climates, and C^* values are nonzero for some common pollutants. Useful guidance on the selection of values is available in Kadlec *et al.* (2000, p. 51), in Kadlec and Wallace (2009, p. 723) and in Rousseau *et al.* (2004). Once the provisional wetland area has been calculated, the effect of precipitation, evapotranspiration and seepage (if applicable) on outflow concentrations can be determined and the wetland area adjusted. If more than one contaminant is to be treated, the wetland area required should be calculated separately for each contaminant and the largest area selected.

The alternative approach to calculating the wetland size required, the loading approach, uses areal contaminant removal rates reported in the literature (e.g. Kadlec *et al.*, 2000, p. 83; Kadlec and Wallace, 2009, p. 717). It is most commonly used for sizing horizontal subsurface flow wetlands for treating domestic wastewater. A widely used sizing in this regard is to allow 5 m^2 of horizontal subsurface flow wetland area per population equivalent of primary-treated domestic wastewater (equivalent to a loading of 8 g BOD/m^2 per day) to produce an outlet BOD concentration of <30 mg/l (Cooper *et al.*, 1996).

Specific guidance has developed in North America and Europe for the sizing of constructed wetlands for management of urban stormwater (e.g. see Woods-Ballard *et al.*, 2007) as part of sustainable drainage systems (SUDS) (see Chapter 22). The permanent water storage capacity of SUDS wetlands is sized to contain the water quality treatment volume (V_t) multiplied by a factor. V_t represents the most polluted runoff resulting from rainfall events and is calculated as the volume of runoff generated when 10–20 mm rainfall depth occurs over the impermeable area of the catchment. The scaling factor for V_t used for a SUDS wetland takes values of <1 to 2–4, depending on the presence of upstream treatment devices and the sensitivity of the receiving watercourse.

25.5.2 Wetland Physical Specifications

Once the required wetland area has been calculated the following aspects of the configuration and composition of the wetland need to be determined:

- *Number of cells*. The total area of a constructed wetland is normally subdivided into three or more cells, the sizes of which are dependent on the topography of the land available. Subsurface flow wetlands treating point sources of pollution commonly contain two parallel treatment trains (see Figure 25.2b) to ensure the greatest flexibility in operation and maintenance.
- *Horizontal and vertical dimensions of the cells*. These should be chosen to minimise preferential and turbulent flow within cells and also to allow drainage by gravity between cells, whilst avoiding flooding of cell inlets.
- *Wetland substrate*. The choice of wetland substrate is influenced by the contaminant being treated, the inflow rate and area available. Coarse sand and gravel substrates are often used in wetlands treating wastewater with high BOD and nutrient concentrations and/or where limited land area is available. More organic-rich substrates (e.g. locally available manures or soils) are more typically used in wetlands treating minewater drainage and other wastewaters with high concentrations of potentially toxic metals.
- *Wetland liner*. Decisions about whether to line the wetland and the type of liner depend on the regulatory requirements and the risk of environmental contamination. Some wetlands receiving water at the end of a treatment train may be unlined, whilst wetlands treating landfill leachate are normally double lined to minimise the risk of groundwater contamination. For wetlands treating lightly contaminated water a compacted soil, clay or bentonite liner may be acceptable if its permeability is less than a specified value, but for treatment of more contaminated wastewater a synthetic liner is often required.
- *Inlet and outlet structures*. The structure for introducing wastewater into the wetland should be designed to minimise scouring and facilitate an even distribution of

wastewater across the wetland. The wetland outlet structure should contain the facility to control water levels in the wetland (e.g. use of a weir or an elbow pipe). Nevertheless, where possible, the use of pipes should be minimised to reduce maintenance needs as the result of pipe blockages.

- *Safety and accessibility for monitoring and maintenance.* Safety is of lesser concern for constructed wetlands, compared to ponds, since water depths are typically shallow. However, where the wetland is part of a WWTW or other industrial facility, or where the inflow water is highly contaminated, it may be desirable to restrict access (e.g. through the use of fencing or the planting of impenetrable barrier vegetation). Nevertheless, it is important that appropriate access to the wetland for monitoring and maintenance (see Section 25.6), particularly of any pre-treatment sedimentation area and the wetland inlets and outlets, is included in the design.
- *Planting and biodiversity considerations.* Decisions about the extent of planting and choice of plants depend on the type of constructed wetland and its purpose. Planting is normally recommended in the following situations: where there is a high risk of erosion within the wetland or of invasion by undesirable plant species, the wastewater is highly contaminated or in gravel-based wetlands treating domestic wastewater to avoid substrate clogging. In other situations, minimal planting may be preferable to allow colonisation of the wetland by a wider range of local plant species and to reduce costs. The plants selected for constructed wetlands should be tolerant of the contaminants in the wastewater, able to grow in the wetland environment and be native plants obtained from sustainable sources. The biodiversity of constructed wetlands, particularly surface flow systems, can be maximised by using a number of plant species, having varying water depths and a mixture of vegetated and open water zones.

25.6 Construction and Maintenance Considerations for Constructed Wetlands

The effectiveness of constructed wetlands for wastewater management requires not only good design but also correct implementation of the design during construction. Furthermore, the long-term performance (and cost) of constructed wetlands is dependent on appropriate operation and maintenance. The typical life expectancy of well-designed constructed wetlands for wastewater management is 50 years (Woods-Ballard *et al.*, 2007; Kadlec and Wallace, 2009). Although constructed wetlands normally have lower capital and maintenance costs compared to alternative methods for wastewater management (see Section 25.7), they are not zero maintenance, and maintenance and inspection are still required (Cooper *et al.*, 2005).

In this section the key considerations for construction and maintenance for constructed wetlands for wastewater management are outlined. More detailed generic information about these topics is available elsewhere (e.g. Woods-Ballard *et al.*, 2007; Kadlec and Wallace, 2009), but the exact requirements for construction and maintenance are site-specific.

25.6.1 Construction and Commissioning Considerations

Since the construction of wetlands for wastewater management is not usually undertaken by the designer it should be ensured that the constructors use the materials and dimensions specified in the design. Any changes to those specified in the design should be approved by the designer. A further key consideration during construction is to minimise the impact on nearby waterbodies. The construction of wetlands in temperate and cold climates should be timed to coincide with the growing season to maximise the time for plant establishment during start-up. Careful water level control is also required at this time to facilitate vegetation establishment. Ideally, mature plants should be planted in shallow water and then the water level maintained at 100–200 mm, without allowing the wetland to dry out, in order to minimise colonisation of the wetland by weeds and grasses (Carty *et al.*, 2008). If the wastewater contains high contaminant concentrations it may be advisable to introduce clean water or dilute wastewater into the wetland during start-up.

25.6.2 Operation and Maintenance Considerations

The operation and maintenance of each constructed wetland for wastewater management should be specified in a manual, since these will vary between systems. An example table of contents for such a manual is shown in Table 25.2.

Table 25.2 Typical table of contents for an Operation and Maintenance Manual for a constructed wetland for wastewater management (adapted from Woods-Ballard *et al.*, 2007, and Kadlec and Wallace, 2009).

Introduction
Summary of how the treatment system works, its purpose and how it can be damaged
Start-up procedures
Treatment wetland
Planting
System operation
Inflow and outflow control
Winter operations (if applicable)
Mosquito control (if applicable)
Sampling procedure and monitoring (if applicable)
Daily/weekly operating logs
Identification of areas where certain activities are prohibited
Maintenance
Maintenance plan
Process control problems
Water control problems
Short-circuiting and deep zones
Vegetation
Embankments
Piping
Explanation of the consequences of not carrying out the specified maintenance
Emergency plans and procedures
Hydraulic overload
Process failure
Safety
Appendices
A. Constructed wetland design drawings
B. Operating log
C. Maintenance log

Maintenance activities can be divided into three categories as follows:

1. *Regular maintenance* tasks that are conducted on a frequent and predictable schedule, such as litter removal, grass cutting, control of weeds and invasive plants, and visual inspections of inlets and outlets, water depths, evidence of contaminants and vegetation health, in particular. Apart from in humid tropical environments where plant growth is prolific, harvesting of plants in constructed wetlands is not normally advised for a number of reasons. Plant material contains only a small percentage of the contaminants removed so harvesting will not increase the wetland treatment efficiency, but does generate material that requires disposal (at extra cost). Furthermore, the presence of plant material in the wetland enhances contaminant removal by providing insulation for the substrate and a carbon supply for microorganisms that degrade contaminants.
2. *Occasional maintenance* consisting of tasks that are likely to be required, but on a less frequent and predictable basis. Examples of these tasks include regeneration of clogged surface gravel in subsurface flow wetlands and removal of sediment from the pre-treatment system upstream of the wetland. Assessment of the extent of contamination of the sediment will be required to select the appropriate disposal method for it. An overview of managing sediment from SUDS, including wetlands, is given in Woods-Ballard *et al.* (2007).
3. *Remedial maintenance* tasks that are intermittent and unpredictable (e.g. repair of inlet after an unforeseen event), although the likelihood of occurrence can be minimised by good design.

25.7 Costs of Constructed Wetlands for Wastewater Management

Studies in which the costs of constructed wetlands for wastewater management are compared with alternative systems have demonstrated that constructed wetlands are cost-effective solutions for wastewater management. An extensive literature exists on this subject and a few selected examples of cost comparisons of constructed wetlands treating different wastewaters are outlined here. The total current cost per population equivalent of using constructed wetlands for urban sewage treatment was estimated to be, at the most, 16% and 61% of the costs of six different conventional treatment units (e.g. activated sludges and drying beds) for systems designed to treat 1000 and 10 000 population equivalents respectively (Gratziou *et al.*, 2005). The total capitalised costs (including construction, operation and maintenance costs) of treating shrimp processing wastewater at a location in the United States using a highly engineered dissolved air flotation system were calculated to be over three times those of using a constructed wetland system (Cardoch *et al.*, 2000). A constructed wetland for urban stormwater management in central Scotland was calculated to have construction, average annual maintenance and wholelife costs 10%, 41% and 17% those of an alternative drainage system (storage chambers) (Duffy *et al.*, 2008). Analysis of actual annual expenditure on maintenance of drainage systems at Hopwood Park motorway service area in central England showed the costs of SUDS maintenance were ~60% of those of highly engineered drainage infrastructure (Heal *et al.*, 2009). Detailed information on the typical capital and maintenance costs of constructed wetlands in the United States are available in Kadlec and Wallace (2009) and for SUDS in the United Kingdom in Woods-Ballard *et al.* (2007) and Wilson *et al.* (2010).

25.8 Conclusions

Constructed wetlands may not be the appropriate solution for wastewater management in all situations, for example in densely populated built environments, due to the land area required. However, the increasing use of constructed wetlands for management of all types of wastewater, coupled with the improved guidance and experience gained, means constructed wetlands will be widely deployed in the future throughout the world. Furthermore, constructed wetlands have been shown to be cheaper than more highly engineered wastewater infrastructure and have other benefits, including reduced flood risk, conservation of water resources, smaller carbon and nitrogen footprints, and, in some situations, enhanced biodiversity and amenity values. In an increasingly changeable world, constructed wetlands are expected to provide a more flexible and resilient means of wastewater management compared to more highly engineered solutions. The benefits of constructed wetlands, particularly the lower costs, make them highly suitable for implementing and improving wastewater management in lesser developed countries (Kivaisi, 2001). Nevertheless, the benefits offered by constructed wetlands for wastewater management are not automatic; they are the product of a treatment train approach to wastewater management and good design, operation and maintenance. Constructed wetlands are not stand-alone solutions for wastewater management but can be deployed in combination with other measures in the built environment.

References

Arias, C.A. and Brix, H. (2005) Phosphorus removal in constructed wetlands: can suitable alternative media be identified? *Water Science and Technology*, **51**, 267–273.

Cardoch, L., Day, J.W., Rybczyk, J.M. and Kemp, G.P. (2000) An economic analysis of using wetlands for treatment of shrimp processing wastewater – a case study in Dulac, LA. *Ecological Economics*, **33**, 93–101.

Carty, A., Scholz, M., Heal, K., Gouriveau, F. and Mustafa, A. (2008) The universal design, operation and maintenance guidelines for farm constructed wetlands (FCW) in temperate climates. *Bioresource Technology*, **99**, 6780–6792.

Chen, G.Q., Shao, L., Chen, Z.M., Li, Z., Zhang, B., Chen, H. and Wu, Z. (2011) Low-carbon assessment for ecological wastewater treatment by a constructed wetland in Beijing. *Ecological Engineering*, **37**, 622–628.

Cooper, P. (2008) The Constructed Wetland Association's database of constructed wetland systems in the UK. In: Vymazal, J. (ed.), *Wastewater Treatment, Plant Dynamics and Management in Constructed and Natural Wetlands*. Springer, pp. 149–160.

Cooper, P. (2009) What can we learn from old wetlands? Lessons that have been learned and some that may have been forgotten over the past 20 years. *Desalination*, **246**, 11–26.

Cooper, P.F., Job, J.D., Green, M.B., Shutes, R.B.E. (1996) *Reed Beds and Constructed Wetlands for Wastewater Treatment*. WRc, Swindon, Wiltshire.

Cooper, D., Griffin, P. and Cooper, P. (2005) Factors affecting the longevity of sub-surface horizontal flow systems operating as tertiary treatment for sewage effluent. *Water Science and Technology*, **51**, 127–135.

D'Arcy, B.J., McLean, N., Heal, K.V. and Kay, D. (2007) Riparian wetlands for enhancing the self-purification capacity of streams. *Water Science and Technology*, **56**, 49–57.

Duffy, A., Jefferies, C., Waddell, G., Shanks, G., Blackwood, D. and Watkins, A. (2008) A cost comparison of traditional drainage and SUDS in Scotland. *Water Science and Technology*, **57**, 1451–1459.

Fuchs, V.J., Mihelcic, J.R. and Gierke, J.S. (2011) Life cycle assessment of vertical and horizontal flow constructed wetlands for wastewater treatment considering nitrogen and carbon greenhouse gas emissions. *Water Research*, **45**, 2073–2081.

Gratziou, M., Ekonomou, S. and Tsalkatidou, M. (2005) Cost analysis and evaluation of urban sewage processing units. *Water Science and Technology: Water Supply*, **5**, 155–162.

Green, M.B. and Upton, J. (1994) Constructed reed beds – a cost-effective way to polish waste-water effluents for small communities. *Water Environment Research*, **66**, 188–192.

Heal, K.V., Bray, R., Willingale, S.A.J., Briers, M., Napier, F., Jefferies, C. and Fogg, P. (2009) Medium-term performance and maintenance of SUDS: a case-study of Hopwood Park Motorway Service Area, UK. *Water Science and Technology*, **59**, 2485–2494.

Kadlec, R.H. (2009) Comparison of free water and horizontal subsurface treatment wetlands. *Ecological Engineering*, **35**, 159–174.

Kadlec, R.H. and Wallace, S.D. (2009) *Treatment Wetlands*. Taylor & Francis, Boca Raton, Florida.

Kadlec, R., Knight, R., Vymazal, J., Brix, H., Cooper, P. and Haberl, R. (2000) *Constructed Wetlands for Pollution Control: Processes, Performance, Design and Operation*. IWA Publishing, London.

Kivaisi, A.K. (2001) The potential for constructed wetlands for wastewater treatment and reuse in developing countries: a review. *Ecological Engineering*, **16**, 545–560.

Langergraber, G., Giraldi, D., Mena, J., Meyer, D., Peña, M., Toscano, A., Brovelli, A. and Korkusuz, E.A. (2009) Recent developments in numerical modelling of subsurface flow constructed wetlands. *Science of the Total Environment*, **407**, 3931–3943.

Liu, D., Ge, Y., Chang, J., Peng, C.H., Gu, B.H., Chan, G.Y.S. and Wu, X.F. (2009) Constructed wetlands in China: recent developments and future challenges. *Frontiers in Ecology and the Environment*, **7**, 261–268.

Nuttall, P.M., Boon, A.G. and Rowell, M.R. (1997) *Review of the Design and Management of Constructed Wetlands*. CIRIA, London.

Onesios, K.M., Yu, J.T. and Bouwer, E.J. (2009) Biodegradation and removal of pharmaceuticals and personal care products in treatment systems: a review. *Biodegradation*, **20**, 441–466.

Radcliffe, J.C. (2010) Evolution of water recycling in Australian cities since 2003. *Water Science and Technology*, **62**, 792–802.

Rousseau, D.P.L., Vanrolleghem, P.A. and De Pauw, N. (2004) Model-based design of horizontal subsurface flow constructed treatment wetlands: a review. *Water Research*, **38**, 1484–1493.

Rousseau, D.P.L., Lesage, E., Story, A., Vanrolleghem, P.A. and De Pauw, N. (2008) Constructed wetlands for water reclamation. *Desalination*, **218**, 181–189.

Uggetti, E., Ferrer, I., Llorens, E. and García, J. (2010) Sludge treatment wetlands: a review on the state of the art. *Bioresource Technology*, **101**, 2905–2912.

Weedon, C.M. (2003) Compact vertical flow constructed wetland systems – first two years' performance. *Water Science and Technology*, **48**, 15–23.

Wilson, S., Bray, R., Neesam, S., Bunn, S. and Flanagan, E. (2010) *Sustainable Drainage Cambridge Design and Adoption Guide*. Cambridge City Council, Cambridge.

Woods-Ballard, B., Kellagher, R., Martin, P., Jefferies, C., Bray, R. and Shaffer, P. (2007) *The SUDS Manual*. CIRIA, London.

Wynn, T.M. and Liehr, S.K. (2001) Development of a constructed subsurface-flow wetland simulation model. *Ecological Engineering*, **16**, 519–536.

Yang, W., Chang, J., Xu, B., Peng, C. and Ge, Y. (2008) Ecosystem service value assessment for constructed wetlands: a case study in Hangzhou, China. *Ecological Economics*, **68**, 116–125.

26

Wastewater Treatment Infrastructure and Design

Joseph Akunna and Joanne Bartie

26.1 Introduction

Wastewater management systems usually comprise collection, treatment and disposal works but only treatment is covered in this chapter. Depending on the characteristics of the wastewater and the preferred final disposal routes, wastewater treatment works can consist of a simple single operation or a series of separate operations, involving multiple units. Each treatment operation or a combination of two or more operation can be carried out in a single treatment unit. The use of separate units for each treatment operation is common in large-scale systems treating wastewaters from urban areas or large industrial sources with high volumetric rates of wastewater production. For small-scale systems handling relatively low and highly variable wastewater flows, for example, wastewaters from small populations and from small to medium sized industries, there is the opportunity for two or more treatment operations to be carried out efficiently in single treatment units.

The degree of treatment that wastewater receives depends on the nature and type of its final destination. In the United Kingdom, as with all EU members, various EU Directives, notably the EU Urban Wastewater Treatment Directive (91/271/EEC), Water Framework Directive (2000/60/EC), Bathing Water Directive (2006/7/EC), Shellfish Directive (79/923/EEC), Dangerous Substances Directive (67/548/EEC) and Groundwater Directive (2006/118/EC), provide guide limits for quality of effluents discharged into natural water bodies. The enforcement and regulation of these directives in the United Kingdom is undertaken by the competent authorities in each country, the Environment Agency (EA) in England and Wales, the Scottish Environment Protection Agency (SEPA) in Scotland and the Northern Ireland Environment Agency (NIEA) in Northern Ireland. In establishing the effluent discharge standards, these competent authorities take into account the relevant EU Directives and other factors including current and desired water quality, current and future use of the receiving water body, downstream users and other sources of water pollutants in

Water Resources in the Built Environment: Management Issues and Solutions, First Edition.
Edited by Colin A. Booth and Susanne M. Charlesworth.

the surrounding catchment. Thus, the degree of treatment required before re-use or direct discharge into a receiving water body (or on land) can vary from one location to another for a given wastewater. The regulating authorities issue permits to each source of wastewater and monitor compliance of the wastewater producer. The permit specifies the allowed quality and discharge rate and requires the producer to notify the authorities in the event of significant permit violations.

Whilst the degree of wastewater treatment is determined by a continuing increase in environmental regulations, other factors also play a major role in how investment into wastewater management infrastructure in the United Kingdom is made. These factors have the potential to become barriers and impact on the ability of the water industry to comply with the regulations; however, they also provide opportunities for innovation.

This chapter deals with identification of appropriate technologies for the treatment and disposal of wastewaters from an urban environment, which originates mainly from domestic and agro-food/beverage processing activities. Management of other types of industrial wastewaters are not covered in this chapter.

26.2 Wastewater Treatment Technologies

Proper consideration of the factors discussed in Section 26.3 and knowledge of the characteristics of wastewater are fundamental in the design and operation of appropriate and efficient wastewater treatment infrastructure. The characteristics of wastewater not only depend on the activities responsible for its production, it also depends on the climatic conditions and methods of collection and transportation. For instance, the composition of domestic wastewater transported over a long distance for treatment will undergo significant changes within the sewer, and these changes will be significantly greater in the summer than in the winter months. Seasonal domestic and industrial water use variations can also bring about significant variations in both the quality and quantity of the wastewater. Consequently, wastewaters from combined sewerage systems tend to be more dilute than those from sanitary sewers during certain periods of the year. By the same token, wastewater characteristics from areas of restricted water use will generally contain higher levels of target pollutants than those from areas with abundant and more readily available water supply.

Depending on the nature and use of the receiving water, both the UK and EU regulations may require the reduction of suspended solids, biological oxygen demand (BOD), chemical oxygen demand (COD), nitrogen, phosphorous and total and faecal coliforms. The levels of these pollutants in domestic wastewater range from 180 to 450 mg/l for suspended solids, 200 to 400 mg/l for BOD, 350 to 750 mg/l for COD, 30 to 85 mg/l for nitrogen, 10 to 15 mg/l for total phosphorus and 10^7 to 10^8 and 10^6 to 10^7 MPN (most probable number)/100 ml for total and faecal coliforms respectively (Butler and Davis, 2010). The treatment options for the reduction of these target pollutants can be categorised as preliminary, primary, secondary and advanced.

26.2.1 Preliminary Treatment Infrastructure

26.2.1.1 Gross Solids Removal

Gross solids consist mainly of sanitary refuse and toilet paper, plus other items such as leaves, dead animals and any other debris that may have been washed down the sewer. In a combined sewerage system, the amount of gross solids will also depend on the frequency of precipitation. The greater the intervals between rainfall events the greater the amount of litter on the top soil and deposition in the sewer, and hence the greater the gross solids load in the sewer during a rainfall event.

Gross solids are generally reduced in wastewater treatment plants using screens. Screens can be coarse or fine depending on the minimum particle size of solids to be removed. Coarse or bar screens have clear openings ranging from 6 to 150 mm, while fine screens have openings less than or equal to 6 mm. The materials removed by fine screening are generally greater than those removed by coarse screens, thereby making the former more prone to clogging. Fine screens are thus usually fitted with mechanical cleaning facilities, whilst it is possible to operate coarse screens with only manual cleaning. An alternative to screening is the reduction of the size of solids in the screen channel, using communitors or macerators. Communitors are installed online in the wastewater flow, intercepting and shredding the solids in the wastewater flow stream. These are generally used in small wastewater treatment systems that generally receive relatively low amounts of gross solids. Macerators are used to shred the solids offline and reintroduce them into the downstream flow.

During the 1990s and 2000s, there was a significant change in the use of the technologies described. The change has been brought about as a result of the financial pressures in meeting the EU Urban Wastewater Treatment Directive. One of the key outcomes was increased automation, to reduce the number of staff, and 'footprint' philosophy where many water companies preferred treatment units that occupied less space. Other pressures also include the need to control odour, which is usually associated with wastewater treatment plants. Consequently, compact technologies, such as fine screens, equipped to also reduce the water content of the screenings and ensure the control of odour have become popular in the United Kingdom. In small wastewater treatment plants, fine screens can replace primary treatment (see Section 26.2.2). The use of fine screens has brought about a significant increase in energy use, and hence high energy bills for the water companies.

One of the challenges in the design and operation of wastewater infrastructure in this 21st century is the need to reduce energy consumption and also to reduce greenhouse gas (GHG) emissions. Taking these factors into consideration, two options for the management of gross solids are likely to be trialled by the water industry in the future. The first is a greater use of the screenings, which are largely organic matter, as co-substrates in the anaerobic digestion of wastewater sludges for the production of biogas, a source of bioenergy that can be re-used on site. The second scenario might involve the removal of the screening stage in wastewater treatment plants. Research is currently being undertaken in the possible use of macerators in households to reduce the size of materials sent to the sewer. If this proves acceptable to the public, the screens will no longer be needed. Furthermore, if current campaigns by some water companies encouraging the public to adopt alternative methods of disposing gross solids items such as sanitary towels, toilet papers and condoms (rather than simply flushing through the toilet) are successful, the use of screens in wastewater treatment plants will be significantly reduced in future wastewater treatment infrastructure.

26.2.1.2 Grit Removal

Grit removal is only necessary in the treatment of wastewater produced from catchments served by combined sewerage systems. Grit consists mainly of materials with specific gravities substantially greater than those of putrescible organic solids such as sand, gravel and cinders. Grit materials are generally heavier and have higher settling velocities than suspended organic solids. Road grit and sand are usually removed early on in the treatment process because they can be highly abrasive to mechanical equipment and can form heavy deposits in pipelines, channels and conduits or set hard in sludge hoppers and in the bottoms of digesters. Grit removal methods tend to separate the grits from the less dense organic solids, so that the latter will be subsequently removed in primary sedimentation tanks located downstream. The methods include the constant velocity grit channel and aerated and vortex flow grit chambers. In constant flow grit channels, the wastewater flow is maintained at a constant velocity, normally 0.3 m/s. This velocity permits only the heavier particles, such as grit to settle, whilst putrescible organic particles remain suspended. In aerated grit chambers, air is introduced from the bottom of the chamber at a velocity, referred to as roll velocity, that will prevent lighter organic particles remaining in the flow, whilst the heavier grit particles settle. The vortex chamber employs a vortex flow pattern to separate grit from organic solids.

Both the aerated and vortex grit removal methods, although having a relatively higher operating carbon footprint and energy use, have been popular in the United Kingdom in the last 20 years, simply because they occupy significantly less space than the constant velocity type. These technologies have also been popular due to the relative ease with which grit processing facilities such as grit washing (to reduce the organic content) and drying and odour control can be incorporated into their operation. These advantages over the less energy consuming and high footprint constant velocity channel are deemed to far outweigh their disadvantages, particularly in large wastewater treatment plants where the high amount and quality of processed grit produced may be suitable for re-use as a construction material.

26.2.2 Primary Treatment

Primary treatment is simply a physical process for organic solids reduction. This can be achieved by gravity settling directly or via chemical addition (i.e. using coagulants such as metallic salts and polyelectrolytes). Chemical addition is usually preceded by sedimentation or flotation. In flotation, solids separation is brought about by introducing compressed air into the raw wastewater basin. The released fine air bubbles then attach to the solids and bring them to the surface where they are collected by a skimming operation. The main advantage of flotation over sedimentation is its (i.e. flotation) greater ability to remove smaller sized particles and in a shorter time. Flotation is not commonly used in domestic wastewater treatment, but is common in the treatment of food processing and beverage wastewaters. Chemical enhanced solids reduction can be carried out in one or a combination of the following ways: (a) when no further treatment is necessary before discharge to receiving waters, since up to 90% and 50% of suspended solids and BOD removals can be obtained respectively; (b) to increase the capacity of a primary sedimentation tank without the need for significant investment; and (c) for phosphorus removal by precipitation, which is necessary when discharging treated effluent into waterbodies prone to eutrophication. Chemical treatment can bring about up to 90% total phosphorus removal; this can lead to microbial phosphorus deficiency if preceded by biological treatment.

An efficiently designed and operated sedimentation tank (without chemical enhancement) can achieve 50–70% of the suspended solids and from 25–40% of the BOD (Tchobanoglous *et al.*, 2002). Although the efficiency of solids removal increases with an increase in the hydraulic retention time, sedimentation tanks are usually designed for a minimum retention time of two hours for the peak design wastewater flow.

Primary sedimentation tanks can be circular or rectangular and have four key features: (i) the inlet zone, which is designed to ensure a uniform distribution of incoming wastewater; (ii) the settling zone, where the actual solids settling takes place, with heavier particles settling closer to the inlet and lighter particles further downstream; (iii) the outlet zone, which is designed to reduce the occurrence of turbulent conditions that can re-suspend the solids; and finally (v) the sludge zone, which holds the settled solids. The volume of the sludge zone is dependent on the frequency of removal of the sludge; hence, the greater the desludging frequency, the smaller the volume of the sludge zone.

Primary treatment reduces the amount of BOD in the wastewater and consequently reduces the amount of organic pollutants that can be removed in the subsequent secondary (or biological) treatment. For small wastewater treatment plants, it is sometimes more cost-effective to omit the primary treatment, because in these types of plant the cost of constructing and operating sedimentation tanks and the treatment of the resulting primary sludge sometimes outweighs the contribution of the primary treatment in the overall treatment train. It is, therefore, necessary to carry out a full cost–benefit analysis for plants involving biological treatment to determine whether or not primary treatment is required.

Primary sedimentation is emerging as an important tool for meeting climate change obligations, as it offers opportunities for energy recovery and helps reduce the carbon footprint of the industry (Palmer, 2011). Biogas production from primary sludge is an important source of energy that the UK Water Industry is now exploiting in order to reduce the impact of the increasing energy bills, and at the same time achieve a significant reduction of GHG emissions. It is known primary sludge provides more than twice more biogas per tonne of dry solids than secondary sludge. Hence future designs of primary tanks will aim to improve on solids capture from the current 50–70% of conventional tanks. Sedimentation efficiencies can be enhanced by the use of some proprietary high-rate clarifiers now being marketed by large water companies and the controlled use of chemical coagulating agents. Some water companies in the United Kingdom are already co-digesting sludge with other organic residuals such as an organic fraction of municipal solid wastes and agricultural residues in order to boost biogas production. Trials are ongoing for co-digestion with other unfamiliar organic materials such as marine macro-algae and screenings from the preliminary treatment stage.

26.2.3 Secondary Treatment

Secondary or biological treatment processes in wastewater treatment consist of aerobic biological oxidation or a combination of anaerobic and aerobic processes.

26.2.3.1 Aerobic Processes

In aerobic processes, aerobic microorganisms require oxygen to oxidise the carbonaceous and nitrogenous compounds present in the wastewater to carbon dioxide and nitrate, with the production of new microorganisms. For carbonaceous (BOD) oxidation, up to 60% of the organic carbon is used in the production of new cells. Hence, aerobic processes produce a significant amount of new microorganisms, known as activated sludge, which must be separated from the treated effluent and disposed of. Where both BOD and nitrogenous oxidation are required, the duration (also known as the aeration or hydraulic retention time) the biological oxidation, is longer, because the nitrogenous oxidation is carried out by microorganisms that grow more slowly than those that break down the carbonaceous BOD.

The supply of oxygen for biological oxidation is one of the major causes of capital and operational expenditure in aerobic treatment processes. In fact, the operational energy requirement for aeration facilities can represent up to 65% of the total operational energy need of a standard activated sludge plant. Oxygen can either be introduced artificially, as in an activated sludge process, or by naturally allowing atmospheric oxygen to diffuse into the wastewater treatment system, as in lagoons and constructed wetlands. Treatment efficiencies achievable in the latter are directly proportional to the surface area of the treatment system; that is the greater the surface–air contact area, the greater the potential for oxygen transfer and consequently the greater the treatment efficiency.

Aerobic microorganisms can either be suspended (as in activated sludge processes) or in the form of biofilms on support media such as stones and plastic materials (as in aerated biofilters and trickling filters). For biofilm systems, the method used for an artificial oxygen supply and the surface area of the support media available for microbial growth are some of the key determinants to process efficiency. Plastic media are more efficient than the stone media because they (i.e. plastic media) have a higher surface area/volume ratios. By the same token, artificially aerated systems are more efficient than nonaerated systems, with the latter requiring large footprints to enhance natural oxygen transfer.

For both suspended and biofilm processes, biological solids produced during treatment (referred to as biomass) must be separated from the treated wastewater before discharge. The solid separation method employed depends on the process type and the quality of treated effluent needed. Aerobic lagoons provide both biological oxidation and solid separation, whilst other processes may require separate sedimentation tanks for solids separation (Figure 26.1).

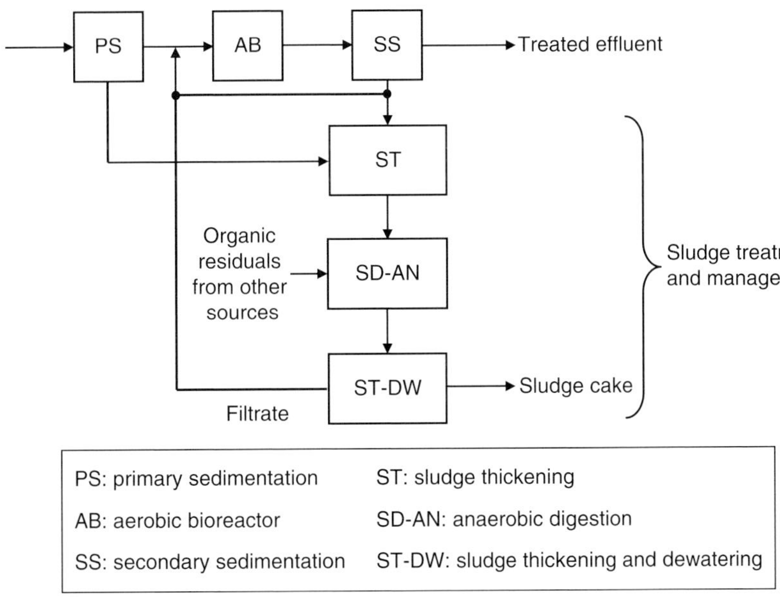

Figure 26.1 Flow diagram for the treatment of domestic wastewater.

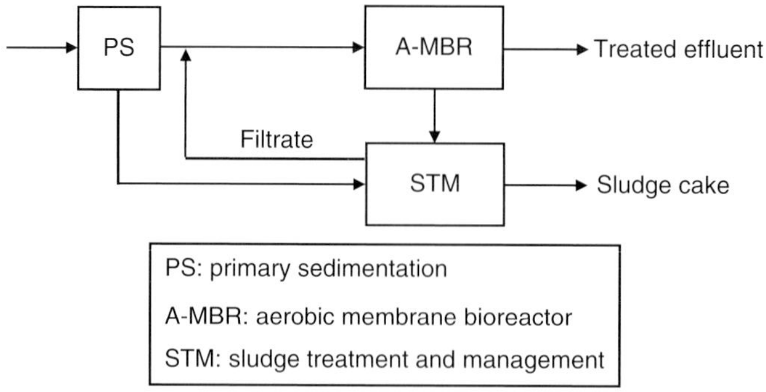

Figure 26.2 Flow diagrams for an MBR plant treating domestic wastewater.

The use of membranes for solids separation is currently gaining acceptance with wastewater treatment plant designers. The addition of membranes in activated sludge systems removes the need for secondary sedimentation tanks and tertiary treatment for suspended solids removal (Figure 26.2). Membrane bioreactors (MBRs) are equipped with either internally immersed or external membrane separation units. The quality of treated effluent is usually high in terms of BOD and suspended solids levels (in most cases less than 10 mg/l for both parameters) and thus suitable for a number of possible re-use applications. Early challenges in the use of the membrane technology lie mainly in the high capital and operational costs of membranes. However, research within the last two decades is now contributing to the development of more cost-effective MBR systems.

Anaerobic digestion of biological (or secondary) sludge can reduce the operating costs and carbon footprint of wastewater treatment operations. However, as stated in Section 26.2.2, in order to maximise efficiency gains from the production of biogas, new wastewater infrastructure will tend to enhance the production of primary sludge (which has higher biogas yield than secondary sludge), and in the process reduce the amount of oxygen required in the secondary treatment.

26.2.3.2 Anaerobic Processes

In anaerobic processes, anaerobic microorganisms convert carbonaceous and nitrogenous compounds to methane gas and carbon dioxide via a series of biochemical steps involving various microbial groups and producing other by-products, notably ammonia and hydrogen sulphide. Anaerobic microorganisms collectively have lower growth rates than their aerobic counterparts, and hence result in the production of smaller amounts of new cells than aerobic processes. Although anaerobic processes do not require an oxygen supply, they require higher temperatures of about 37°C for effective operation, unlike aerobic processes that can function effectively at lower temperatures. Anaerobic systems are usually employed where either the amount of methane gas produced will be sufficient to raise the operational temperature to the appropriate level (with or without net energy gain), such as in the treatment of sludge and high-strength wastewaters or where the raw wastewater and/or natural ambient temperature is such that little or no additional heating is required to bring the digestion temperature to the appropriate level. Anaerobic systems are commonly used to treat primary and secondary sludge produced in aerobic systems, as shown

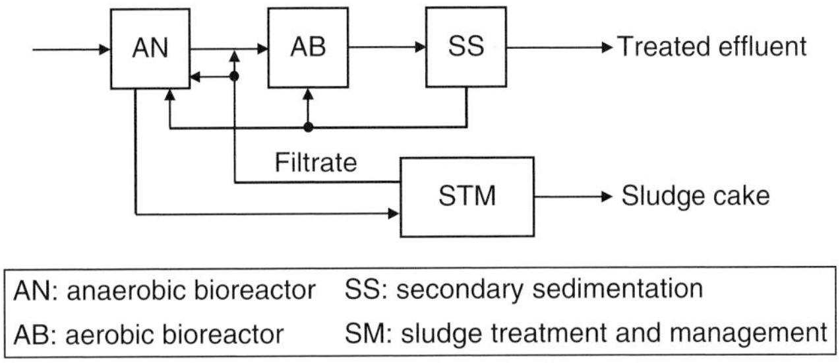

Figure 26.3 Anaerobic–aerobic treatment for carbonaceous pollutant reduction.

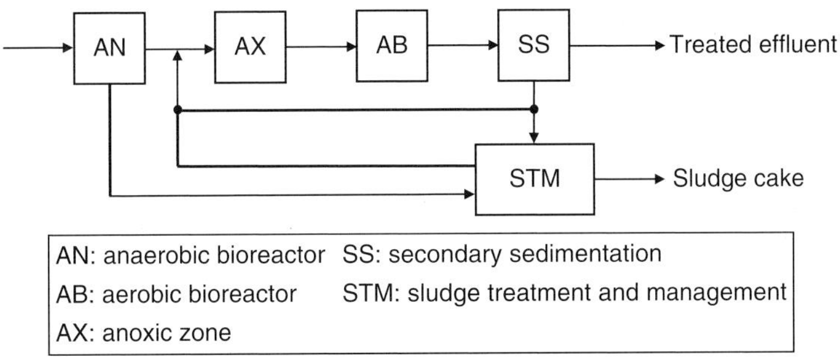

Figure 26.4 Anaerobic–aerobic treatment for carbonaceous and nitrogenous pollutant reduction.

in Figures 26.1 and 26.2, since the amount of methane gas obtainable from anaerobic sludge digestion is generally greater than the amount needed to maintain the digestion temperature at an appropriate level.

26.2.3.3 Combined Anaerobic–Aerobic Processes

In general aerobic systems can ensure good-quality treated effluent that can either be discharged directly to surface waters or re-used for irrigation. On the other hand, anaerobic treatment represents a pre-treatment step and, as such, the treated effluent usually requires 'aerobic polishing' before discharge into receiving waters.

For the treatment of domestic wastewater, the aim of employing anaerobic pre-treatment is not usually for biogas production, but for the breakdown of complex organic molecules into simpler organic acids, which will require lower oxygen demand for the final aerobic post-treatment, and hence bring about savings in both capital and operational costs. Anaerobic pre-treatment also encourages the use of less energy intensive aerobic post-treatment methods, such as constructed wetlands. Process options for anaerobic–aerobic treatment of domestic wastewater for the oxidation of carbonaceous compounds only and with nitrogenous pollutants are shown in Figures 26.3 and 26.4 respectively.

26.2.4 Advanced treatment

Advanced treatment of domestic wastewater may be required where treated effluent is to be discharged into: (i) 'sensitive' waters, meaning that the receiving water is either already eutrophic or about to become so, thus necessitating nitrogen and phosphorus removal; (ii) water used for recreational purposes or for shellfish farming, in conformity with EU Bathing Water and Shellfish Directives, thus requiring the need for the reduction of microbial pollutants; and (iii) relatively small water bodies, where the organic load (particularly from particulate sources) in the secondary effluent may result in excessive oxygen demand in the water body.

There is currently interest in recovery of phosphorus from secondary effluents. Phosphorus is a strategic commodity in the production of fertilizer, and its price has seen a significant increase during the last ten years. There are fears that this increase might impact on global food production. There may also be a need to remove specific pollutants that can pass through the biological treatment unaffected, such as certain organic compounds originating from household chemical and pharmaceutical products. These compounds are collectively referred to as organic micropollutants, the most common being the endocrine-disrupting compounds (EDCs). The common treatment methods used in advanced treatment are described in the following sections.

26.2.4.1 Solids Removal

Post-secondary treatment for solids removal is usually carried out by filtration. As explained in Section 26.2.3.1, membrane bioreactors (MBRs) are equipped with membrane filters and therefore do not need additional treatment for liquid–biomass separation. For effluents from secondary sedimentation tanks, filtration can be carried out using grass plotting, constructed wetland, sand filtration or mechanical devices such as upflow and drum filters. In grass plotting or irrigation over grassland, treated effluent is allowed to flow over a plot of land, during which some of its suspended solids will be retained by the vegetation. Constructed wetland systems, such as overflow, vertical and horizontal flow systems utilise the combined effects of biological degradation, soil and vegetation filtration (the individual contributions of each of these processes depend on the type of wetland system) to 'polish' the effluent. All the above ecological systems are suitable for small- to medium-sized wastewater treatment plants. For large plants, more energy-dependent technologies such as rapid sand gravity filters and upflow clarifiers are commonly used. A detailed description of these technologies can be found in standard textbooks.

26.2.4.2 Disinfection

The aim of disinfection is to reduce pathogenic organisms before discharging the effluent. While specific organisms are not routinely monitored, the indicator organisms of faecal and total coliforms are used. Disinfection is required if the effluent discharges into waters used for bathing, shellfish farming or for potable water abstraction, as stipulated in relevant EU Directives. The Directives and their guide limits for faecal coliforms are shown in Table 26.1. Common procedures for disinfection of wastewater include chlorination, ozonation, ultraviolet(UV) radiation and microfiltration.

Table 26.1 Faecal coliforms standards in the European Union (EU).

EU Directive	Maximum guide limit Faecal coliforms MPN/100 ml[a]
Surface water for potable abstractions (Directive 75/440/EEC)	20000
Bathing water (Directive 76/160/EEC)	2000
Shellfish (Directive 2006/113/EC)	≥300
Drinking water (Directive 98/83/EC)	0 (0)

[a] Untreated domestic wastewater has a faecal coliform count in the range of 10^6 to 10^7 MPN/100 ml (Butler and Davis, 2010).

Disinfection with oxidising agents, such as chlorine and ozone, will firstly oxidise any residual organic compounds in the effluent before being effective as a disinfectant. As a result, it is usually advisable to firstly carry out advanced treatment to reduce any organic solids remaining in the effluent after secondary treatment. Chlorination is usually recommended in the United Kingdom mainly for temporary use, because of the possible formation of halogenated compounds from its reactions with organics. Some halogenated compounds have been known or suspected to be carcinogenic and mutagenic.

UV treatment kills the vegetative and spore cells of microorganisms. The lethality of UV light varies greatly with wavelength. Bacterial wavelength ranges from 200 to 295 nm, with a maximum effect at 254 nm. Suspended solids in wastewater can reduce the transmission and absorption of UV rays. They can also shield microorganisms from exposure, particularly bacteria within suspended particulates. Solids reduction using some of the techniques described in Section 26.2.4.1 therefore usually precedes UV treatment.

Pathogen reduction can also be achieved using membrane filtration techniques, with microfiltration being the most commonly used technique. With a pore size ranging from 0.1 to 1 μm, microfiltration can remove bacteria and most viruses. MBRs (see Section 26.2.3.1 above) can be designed to carry out disinfection.

26.2.4.3 Nitrogen Removal

Nitrification

Human wastes contain appreciable amounts of protein and urea, which are converted to ammonia during treatment by both aerobic and anaerobic processes. Part of the ammonia formed is used for bacterial growth, to produce new cells, and the excess will be discharged in the effluent. Ammonia can bring about excessive oxygen demand in the receiving water, since ammonia oxidises naturally to nitrate in a process called nitrification and is also toxic to fish at elevated levels.

Nitrification is an aerobic two-step biological process, carried out by two different microbial groups: one converting ammonia to nitrites and the other converting nitrite to nitrates. More details of these processes can be found in standard books.

Combined organic carbon oxidation and nitrification can be accomplished in aerobic processes (see Section 26.2.3.1), by increasing the aeration or hydraulic retention times. An alternative approach, commonly used to upgrade existing plants that have not been

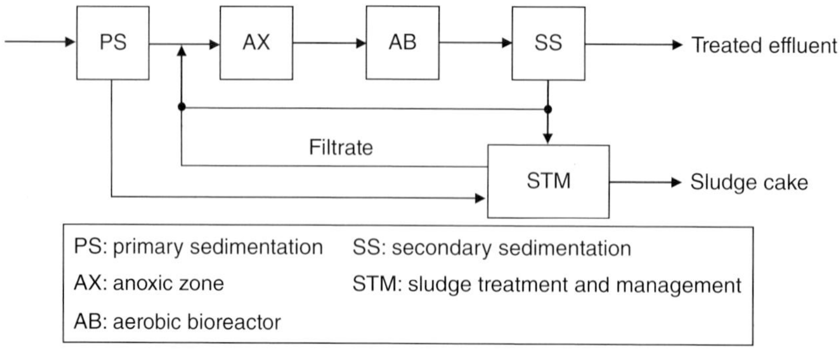

Figure 26.5 Typical flow diagram for the treatment domestic wastewater for carbon and nitrogen removal.

originally designed to carry out nitrification is via a separate nitrification unit following a conventional secondary aerobic treatment process.

Denitrification

Complete nitrogen reduction is required where there is a need to control eutrophication in the receiving water or in water re-use applications such as downstream abstraction for water supply or groundwater recharge. As a reference, the EC Directive for Drinking Water Quality (98/83/EC) stipulates the maximum concentration for nitrate as 50 mg/l or 11 mg/l as nitrogen.

In the absence of oxygen, nitrates and nitrites can be converted to nitrogen gas in a process called denitrification by microorganisms. Biodegradable organic compounds must be available for this reaction to occur. Nitrates and nitrites can also be converted back to ammonia by another group of microorganisms, although denitrification is a more dominant reaction pathway.

Denitrification can be achieved when there is a contact between nitrified effluent and a source of biodegradable organic carbon and in the absence of molecular oxygen. External or internal sources of carbon can be used. The use of external carbon sources (such as methanol) entails handling and storage of chemicals. Denitrification using internal carbon sources (i.e. organic carbon present in the raw wastewater) is the most widely method practised in the United Kingdom. Figure 26.5 shows an example of the process configuration for the treatment of domestic wastewater for the reduction of carbonaceous and nitrogenous pollutants. Other process options can be found elsewhere.

26.2.4.4 Phosphorus Removal

Phosphorus removal is essential for the control and prevention of eutrophication as stipulated in the EC Urban Wastewater Treatment Directive (91/271/EEC). Phosphorus is also an important raw material for the production of fertiliser, as explained in Section 26.2.4. Domestic wastewater contains phosphorus compounds, which come from the breakdown of urine and proteins in human waste. Phosphorus is also contained in synthetic detergents used for household cleaning and washing. Phosphorus is an essential element for the growth of microorganisms, hence, a limited amount of phosphorus in the wastewater will adversely affect the efficiency of biological processes. Domestic wastewater contains phosphorus in excess of the amount necessary for the breakdown of the organic and nitrogenous pollutants present. However, some industrial wastewaters may require phosphorus addition for effective biological treatment. Phosphorus in wastewater can be removed by biological or chemical methods, or by a combination of both.

Phosphorus Removal: Chemical

Chemical phosphorus removal is carried out by adding chemicals, or coagulants, that can convert soluble inorganic phosphorus compounds into insoluble phosphate, which is then allowed to settle out in a sedimentation tank. Commonly used coagulants include metallic salts such as ferric or aluminium chlorides and sulphates, and lime. Polyelectrolytes (or polymers) have often been used in conjunction with the metallic salts to enhance the stability of the precipitates.

Precipitation of phosphorus can be accomplished at three locations in the treatment plant, classified as: (a) primary or pre-precipitation, (b) simultaneous or co-precipitation and (c) tertiary or post-precipitation.

In primary precipitation, the salts are added to raw wastewater or to the primary sedimentation tank and the precipitated phosphate is removed with the primary sludge. It is suitable for application in both suspended and attached growth aerobic processes. Since the coagulants will also enhance the removal of suspended solids in the sedimentation process, this method will require a higher chemical dose and will lead to the production of greater quantities of primary sludge. As explained in Section 26.2.2, enhanced primary sedimentation can bring about a net energy gain through the production of higher amounts of primary sludge and consequently lesser amounts of secondary sludge and through the resulting reduction of oxygen and energy demand for secondary treatment.

Simultaneous precipitation involves the addition of the coagulants directly into the aeration tank of an activated sludge plant for subsequent removal in the secondary clarifiers. This method improves the efficiency of solids separation in the secondary clarifier, thereby lowering the suspended solids content of the final effluent.

Tertiary precipitation involves the addition of coagulants to the effluent from the secondary clarifier. Substantially lower amounts of chemicals are required, but an additional clarifiers is necessary for the removal of the precipitates. This option will be preferable for phosphorus recovery. Recovery of phosphorus from waste streams as struvite for agricultural use is increasingly as an attractive option for the water industry due to the rising cost of phosphorus.

Phosphorus Removal: Biological

Biological phosphorus removal is a complex process combining both anaerobic and aerobic biochemical reactions and involving special groups of microorganisms, which can accumulate intracellularly large quantities of phosphorus in the form of insoluble polyphosphate. In an anaerobic condition and in the absence of nitrates, these organisms are able to absorb short-chained volatile fatty acids (mainly acetic acid) and release polyphosphate in the form of soluble orthophosphate into the effluent being treated. In a subsequent aerobic condition, the absorbed short-chained volatile fatty acids are oxidised, thereby providing energy within the cell for the uptake of available soluble orthophosphate in the effluent (i.e. both orthophosphates released from the cell in the anaerobic zone and those existing in the effluent being treated). These reactions will result in a net reduction of phosphorus from the solution and its accumulation in the microbial cells that make up the waste surplus sludge. A more detailed consideration of biological phosphorus removal processes can be found elsewhere (Chartered Institute of Water and Environmental Management (CIWEM), 1994; Tchobanoglous *et al.*, 2002).

Biological phosphorus removal plants are not as common as chemical phosphorus removal plants in the United Kingdom. Plant operators prefer the latter for its relative simplicity in operation and process control.

26.2.4.5 Removal of Organic Micropollutants

There is currently a growing concern worldwide regarding the potential dangers of organic micropollutants such as endocrine-disrupting compounds (EDCs), pharmaceuticals and other synthetic organic compounds in domestic wastewater, with levels ranging from 10 μg/l to 1 mg/l. The concerns have grown since the discovery of the feminisation of male fish and other aquatic organisms exposed to wastewater treatment plant effluents and the more recent reports on their potential genetic effects on humans (Wu *et al.*, 2011). Although removal of these compounds is not currently part of the mission of domestic wastewater treatment, it is likely that future wastewater treatment infrastructure designers may be favourable to technologies that will ensure the treatment of these pollutants. Current research in this field has seen the emergence of various treatment options, involving biological and physicochemical treatment processes. One of the most promising techniques involves the use of MBR systems (see Section 26.2.3.1). The long solid retention times and low food-to-microorganism ratio obtainable in MBR systems are favourable to enhanced biological degradation of nonreadily biodegradable (or 'hard') organic compounds such as some of the EDCs. MBR can also be equipped with processes, such as ultrafiltration or activated carbon adsorption, which are also effective in separating these compounds from the effluent. These treatments can achieve solids and pathogen reduction, and produce effluents with wider re-use options.

Another feasible option, suitable for non-MBR plants, is ozone treatment. Ozonation can oxidise all residual organic compounds, both biodegradable and nonbiodegradable, without resulting in solids production. Ozone treatment is commonly used for the reduction of pathogens, solids, colour and odour in the treated effluent.

26.2.5 Sludge Treatment

Land application is the most common ultimate disposal outlet for sewage sludge in the United Kingdom. Since 1999, sludge destined for land application must receive advanced treatment. Advance treatment comprises a collection of treatment processes and operations aimed at virtually eliminating any pathogens that may be present in the original sludge. An example of a typical sludge treatment process configuration is shown in Figure 26.1. It generally include water reduction (by thickening), organic solids and pathogen reduction by anaerobic (aerobic) digestion followed by a further water reduction, dewatering and drying to produce sludge cakes or pellets which can be used as organic fertiliser. Lime treatme can be used for pathogen reduction. Detailed coverage of sludge treatment and management can be found elsewhere (CIWEM, 1996; Tchobanoglous et al., 2002).

26.3 Factors Affecting Investment in UK Wastewater Treatment Infrastructure

Globally there are a range of challenges currently facing the water industry that may affect asset investment. These challenges include energy price volatility, climate change and its regulation, asset capital cost and strategic resource considerations (Palmer, 2010). Global macroeconomics, whilst in some respects appears far removed from the UK Water Industry, has a significant part to play in investment decisions made locally. The water cycle is global; the availability, use and security of water therefore transcends local, national and even continental boundaries (Royal Academy of Engineering (RAEng), 2010).

A salient feature of the water and sewerage services sector in the United Kingdom compared to other developed countries is the diversity of institutional arrangements between its constituent parts. In England and Wales water and sewerage services are provided by private companies, whilst in Scotland and Northern Ireland these services remain government owned. This allows for an interesting and diverse range of approaches to be viewed for tackling these challenges within one country, the United Kingdom.

26.3.1 Economic Factors

26.3.1.1 Global Macro Economics

Population is the overarching factor driving current macro economic factors. Monday, 31 October 2011 was designated by the United Nations as the day that the World would likely meet the new population milestone of 7 billion. The United Nations have predicted that the World's population will continue to increase, albeit at a lesser rate, to more than 9 billion by 2050 (see Figure 26.6).

As the global population grows the pressure on resources including water, food, fuel, chemicals and other resources also increases (United Nations (UN), 2008). Imbalances in supply and demand result in price changes so when the rate of demand does not match the rate of supply, prices increase. Population growth also results in increased greenhouse gas (GHG) emissions, which drives the level of climate change. If climate change predictions are correct that increasing temperatures will result in the migration of population to regions where life is more sustainable (Stern, 2006) then the pressure on resources from the population will become even more pronounced.

The potential impacts of global macroeconomic risk factors on the water industry are:

- Lack of water security
- Increased pressure on water resources:
  - Drinking
  - Agriculture (food and biofuels)
  - Wastewater treatment
  - Electricity generation (hydro, nuclear, thermal)

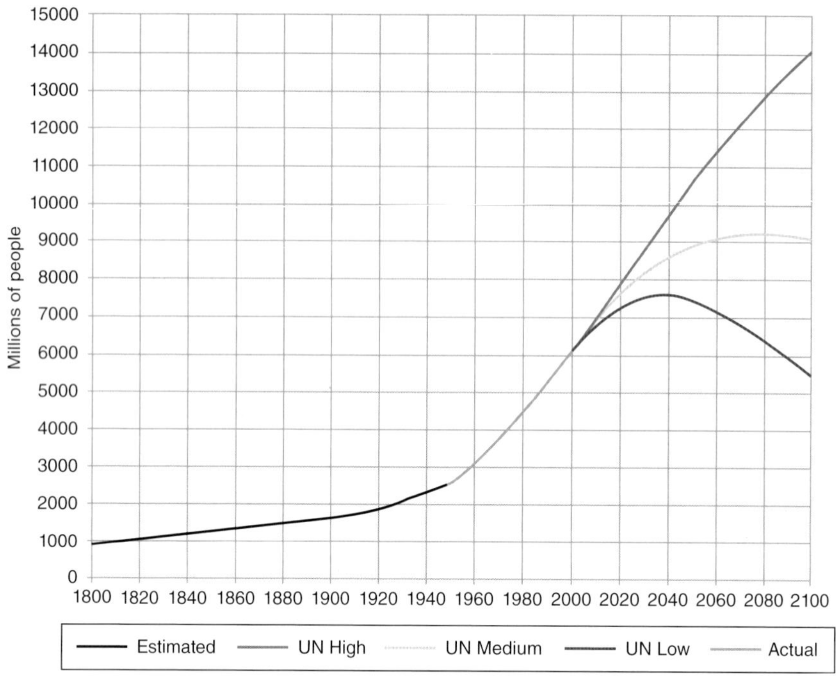

Figure 26.6 World population, estimated, actual, and projected (UN high, medium and low 2004 growth rate projections): 1800–2100 (source: http://globalconsensus.wordpress.com/2012/02/08/world-population-growth-projections/)

- Increased water pollution
- Increased cost of commodities associated with water treatment such as fuel and chemicals
- Increased production of GHGs associated with increased treatment standard requirement
- Increased cost in meeting challenges associated with climate change such as the cost of carbon
- Transboundary conflict

(Pittock, 2008; RAEng, 2010; Palmer and Nair, 2011) (http://www.water.org.uk/home/policy/climate-change/mitigation)

26.3.1.2 Microeconomics

Political

The UK Water Industry follows a 5-year price-setting and investment cycle. The duration of the investment cycle will influence the investment choices made. The review of prices that consumers pay for water services, which is a major source of funding for infrastructural improvement and maintenance, is made by the economic regulators who independently set the cost of the services. These are the Water Services Regulation Authority (OFWAT) (England and Wales), the Water Industry Commission for Scotland (WICS) (Scotland) and The Utility Regulator – electricity, gas, water (Northern Ireland). An historic overview shows that the water industry has found it difficult to ensure smooth expenditure patterns across the 5 years, due mainly to the following reasons:

- Nonflexible 5-year price-setting cycle itself, and its impact on investment planning
- Need to meet demanding efficiency targets
- Lack of clarity in what is required by directives or in government guidance
- Changes made to delivery of improvement and maintenance projects in agreement with regulators
- Inevitable changes to priorities, for example to cope with drought
- Projects held up by planning delays

(http://www.water.org.uk/home/news/press-releases/response-to-ofwat)

The 5-year planning cycles do not fit well with macroeconomic challenges, such as climate change and reducing resources. Significantly longer term strategic investment planning is required to action these effectively. In recognition of this, OFWAT's approach to the 2009 Price Review went some way to address this. It required companies to place their 5-year plans in a 25-year context, for the first time, through the production of strategic direction statements (OFWAT, 2008a). Within these documents, companies are required to demonstrate that they have a sound understanding of the risks from climate change that may affect their ability to provide essential water and sewerage services. They then needed to show that they have arrived at robust and appropriate solutions for managing these risks. The requirement enabled companies to consider both their long-term and short-term activities (http://www.ukcip.org.uk/business/business-case-studies/ofwat/).

Indirectly these short-term cycles mean views and political ideas can influence investment choices and priorities, thereby creating uncertainty over the future. Whilst the economic regulators are independent of government, they represent the interests of customers so will indirectly take into account the economic and political climate on the customers' ability to afford the service.

Funding

The diversity of UK Water Company models means there are a number of funding models applied across the industry: private, public and mutualised. Investment is delivered in 5-year cycles and is funded in part by water charges, other revenue streams and, depending on the model, investors or government funding. All models have limited budgets and lead to investment constraints; for example, the recent global economic downturn may, in some cases, have necessitated a reduction of government funding, with shareholders expecting a return on their investment and water charges being controlled by the water industry's economic regulators (OFWAT, WICS and The Utility Regulator) in conjunction with increasing chemical and energy prices driven by the global markets.

Water companies have to find ways to balance the expectation of customers and shareholders, the risk of maintaining investment and the potential consequences to the water environment of not meeting regulatory targets against these financial constraints. These models drive how choices are made and investment is balanced.

Loss of Manufacturing Industry – Economic Downturn

Over the last century the United Kingdom has seen a shift in industrial composition, with the decline of agriculture and manufacturing's share of total employment and the rise of services. In the United Kingdom, manufacturing's share fell from 28 to 14% of employment, and agriculture's share from 11 to 2% (Office for National Statistics (ONS), 2003). The consequent loss of industrial wastewaters from the manufacturing industry has resulted in wastewater treatment plants originally designed to accommodate them becoming underloaded. Underloading brings treatment challenges as it can be 'harder to run an overdesigned plant than an underdesigned plant' (http://www.environmentalleverage.com). Additionally, loss of revenue from industrial wastewater charges means less money is available for investment in improving wastewater treatment infrastructure.

Investment Risk

Risk is assessed based on a company's criteria. The level of risk that an organisation is willing or directed to take will drive their infrastructure and services investment profile. For example, it may be deemed better value to not invest in staff to maintain equipment if the cost of replacement is more effective. It could be that the civil structure of a wastewater treatment infrastructure has deteriorated but that the final effluent quality is still excellent and therefore will not be invested in at this time. It could be that reputational issues (such as corporate social responsibility) inform investment or that it is more cost-effective to outsource to third parties. Decisions made regarding risk are generally driven by resource availability.

Changes in Mandatory Regulatory Requirements

There are global, regional and individual protocols, directives and legislation that drive the reduction of greenhouse gas (GHG) emissions, for example the Kyoto Protocol, the European Union Emission Trading Scheme (EU ETS), which is the largest multinational GHG emissions trading scheme in the world, and the UK's Climate Change Bill, which has set a target of 80% reduction in GHG emissions by 2050.

There are environmental protocols, directives and legislation, such as the EU Water Framework Directive, that are driving water quality, protection and improvement of river basin areas, air quality, habitats, etc.

Delivering these improvements requires significant capital expenditure. For the water industry, this may mean expanding, enhancing and building increasingly energy-intensive treatment processes that in turn will release more GHG and carbon emissions. The regulatory investment drivers can therefore exist in conflict, where one set of legislation is being addressed at the expense of another. More synergy between branches of environmental regulation is required to ensure that desirable outputs in one area do not result in undesirable outcomes for another. The water industry in the United Kingdom recognised this conflict and has committed to working with the government and European Union to ensure carbon implications of environmental regulations are fully considered (http://www.water.org.uk/home/policy/climate-change/mitigation/mitigation-table-1-may09.pdf and http://ec.europa.eu/governance/impact/consultation/docs/contributions/organisations/water_uk_en.pdf). The European Union have developed impact assessment guidelines, which means that before the Commission proposes a policy initiative, it shall assess the economic, social and environmental consequences in an impact assessment to ensure there is a holistic approach to European policy decisions (http://www.water.org.uk/home/news/press-releases/eu-impact-assessment-guidelines?s1=consultations).

Energy Costs

The water industry in the United Kingdom is energy-intensive and contributes about 1% of national GHG emissions (http://www.water.org.uk/home/policy/climate-change/mitigation; OFWAT, 2008b). As companies look to reduce their fuel bills and meet their climate change obligations (reduction in greenhouse and carbon emissions, see Section 26.3.2) a shift in thinking is occurring within the water industry. By looking at the wastewater stream as a commodity rather than a 'waste', an opportunity arises to recover energy and move towards energy security and carbon neutrality.

By making use of energy from the water cycle through methods such as biogas production from anaerobic digestion of sludge, gasification, heat exchange, incineration of screening, using its land portfolio for renewable energy generation, water companies can compensate for their CO_2 emissions. Wastewater sources of renewable energy are becoming increasingly interesting to the water industry in the United Kingdom as prices of grid electricity increases and climate change taxes become more expensive. Additionally, greater incentives are being provided by the UK government in the form of favourable tariffs for renewable sources, in order to meet the ambitious renewable energy target of 15% of the UK energy sources by 2020 (supported by the devolved administrations' targets of 100% renewable electricity in Scotland, 40% in Northern Ireland and double the renewable capacity in Wales by 2025) (Department of Energy and Climate Change (DECC), 2011).

26.3.2 Climate Change Factors

Climate change has the potential to cause an unlimited and immeasurable impact on water availability and treatment. It brings two serious challenges for water treatment infrastructure investment, one being mitigation and the other adaptation. The ability to protect and invest in water treatment whilst taking these factors into account is further complicated by the deeply embedded uncertainty surrounding climate change impact prediction.

26.3.2.1 Mitigation

Mitigating the impact of future climate change by reducing GHG emissions is a generally accepted obligation with shared responsibility in the United Kingdom. The Scottish government led the way by setting the most ambitious and comprehensive piece of climate change legislation in the world, closely followed by the UK government. The legislation established legally binding targets of at least an 80% reduction in GHG emissions by 2050 and a reduction in carbon dioxide (CO_2) emission of at least 42% (34% in the United Kingdom) by 2020, using 1990 as the emission baseline (http://www.legislation.gov.uk/asp/2009/12/pdfs/asp_20090012_en.pdf and http://www.legislation.gov.uk/ukpga/2008/27/pdfs/ukpga_20080027_en.pdf).

The manner in which the United Kingdom has chosen to legislate and regulate climate change mitigation affects UK businesses, including the water industry. The UK water industry provides an example of how climate change legislation is a factor that must be taken into account when making investment decisions. Global legislation on climate change mitigation is evolving and water industries in other countries are also likely to face the same challenge of factoring climate change legislation into investment decisions in the near future (Palmer, 2010).

26.3.2.2 Adaptation

The uncertainties around climate change are propagated with increasing magnitude through all stages of climate change impact assessment methodology from emission scenarios through global climate model (GCM) modelling, downscaling and finally engineering modelling (Schneider and Kuntz-Duriseti, 2002). This makes it extremely difficult to determine, with any degree of certainty, how to balance the investment in climate change adaptation against risk. This raises some questions:

- How can water infrastructure be climate 'proofed' (or adapted)?
- How can the 'right' investment decisions be made based on predicted models?
- How can investment be justified for things that are predicted but have not happened?
- Is climate proofing a justifiable investment?
- What is the appropriate level of climate change impact to be invested for – minimum, middle of the road or worst-case scenario?

The water industry is at the forefront of climate change. It will impact all areas of the water industry as its raw material is directly dependent on the natural environment (Water UK, 2007). Appropriate adaptation investment strategies can only be determined by identifying potential climate change impacts and assessing the likelihood and severity of these utilising climate change impact scenarios. A comprehensive assessment of potential climate change impacts facing the UK water industry was commissioned by Water UK in conjunction with the UK Water and Sewerage companies and can be found at: http://www.water.org.uk/home/policy/publications/archive/industry-guidance/asset-management-planning/water-uk-climate-change-adaptation-approach-for-asset-management-planning---information-tables.xls. The data given in this study provide a tool to enable assessment of climate change impacts of precipitation, flooding, temperature and sea level rise on infrastructure. It scores the impacts and assesses them for severity and urgency, indicating the industry's ability to secure services and meet the needs of their customers. The tool also provides guidance for the water industry on where to focus adaptation investment in response to climate change.

26.4 Conclusions

Wastewater treatment, infrastructure and design in the United Kingdom have been historically determined by environmental regulations. Tightening environmental quality standards have been achieved through more complex and energy-intensive treatment technologies. Other equally important factors now influence the investment decisions and risks that water companies are willing to take. These include macroeconomics, microeconomics and climate change and necessitate the need for the industry to take a holistic view of these drivers for infrastructure selection and design. The industry, long known for its aversion to risk, will now have to become innovative, finding new ways, processes and technologies to secure the environmental quality whilst taking account of these new challenges.

Existing treatment technologies will not become redundant but will be utilised in a new manner. It may mean a shift backwards to the turn of the century, when less energy intensive forms of wastewater treatment were employed. It is more likely, however, that wastewater will be looked upon and exploited as a resource.

New areas of concern, such as the presence of micropollutants originating from domestic wastewaters in the water environment, will increase the need for new advances in treatment technologies. There is also increased pressure on the availability of water resources in various parts of the country, leading to considerations for greater re-use of treated wastewater for domestic purposes. These two factors will greatly influence future regulations on the quality of discharges and may, as a consequence, increase the use of certain technologies.

Investment in wastewater infrastructure in the future is likely to continue to be determined by environmental regulations. Since 2008, however, a new financial era has dawned globally and as a result the water industry faces a paradigm shift in how it must now approach wastewater treatment, infrastructure and design. Global macroeconomics, whilst in some respects appears far removed from the UK water industry, has a significant part to play in investment decisions made locally.

References

Butler, D. and Davies, J.W. (2010) *Urban Drainage*. Spon Press.

Chartered Institution of Water and Environmental Management (CIWEM) (1994) *Tertiary Treatment*. Handbooks of UK Wastewater Practice.

Chartered Institution of Water and Environmental Management (CIWEM) (1996) *Sewage Sludge*. Handbooks of UK Wastewater Practice.

Department of Energy and Climate Change (DECC) (2011) *UK Renewable Energy Roadmap*. Crown Publishing, URN 11D/698.

Office for National Statistics (ONS) (2003) *Labour Market Trends – A Century of Labour Market Change: 1900 to 2000*, March 2003, Volume 111, No. 3.

OFWAT (2008a) *Setting Price Limits for 2010–15: Framework and Approach*, Birmingham. http://www.ofwat.gov.uk/pricereview/pap_pos_pr09method080327.pdf.

OFWAT (2008b) *Preparing for the Future- OFWAT Climate Change Policy Statement*, Birmingham.

Palmer, S.J. (2010) Future challenges to asset investment in the UK Water Industry: the wastewater asset investment risk mitigation offered by minimising principal operating cost risks. *Journal of Water and Climate Change*, **1** (1), 17–35.

Palmer, S.J. (2011) Primary sedimentation: from neglected common asset to future keystone technology. In: *Proceedings, 5th European Water and Wastewater Management Conference*, 26–27 September 2011.

Palmer, S.J. and Nair, A. (2011) Designing for the future: optimum resource recovery process design and long term value analysis of recovered commodities. In: *Proceedings, 5th European Water and Wastewater Management Conference*, 26–27 September 2011.

Pittock, J. (2008) Climate change and energy: the water dilemma. In: *Water 21:IWA*, 13–16 August 2008.

Royal Academy of Engineering (RAEng) (2010) *Global Water Security – An Engineering Perspective. Engineering the Future*, April, London. Available at: http://www.raeng.org.uk/news/publications/list/reports/Global_Water_Security_report.pdf; http://www.bis.gov.uk/assets/cst/docs/files/whats-new/09-1632-improving-innovation-water-industry.pdf.

Schneider, S.H. and Kuntz-Duriseti, K. (2002) In: Schneider, S.H., Rosencranz, A. and Niles, J.O. (eds), *Climate Change Policy: A Survey*. Island Press, Washington, DC, pp. 53–87.

Stern, N. (2006) *The Economics of Climate Change: The Stern Review*. Cambridge University Press, Cambridge, UK.

Tchobanoglous, G., Burton, F.L. and Stensel, H.D. (2002) *Wastewater Engineering*, 4th Edition. McGraw-Hill.

United Nations (UN) (2008) *Demographic Aspects of Climate Change, Environmental Degradation and Armed Conflict*. Department of Economic and Social Affairs – Population Division. UN, New York (UN/POP/EGM-URB/2008/18). Available at: http://www.un.org/esa/population/meetings/EGM_PopDist/P18_Urdal.pdf.

Water UK (2007) *A Climate Change Adaptation Approach for Asset Management Planning*, MWH 41414874 V1.0. Available at: http://www.water.org.uk/home/policy/publications/archive/industry-guidance/asset-management-planning/water-uk-climate-change-adaptation-approach-to-asset-management-planning.pdf.

Wu, C., Xue, W., Zhou, H., Huang, X. and Wen, X. (2011) Removal of endocrine disrupting chemicals in large scale membrane bioreactor plant combined with anaerobic-anoxic-oxic process for municipal wastewater reclamation. *Water Science and Technology*, **64**, 1511–1518.

Websites

www.sepa.org.uk
www.environment-agency.gov.uk
http://www.guardian.co.uk/environment/series/crowded-planet-population
http://en.wikipedia.org/wiki/World_population
http://www.legislation.gov.uk/asp/2009/12/pdfs/asp_20090012_en.pdf
http://www.legislation.gov.uk/ukpga/2008/27/pdfs/ukpga_20080027_en.pdf
http://www.ukcip.org.uk/business/business-case-studies/ofwat/
http://www.water.org.uk/home/policy/climate-change/mitigation
http://www.water.org.uk/home/policy/climate-change/mitigation/mitigation-table-1-may09.pdf
http://ec.europa.eu/governance/impact/consultation/docs/contributions/organisations/water_uk_en.pdf
http://www.water.org.uk/home/news/press-releases/eu-impact-assessment-guidelines?s1=consultations
http://www.water.org.uk/home/policy/publications/archive/industry-guidance/asset-management-planning/water-uk-climate-change-adaptation-approach-for-asset-management-planning—information-tables.xls.

Section 7 International Case Studies

27

An Overview of Management Issues in Developing a Sustainable Water Supply, Sanitation and Hygiene (WASH) Service Delivery in Nigeria

David Oloke and Dayo Olugboye

27.1 Introduction

Water supply, sanitation and hygiene (WASH) is described as all works related to water, sanitation and hygiene, including the provision of safe and affordable access to a clean water supply and methods of disposing of the waste. This also involves the provision of services and training on how to manage them (Bunclark *et al.*, 2011). Over the last three decades, huge investments have gone into delivering WASH infrastructures in many developing countries across Africa. However, in Sub-Saharan Africa, only about 40% of such intervention's shows an element of sustainability. The Millennium Development Goal (MDG) agreed at the United Nations in 2000 is to halve the proportion of people without sustainable access to adequate and affordable safe drinking water by 2015 (UN Millennium Project, 2005). This goal will be much harder to achieve in Africa than in the rest of the developing world due to the low levels of existing coverage coupled with high population growth rates in some areas. This is compounded by the fact that existing services demonstrate limited sustainability throughout the continent (Harvey and Reed, 2004). Major changes in policy and planning are required if past, ongoing and future investment in WASH is not to be wasted. Sustainable water supply and sanitation development does not result simply from the installation of new infrastructure (Water, Engineering and Development Centre (WEDC), 2010). It is now widely accepted that water and sanitation projects need to address a range of concerns, only some of which are purely technical. Successful water and sanitation projects need to integrate perspectives from a wide range of disciplines such as: social, health and hygiene, technical economics, financial, institutional and environmental (WEDC, 2010).

The lack of clean water and basic latrines positioned close to people's homes hampers livelihoods, places a burden on people's time and reduces quality of life (WaterAid, 2011). The World Health Organization (WHO) and United Nations Children's Fund (UNICEF)

Water Resources in the Built Environment: Management Issues and Solutions, First Edition.
Edited by Colin A. Booth and Susanne M. Charlesworth.

2012 report on progress on drinking water and sanitation update states that, at the end of 2010, 89% of the world's population, that is 6.1 billion people, are using improved water sources. However, in Sub-Saharan Africa only 61% of the people have access to improved water supplies, compared to 90% in Latin America, the Caribbean, Northern Africa and large parts of Asia. It was also noted that >40% of all people globally who lack access to drinking water live in Sub-Saharan Africa. Also in those areas, of the 1.1 billion people who still practice open defecation, the vast majority live in rural areas (WHO/UNICEF, 2012).

Inadequate water services cause an estimated 500 million people to fall ill each year, 15 million to become economically inactive, and a further 12 million to die (Delmon, 2001). Insufficient water supplies and sewerage services represent a major cause of illnesses and death. Poor infrastructure impedes a nation's economic growth and international competiveness (Malhotra, 1997). Also, certain historical analyses demonstrate that global water consumption doubles every 20 years, at twice the rate of human population growth (Barlow, 1999). Efforts by the central government, local authorities, external support agencies, nongovernmental organisations, communities as well as individual efforts have led to an appreciable impact on the livelihoods of many. However, initial gains in WASH infrastructure service provision do not normally translate into continuous services benefit to end users in the long term, due to significant management deficits at national, state, local government and community levels in Nigeria. Measuring the actual sustainability of both water and sanitation facilities remains an area that could benefit from further attention (WHO/UNICEF, 2012).

Sustained improvement in water and supply and sanitation can bring about significant health improvement – notably the reduction of diarrhoea – however, the benefits reach beyond the following:

- Accessible water supplies reduce drudgery and increase time for other activities such as education, income-generation or an essential extra hour of sleep.
- Improved sanitation facilities afford privacy and assist in retaining human dignity. They can increase school attendance by girls, reduce vulnerability of women to attack and rape, and make living conditions more pleasant.
- Adequate drainage can reduce the cost of controlling mosquitoes, decrease flooding and sewerage contamination.
- Facilitating basic services in a participative manner can increase the confidence of socially excluded groups and, especially of women, can be a catalyst for economic activities.

Though hard to measure, these benefits are among those most commonly cited by people in low-income communities (Water and Environmental Health – London and Loughborough (WELL), 2004).

27.2 Background on WASH Development

27.2.1 International Decade for Clean Drinking Water (1981–1990)

In the 1980s, there were 1.8 billion people living in rural areas of developing countries. Only one person in five had access to clean water and 590 million (41%) of children under 15 years did not have clean water. In developing countries, one hospital patient in four suffered from an illness caused by polluted water. Daily, millions of women and children had the chore of fetching water, taking up half a day and using energy that would have been better spent on education, training or simple survival. Even then, the water may not be clean, yet contaminated water meant sickness and death. In 1977, the United Nations 'Water Conference' at Mar del Plata set up an International Drinking Water Decade (1981–1990). Its aim was to make access to clean drinking water available across the world. Since then, there has been a concerted effort to improve the domestic water supply and sanitation coverage in developing countries (WEDC, 2010).

The decade focused on safe water and sanitation for everybody by 1990. Among the obstacles were the following: whether developing countries would give water and sanitary disposal high enough priority to get results; whether an effective organisation could be created in countries to carry out a water and waste programme; how manpower training and financing could be accomplished; and whether or not appropriate technology would be used. This first water decade brought water to more than 1.2 billion people and sanitation to about 770 million; however, growth and rapid urbanisation, together with the low level of public awareness about health, drastically reduced many countries' abilities to keep up with demand; today, there are still about 1.1 billion people who have inadequate access to water and 2.4 billion without appropriate sanitation. Since the decade ended in 1990, hopes for improvement are centred on the World Water Assessment Programme, a joint effort of the United Nations and its member states, which includes a biennial assessment of the state of global fresh water resources (http://www.gdrc.org/uem/water/decade_05-15/first-decade.htmlater).

In 2004, more than three out of every five rural people, over 2 billion, did not have access to basic sanitation facilities. If the current trend persists, nearly 1.7 billion rural dwellers will still not have acess to improved sanitation by 2015. Similarly, urban sanitation coverage more than doubled the rural sanitation coverage, according to the Joint Monitoring Programme (JMP) 2010 updated Report. Of the approximately 120 million children born in the developing world each year, half will live in households without access to improved sanitation, at grave risk to their survival and development. Poor hygiene and lack of access to sanitation together contribute about 88% of deaths from diarrhoeal disease (Federal Ministry of Water Resources (FMWR), 2004).

27.2.2 United Nations International Water Decade (2005–2015)

At its 58th Session, the UN General Assembly adopted a draft resolution, without a vote (A/RES/58/217), proclaiming 2005–2015 as the International Decade for Action. This recommendation, which comes at the close of the International Year of Freshwater 2003, calls for a greater focus on water-related issues and for actions to ensure the participation of women in water-related development efforts. It also recommits countries to achieving the water-related goals of the 2000 Millennium Declaration, the 2002 Johannesburg Plan of Implementation and of Agenda 21. The goal of the decade is a greater focus on water-related issues, with emphasis on women as managers of water to help to achieve internationally agreed water-related goals, which are to halve by 2015 the proportion of people who are unable to reach or afford safe drinking water and who do not have access to basic sanitation (http://www.gdrc.org/uem/water/decade_05-15/index.html).

A summary from the WHO/UNICEF 2012 report highlights the fact that an estimated 89% of the global population now use improved drinking water sources. Despite this enormous accomplishment 780 million people remain unserved; 4 out 10 people without access to improved drinking water live in Sub-Saharan Africa. While coverage of improved water supply sources is 90% or more in Latin America and the Caribbean,

Northern Africa and large parts of Asia, it is only 61% in Sub-Saharan Africa. The number of people in rural areas using unimproved water sources is five times greater than in urban areas. Eight out of 10 people living in urban areas have piped water connection on their premises, compared to only 3 in 10 people in rural areas. In Sub-Saharan Africa, almost 90% of the population in the richest quintile use improved drinking water sources, compared to only 35% of people in the poorest quintile.

27.3 Situation of WASH in Nigeria

Water is a necessity for life. Provision of an improved water supply is one of the basic social responsibilities of government. Within WASH programming in Nigeria, safe water provision is one of the core components of WASH delivery. Nigeria, with a land area of 924 000 km is endowed with about 267 billion m^3 of groundwater annually (FMWR, 2004). Table 27.1 shows population distribution in Nigeria. It is clear that people living in rural and small towns combined together constitute more than 70% of the entire population.

In Nigeria, rapid population growth has not been accompanied by an increase in the delivery of essential urban services such as water supply, sewerage and sanitation, and collection and disposal of solid waste. It is estimated that currently only about 50% of the urban and about 20% of the semi-urban population have access to a reliable water supply

Table 27.1 Population distribution in Nigeria.

Population distribution type	Community size	Population (million)	% of total
Urban	>20 000	45.0	38
Small towns	5000 to 20 000	40.0	33
Rural	<5000	35.0	29

Source: Federal Government of Nigeria (FGN) (2000).

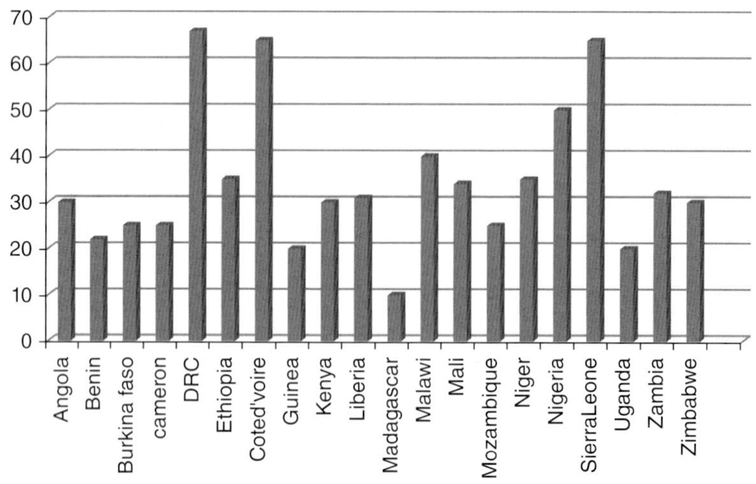

Figure 27.1 Tracking functionality for sustainability (Source: Harvey, 2009).

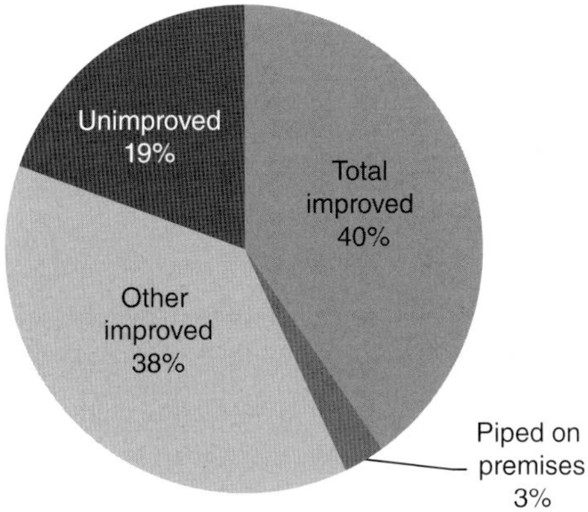

Figure 27.2 Sanitation facilities in Nigeria (Source: WHO/UNICEF, 2012).

of acceptable quality (i.e. one that is better than a traditional source). Overall, effective urban water supply coverage may be as low as 30% of the total population due to poor maintenance and unreliability of supplies. Rural coverage is estimated at 35%. Except for Abuja and limited areas of Lagos, no urban community has a sewerage system, with the result that sewage and sullage either lie stagnant or are disposed of through the stormwater drainage system. According to the WHO/UNICEF (2012), the situation of the sanitation and drinking water sources is illustrated in Figures 27.2 and 27.3.

27.4 Sanitation and Hygiene

Sanitation encompasses a wide range of challenges, including excreta disposal, hygiene, solid waste (rubbish/garbage) disposal, drainage, amongst others; this section focuses on excreta disposal. In Nigeria the minimum standard for household excreta disposal is a safe, hygienic and conveniently located facility (FMWR, 2004). However, lack of safe, private toilets and hand-washing facilities in schools affects educational enrolment, retention and performance. Girls are particularly affected, and poor sanitation is a contributing factor in Nigeria's low female enrolment rates in school (WHO/UNICEF, 2008). Global statistics on sanitation hide the dire situation in some developing regions. With an average coverage in developing regions of 50%, only one out of two people has access to some sort of improved sanitation facility. The regions presenting the lowest coverage are Sub-Saharan Africa (37%), Southern Asia (38%) and Eastern Asia (45%). Western Asia (84%) has the highest coverage among developing regions (WHO/UNICEF, 2008).

Nigeria lacks a comprehensive strategy on sanitation as a whole, including excreta disposal, solid waste disposal, wastewater disposal, drainage and treatment of wastewater. The new Water Supply and Sanitation Strategy document links sanitation development to water supply under the Ministry of Water Resources. However, sanitation units from the Ministries of Health and Works and Housing have recently been transferred to the Ministry of Environment. Currently, individual solutions are adopted at the household level, for example pit latrines, septic tanks and storage (Table 27.2).

Table 27.2 National definitions of acceptable access to water supply and sanitation in Nigeria.

Settlements	Water supply	Sanitation facility
Rural	Rural water supply: guaranteed minimum level of service 30 litres per capita per day within 250 m of the community of 150 to 5000 people serving about 250–500 persons per water point	Each household in rural areas (population <5000) must own and have access to at least an upgraded pit latrine
Semi-urban	Semi-urban (small towns) water supply: represents settlements with population between 5000 and 20 000 with fair measures of social infrastructure and some level of economic activity with minimum supply standard of 60 litres per capita per day with reticulation and limited or full house connections as determined by beneficiaries/ government	Each household in semi-urban areas (population 5000–20 000) must own and have access to safe sanitary facility of at least a Sanplast latrine
Urban	Urban water supply: 120 litres per capita per day for urban areas with population >20 000 inhabitants to be served by full reticulation and consumer premises connection	Each household in urban areas (population >20 000) must own and have access to safe sanitary facility of at least a pour flush toilet

Source: FGN, *National Rural Water Supply and Sanitation Project (NWSSP)* (2001).

There is very little sewerage in urban Nigeria. Regarding solid waste, while there is some level of public and private solid waste collection, the frequency of collection is poor. The stormwater drainage system is frequently a disposal point for solid waste. Moreover, disposal, when waste is collected, is by dumping rather than sanitary landfill and is a major cause of water pollution, either through the stormwater drainage system or seepage into the groundwater. Wastewater disposal pollutes the surface water. Being in an embryonic stage, the sanitation subsector requires better-formulated policies and a massive injection of well-formulated investments, designed specifically for African conditions, combined with institutional reforms. The Bank has been the only donor in the subsector with three projects to address this situation, but these efforts need to be multiplied significantly (Federal Government of Nigeria (FGN), 2000).

The benefits of good sanitation far outweigh the costs, including health care costs and loss of productivity. Waterborne sewerage systems are rare in Africa. Only half of Africa's large cities have sewerage networks, and only Namibia, Senegal and South Africa provide universal sewerage access. Sewerage networks reaching approximately 10% of the population, such as those in Côte d'Ivoire, Kenya, Lesotho, Madagascar, Malawi and Uganda, are more typical. Little more than half of the households with piped water also have flushing toilets, which are often connected to septic tanks rather than to sewers. This is not surprising given that development of waterborne sewerage networks generally lags substantially behind the evolution of the piped water networks on which they depend. In low-income countries of Africa, only 15% of the population enjoy private connections to piped-water networks, and this already places a low ceiling on the potential for waterborne sewerage (World Bank, 2004). Sanitation is predominantly on-site and typically takes the form of traditional pit latrines. Half of the population use these, and the rate of use is approximately equal in both urban and rural areas. Curiously, the number of improved latrines is not much greater than that of septic tanks, despite a significant cost difference between the two. An urban–rural divide emerges when access to improved sanitation is considered (World Bank, 2004).

Also, a current report on sanitation from the WHO and UNICEF states that, globally, 63% of the population use improved sanitation facilities. Since 1990, 1.8 billion people have gained access to improved sanitation. An estimated 0.25 billion people are still without improved sanitation; almost three-quarters of them live in rural areas. In urban areas, 8 out of 10 people use an improved sanitation facility, compared to only half of the rural population. However, the number of people without improved sanitation in urban areas has grown by 183 millionsince 1990, during a time of rapid urbanisation. The number of people resorting to open defecation globally has decreased by 271 million since 1990. Still, open defecation is practised by 1.1 billion people – 15% of the global population (WHO/UNICEF, 2012).

27.5 New Approach to Rural Sanitation

Community-led total sanitation (CLTS) is a concept introduced by Kamar Kar to mobilise communities to take on responsibilities of proper faeces disposal, considering the menace caused by faeces. It revisits all the past approaches, particularly the promotion of household sanitation within the context of basic human dignity (Sah and Negussie, 2009). Community-led total sanitation initiatives are not emphatic on latrine construction as such and they therefore avoid the use of hardware subsidies. Instead, mobilisation efforts focus on helping communities and individuals understand the health risks of open defecation and use disgust and shame as 'triggers' to promote action. This ultimately leads to the construction and exclusive use of locally built low-cost household latrines. The ultimate goal of community-led total sanitation is for communities to achieve and maintain open defecation-free status and improved hygiene practices (Kar and Chambers, 2008).

The main difference between community-led total sanitation and other approaches are the rejection of hardware latrine subsides and the use of shame and disgust to trigger as the key motivation for behaviour change and involve the entire community in the process (as opposed to individual households). Other participatory approaches such as the Participatory Hygiene and Sanitation Transformation (PHAST) have led to significant change in some countries; in other countries, such as India and South Africa, even subsidy-based approaches have been successful. Nevertheless, there is mounting evidence that community-led total sanitation can lead to more rapid and sustainable behavioural change than other approaches and with a significantly lower programme cost (Kar, 2003).

27.6 The Sanitation Ladder Concept

The WHO and UNICEF Joint Monitoring Programme (2008) adopted the concept of a ladder in developing a global monitoring framework for the achievement of water and sanitation Millennium Development Goals (MDGs) by distinguishing between 'improved' and 'unimproved' sanitation facilities (WHO and UNICEF, 2008, p. 6). The focus has recently shifted from the facilities themselves to the 'use of facilities', but in the WHO/UNICEF (2010) report, the emphasis remained on types of latrines or technology options and therefore on the 'containment' part of the sanitation service delivery chain, rather than on disposal, treatment and reuse, or on solid and liquid waste. Hygiene covers a range of health and environmental issues, including the use of water and sanitation to block the transmission of related diseases and improve health. As a central component in both water and sanitation services, hygiene cannot simply be an add-on to either the water or sanitation service ladders.

It is widely accepted that effective, sustainable hygiene promotion cannot be achieved through a once-only intervention and requires ongoing activities from multiple sources. Hygiene promotion can be seen as a public or environmental health function and therefore 'a service', either undertaken by public or environmental health departments or by the sanitation provider or utility. However, water and/or sanitation infrastructure-related hygiene promotion is usually 'an intervention' that happens between once and five times in a project cycle, and is unlikely on its own to result in sustainable improvement in hygiene practices. Arguably, hygiene promotion will only result in sustainable behaviour change if it is an ongoing, integrated service management (IRC, 2011).

27.7 Management and Sustainability of WASH facilities

Sustainability of water and waste management means a beneficial change in access to services, leading to corresponding lasting outcomes and impacts on people's lives. The time dimension implied in the idea of sustainability is not finite. Once change for the better has been brought about, that trajectory of change must be maintained and enhanced. If communities slip back into a situation where they have to rely on unimproved water and sanitation services then investment has effectively been wasted (WaterAid, 2011). Achieving sustainability of the water supply and sanitation facilities has remained a daunting challenge for stakeholders in the WASH sector.

The concept of sustainability is generally viewed as implying economic, social, institutional, financial and technical viability of specific development endeavours. Sustainability can also be assessed in terms of development activities embedded in national and sectoral priority goal(s) supported by an environment able to maintain them. Sustained human and institutional development efforts could greatly contribute to long-term sustenance of the development effectiveness of programmes and projects as well as overall sector development.

Figure 27.1 shows a cost of investment loss of between US$1.2 to 1.3 billion in 20 countries in Sub-Saharan Africa over 20 years. It specifically shows the percentage of nonfunctioning hand pumps (HP). The adverse impact of these factors on sustainability is exemplified by the significant deterioration in the performance of a number of projects in many rural, small town and urban communities across Africa.

The WASH sector has made important progress in bringing initial infrastructures to many, but the reality of short-lived investment has not met expectations. This has often resulted in broken systems and wasted resources. The sector has seen progress in recent decades, yet outputs of infrastructure and decentralisation have not brought lasting service outcomes. Billions of investment dollars and the evolution of WASH approaches have brought new services to nearly 720 million people in the last 20 years. However, the reality is that unsustainable services are undermining progress due to nonfunctioning infrastructures. Typical water point failure rates of 30–40% in Sub-Saharan Africa equates to billions of dollars of wasted financial investment over the years and a subsequent regression in health and quality of life. Figure 27.3 represents an average 36% investment loss in Sub-Saharan Africa of between US$1.2 to 1.3 billion over 20 years.

The responsibility of communities to manage their water and sanitation services forms a central component of much WASH sector policy and strategy. However, subscribing to this

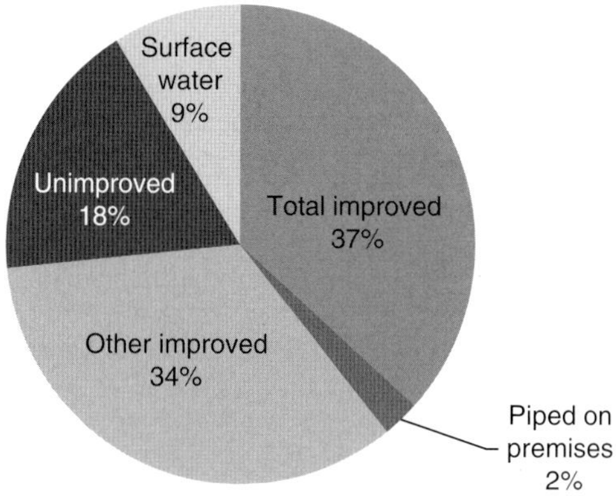

Figure 27.3 Current drinking water situation in Nigeria (Source: WHO/UNICEF, 2012).

Figure 27.4 A typical picture of an abandoned hand pump borehole constructed between 2005 and 2010 in Mopa Kogi state (Nigeria).

principle has not delivered the results expected. In some cases this is due to poor implementation; in other cases the principles are simply inadequate. The community management model has sometimes been presented as a panacea for achievement of lasting services, but in the absence of external support, there is extensive evidence of its weaknesses. The evolution of community management as a pragmatic response to weaknesses in public service provision, and its subsequent promotion as an ideal model of service delivery, was a triumph of hope over realism (Feachem, 1980). Water supply and sanitation development of any nation are continuing long-term processes, which require careful planning and implementation geared towards achieving improved quality of life (Babalola, 1997). Consequently, there is a need to find and identify an integrated approach specific to each community in planning WASH programmes. Figure 27.4 is an illustration of an abandoned facility in Kogi State Nigeria – typical of such failed projects. Several thousands of such nonfunctional abandoned hand pump boreholes exist across the country. However, government and other external intervention agencies at the initial installation would have taken user numbers from the community towards achieving the MDG's goal on water supply and sanitation, but in reality the people are still lacking adequate access to a safe water supply and improved sanitation as a result of management and sustainability issues.

For example, despite the fact that one-third of Malawi is covered by water, 47% of the population, estimated at 11 million, have no access to safe drinking water due to both nonfunctional water facilities and their geographical distribution (Ministry of Water Development (MWD), 2003). With widespread poverty and the high prevalence of HIV and AIDS, the country is facing enormous challenges that the government cannot resolve alone. The majority of rural people continue to rely on unprotected and unsafe water sources such as rivers, open wells, springs and lakes for their water supply.

During the early 1980s to 1990, the water supply sector received considerable financial support and investments allied to the UN Water Decade (International Drinking Water Supply and Sanitation). Despite such heavy investment, the water facilities provided were not sustainable because they lacked effective community participation from the water users according to the Water Services Sector Study (WSSS) carried out in 1994 (MWD, 2003). The WSSS noted that lack of community involvement and ownership of the water facilities were critical issues to achieving sustainability. The Government of Malawi has subsequently revised its Water Development Policy (2003) and is now advocating demand–response approaches to rural water supplies using community-based management of water and water facilities, on the premise that users are the best managers of the resources upon which their lives depend. Community ownership of the water schemes and points can also reduce government responsibility for repair and maintenance since this rests in the hands of the rural people.

27.8 Rural WASH in Nigeria

Rural Nigerian communities with populations less than 5000 usually do not have access to electricity, piped water or tarred roads. The National Standards of water consumption for rural areas is currently 30 litres of safe water per capita per day delivered within 250 m of the community and serving about 250–500 persons per water point (Multiple Indicator Cluster Survey (MICS), 1999). 'Safe water' means that it meets the National Drinking Water Quality for Nigeria. However, rural water supply coverage is low, much lower than the Sub-Saharan average of 42%. Nigeria is one of only eight countries in the world where rural water coverage had been assessed to be dropping (FGN/UNICEF/WaterAid, 2009).

In rural water supply and sanitation, demand for community water supply services are localised demands. Therefore, managerial decisions about levels of service, locations of facilities as well as cost sharing are made locally. The main role of the higher government agencies should be to establish institutional rules, regulations and processes that encourage such local decisions.

27.8.1 Small Towns and Urban WASH

Evaluations carried out by the African Development Bank and other donors show that many of the urban entities providing water supply-related services were weak and unable to achieve financial viability (African Development Bank (ADB), 2004). In 1980, about 44% of people in developing countries had access to safe water supplies, rising to only 61% by 1998. Similarly, in 1980, only 46% of people in developing world had access to sanitation, rising to 56% by 1998. In developing countries, the expansion of water and sanitation services and associated infrastructure in towns and cities is not keeping pace with the growth of the urban population and their communities of slums and informal settlements (WaterAid, 2009).

In an attempt to increase viability and sustainability of water supply coverage, urban and small town utilities are increasingly looking for win–win approaches to reaching the urban poor. Research carried out by WaterAid (2011) showed that there is a vast potential market for potable water in the poorer urban areas of Africa and Asia. It also showed that 65% of urban residents in Sub-Saharan Africa are presently paying for water at exorbitant prices, but remain unconnected to the public utilities pipe network (WaterAid, 2011). According to WHO/UNICEF (2008), 40–46% of urban water supply is through an off-grid water supply system, with the more formal urban water and sanitation provision tending to prioritise wealthier areas, neglecting the vast populations of informal settlements with no access to WASH facilities. Policy-makers, sector regulators and utility managers often have perceived poorer households as not being viable customers; however, recent evidence suggests that, by meeting the needs of poor people, utilities can grow and increase profitability (WaterAid, 2011). Core management issues facing many urban water utilities around the world, particularly in Asia and Sub-Saharan Africa, include:

- Lack of stable, predictable and sufficient finance leading to underinvestment.
- High population growth in urban areas without proportionate expansion in the services provided by utilities.
- Public utilities being treated as social services.
- Insufficient operational and management autonomy from government.
- Poor management with no regulatory mechanisms to monitor public utilities or making them accountable for performance.
- Inefficiency whereby government is often forced to functionally 'prop-up' failing and bankrupt utilities. The utilities cannot expand to serve new customers and would normally use state funds inefficiently to serve existing ones, almost always the nonpoor (WaterAid, 2011).

Therefore it is clear that there are obvious management issues that particularly remain as considerable challenges to delivering efficient and sustainable WASH infrastructure in rural, small towns and most urban areas in Nigeria. These are briefly discussed as follows:

1. *Capacity gap*. A capacity gap is generated by insufficient scientific and technical expertise and infrastructures for designing and implementing water policies. If there is a difference between the capacity needed to shoulder water responsibilities and the local authority's technical, procedural, networking and infrastructure capacity, consequences for implementation of national water policies are unavoidable. The local authority may not have the funding to operate and maintain services effectively. This may lead to the deterioration and potential failure of services and infrastructure, which in turn threaten the quality of water resources (Organisation for Economic Co-operation and Development (OECD), 2012). Addressing capacity gaps assumes significant importance with the growing complexities around provision and management of water and sanitation facilities. There is also a greater need for promoting aspects of social engineering, both at implementation and educational

levels, to address people-friendly sustainable solutions. Unlike in India and some other developing countries, universities in the West (Water Engineering Development Centre (WEDC), Surrey and Grandfield in the United Kingdom and the Institute for Water Education in the Netherlands) have several tailor-made courses focused on these issues, both in the short and long term. These courses are unaffordable by many due to their financial implications (http://www.washinstitute.org/about_washi.php).

2. *Poor community participation.* The lack of community participation has led to poor operation and maintenance of WASH projects. Many such projects were executed in the form of a top–down approach, where a central government agency or an external support agency intervened in a community without the necessary consultation with the community. In some cases, the facilities were unused, resulting in a lack of sense of ownership. Where the facility suffered a breakdown, the community tended to rely on, and expect, external solutions. This has long been identified as a major management challenge to WASH service provision and sustainability.
3. *Lack of reliable data, especially at subnational and local levels.* This constitutes a serious management challenge and a barrier to developing and implementing investment plans in sanitation and drinking water (WHO/UNICEF, 2010). Furthermore, accurate and reliable WASH data are not readily available to support effective and timely decision-making in African countries (WHO/UNICEF, 2008). Generally, data streams are not harmonised and data literacy across the sector is low. Different institutions, both national and international, provide different data on water and sanitation as a result of different definitions and indicators (Sarkolie and Addai, 2009). In many countries, at least three sector data collection activities run in parallel. These include: nationally consolidated data (taken from regular government agencies, such as ministry of water resources, census data and data from household survey information (WHO/ UNICEF, 2010). These are analysed to satisfy global reporting requirements and, in addition, community-based qualitative data are often also facilitated by local government and nongovernmental officer (NGO) partners (Alana, 2011). At the subnational level, invalid, biased or unrepresentative data lead to repeated failures to target the vulnerable and extend services to the poor (WELL, 2004).
4. *Poor policy formulation and implementation.* The poor state of affairs with regard to access to safe water, hygiene and sanitation in Nigeria is thus partly as a result of the lack of harmonised policies on WASH data development (WHO/UNICEF, 2010). Without major policy changes and considerable improvements in water management processes and techniques, by 2050 the situation is likely to deteriorate, and will be compounded by increasing competition for water and increasing uncertainty about water availability. Policy solutions are often apparent, with key challenges lying in implementing those reforms to existing water policy (OECD, 2012). The policy context within which a project operates has a major influence on the prospects for sustainability. The existence of a well-formulated policy does not guarantee sustainability, especially when it is not backed by political will, but it can at least provide the basis for a common understanding among stakeholders (Parry-Jones *et al.*, 2001). In the absence of a coherent policy, implementation approaches can become fragmented and inconsistent, and use of a wide variety of technologies for the same purpose exacerbates the maintenance problem (Harvey and Reed, 2004).

5. *Environment factors*. These include issues such as fast depletion of groundwater, water quality issues, poor sanitation coverage, poor focus on hygiene, problems in dealing with solid and liquid waste, rapid urbanisation and emerging issues due to climate change, amongst others. These all demand adequate human resources with necessary capacity. The present business environment in Nigeria does not give confidence to private sector participation. There are models such as public private partnership (Harvey and Reed, 2004) promoting private sector participation, but in situations where government policies and disposition keep changing with subsequent changes in regime it creates an unsettled environment for investment consideration in the WASH sector.
6. *Financial viability*. Raising and maintaining adequate funding for water supply and sanitation facilities is of critical importance for sustainability. Insufficient financing is a major factor for poor maintenance, which is often cited as the main reason for failure. This idea is more specifically developed by the OECD in a policy brief about a new financing strategy in water supply and sanitation (http://www.oecd.org/environment/39157644.pdf). According to OECD experts, 'an important obstacle to achieving water supply and sanitation goals in many countries has been the failure to address financial issues' (Diane, 2008).
7. *Undue political interference*. Water supply and sanitation infrastructure scores very high political points in Nigeria and reforms are being introduced to improve services and ensure sustainability of infrastructures. Some politicians keep promising a free water supply for all. There are instances where a newly elected political office holder or the opposition discourage the people from participating in community water management activities or even making water tariff payments, resulting in the inability to meet basic operation and maintenance of facilities.

27.9 Conclusions

Over the last three decades, huge investments have gone into delivering WASH infrastructures in many developing countries across Africa. However, in Sub-Saharan Africa, only about 40% of such intervention's shows an element of sustainability. The Millennium Development Goal (MDG) agreed at the United Nations in 2000 is to halve the proportion of people without sustainable access to adequate and affordable safe drinking water by 2015. However, major changes in policy and planning are required if past, ongoing and future investment in WASH is not to be wasted, as a sustainable water supply and sanitation development does not result simply from the installation of new infrastructure. The benefits of good sanitation far outweigh the costs, including health care costs and loss of productivity. Waterborne sewerage systems are rare in Africa. Only half of Africa's large cities have sewerage networks, and only Namibia, Senegal and South Africa provide universal sewerage access. Sewerage networks reaching approximately 10% of the population, such as those in Côte d'Ivoire, Kenya, Lesotho, Madagascar, Malawi and Uganda, are more typical (Tables 27.3 and 27.4).

It has been clearly shown in the foregoing, therefore, that core management issues in rural, small towns and urban WASH sectors in the developing countries centre on core policy issues and the proper orientation about water supply and sanitation infrastructure services delivery. This is hinged on the long-term tradition of treating water as a social

Table 27.3 Water supply assessment.

Improved water supply	Not improved water supply
• Household connections • Public standpipes • Borehole (motorised and hand pump) • Protected dug well • Protected spring • Rainwater harvesting	• Unprotected well • Unprotected spring • Vendor-provided water • Bottled water • Tanker truck-provided water • Streams and ponds

Source: FGN, *National Rural Water Supply and Sanitation Project (NWSSP)* (2001).

Table 27.4 Sanitation assessment.

Improved sanitation facility	Not-improved sanitation facility
• Connection to public sewer • Connection to septic system • Pour flush latrine • Simple pit latrine • Ventilated improved pit latrine • Latrine in public places • Shared latrines	• Service or bucket latrines, where excreta are manually removed • Latrine with an open pit • Defecation in bushes

Source: FGN, *National Rural Water Supply and Sanitation Project (NWSSP)* (2001).

good, capacity gaps at the technical and management level, and an inability to meet financial obligations of operation and maintenance. Making progress towards attaining and ideally surpassing MDG targets for water and sanitation will require a shift from singularly focusing on expanding infrastructure in areas without service to concentrating on both achieving long-term functionality goals through improved operation and maintenance of existing supplies.

While government at the state levels are forced into making attempts to keep utilities running, it will be necessary to also emphasise the need to keep up environmental sanitation in the urban areas. Keen attention must be given to managing WASH infrastructures in semi-urban and rural communities in Nigeria. Further research and studies are required to understand in more detail community-based participatory cross-sectorial approaches that have a sustained service delivery for possible adaptation.

References

African Development Bank (ADB) (2004) *Evaluating Bank's Assistance for Capacity Strengthening of Urban Water Supply and Sanitation Entities in Regional Member Countries Operations*. Operations Evaluation Department, ADB.

Alana, P. (2011) *Assessing Sanitation Service Levels*, Second Edition. IRC International Water Sanitation Centre.

Babalola J.O. (1997) Rural water supply: issues, problems and prospects, water resources. *Journal of Nigerian Association of Hydrologists (NAH)*, 8, 19–25.

Barlow, M. (1999) The global water crisis and the common definition of the world's water supply. In: *International Forum on Globalization.*

Bunclark, L., Carter, R., Casey, V., Day, St J. and Guthrie, D. (2011) *Managing Water Locally: An Essential Dimension of Community Water Development.* Institution of Civil Engineers, London.

Delmon, J. (2001) *Water Projects: A Commercial and Contractual Guide, International and National Water Law and Policy Services.* Kluwer Law International, Waterlands.

Diane, B. (2008) *Service Delivery in Rural Environment: Past Approaches and the Way Forward Review of Literature.* Emerging Market Group Ltd.

Feacham, R.G. (1980) Community participation in appropriate water supply and sanitation technologies: the mythology for the decade. *Proceedings of the Royal Society, Biological Sciences*, **209**, 15–29.

FGN (2000) *Federal Republic of Nigeria, Water Supply & Sanitation Interim Strategy*, FGN, Nigeria.

FGN (2001) *National Rural Water Supply and Sanitation Project (NWSSP)*, FGN, Nigeria.

FGN/UNICEF/WaterAid (2009) *Sanitation Fact Sheets Nigeria.* Federal Ministry of Agriculture and Water Resources, Abuja, Nigeria.

FMWR (2004) *National Rural Water Supply and Sanitation Programme a Strategic Framework.* Draft Final prepared by The Department of Water Supply and Quality Control, Federal Ministry of Water Resources.

Harvey, P. (2009) *Water Point Mapping for the Analysis of Rural Water Supply Plans: Case Study from Tanzania.* UNICEF, Geneva.

Harvey, P.A. and Reed, R.A. (2004) Rural Water Supply in Africa: Building Blocks for Hand Pump Sustainability. WEDC, Loughborough University. Available at: http://www.oecd.org/environment/39157644.pdf cited 28/08/2012.

IRC (2011) *Assessing Sanitation Service Levels*, Second Edition, July. Working Paper 3. IRC International Water and Sanitation Centre, IRC.

Kar, K. (2003) *Subsidy or Self-Respect? Participatory Total Community Sanitation in Bangladesh, Working Paper 104.* Institute of Developmental Studies, University of Sussex.

Kar, K. and Chambers, R. (2008). *Handbook on Community Led Total Sanitation.* Institute of Development Studies/PLAN International, 109 pp.

Malhotra, A.K. (1997) *Private Participation in Infrastructure: Lessons from Asia's Power Sector on Finance and Development*, December 1997, 33 pp.

Ministry of Water Development (MWD) (2003) *National Water Policy.* Malawi Government.

Multiple Indicator Cluster Survey (MICS) (1999) *Multiple Indicator Cluster Survey Results*, MICS3. National Planning Commission, Nigeria.

OECD (2012) *Meeting the Water Reform Challenges.* OECD Studies on Water, OECD Publishing, pp 102–103. Available at: http://dx.doi.org/10.1787/97892641700001.

Parry-Jones, S., Reed, R. and Skinner, B.H. (2001) *Sustainability Handpump Project in Africa: Draft Guidelines for Field Evaluation Of Handpump Projects.* WEDC, Loughborough University, UK.

Sah, S. and Negussie, A. (2009) Community led total sanitation (CLTS): addressing the challenge of scale and sustainability in rural Africa. *Desalination*, 248 (1), 666–672.

Sarkolie, Y.A and Addai, E. (2009) Making use of available water and sanitation data. In: *Proceedings of the 34th WEDC International Conference on Water, Sanitation and Hygiene. Sustainable Development and Multisectoral Approaches*, UN Conference Centre, Addis Ababa, Ethiopia. WEDC UK.

UN Millennium Project (2005) *Investing in Development: A Practical Plan to Achieve the Millennium Development Goals.* United Nations Development Programme, New York.

WaterAid (2011) *Reaching the Urban Poor: Supporting Utilities to Engage Communities in Service Delivery.* WaterAid UK.

WEDC (2010) *Management of Water and Sanitation.* A WEDC Postgraduate Module, Loughborough University.

WELL (2004) *Water Supply and Sanitation Sector Monitoring: A New Decade to Measure.* Notes. WEDC UK, pp. 5–8.

WHO/UNICEF (2008) *Progress on Drinking Water and Sanitation: WHO/UNICEF Joint Monitoring Programme for Water Supply and Sanitation*. Geneva.

WHO/UNICEF (2010) *Progress on Drinking Water and Sanitation: 2010 Update, Joint Monitoring Programme for Water Supply and Sanitation (JMP)*. UNICEF, New York, and WHO, Geneva.

WHO/UNICEF (2012) *Progress on Drinking Water and Sanitation: WHO/UNICEF Joint Monitoring Programme for Water Supply and Sanitation*. Geneva.

World Bank (2004) Africa's Water and Sanitation Infrastructure. In: Banerjee, S.G., Morella, E., Foster, V. and Briceño-Garmendia, C. (Series eds), *Access, Affordability, and Alternatives*, Volume 1.

World Health Organization (WHO) (2010) *Global Annual Assessment of Sanitation and Drinking Water (GLASS). Targeting Resources for Better Result*. WHO, USA, 8 pp.

28

Balancing Flood Risk and Water Scarcity of the Asian Delta Regions

Jessica E. Lamond

28.1 Introduction

Flood disasters and droughts cause devastation and hardship in some part of the world each year. The annual average reported cost of flooding has been $54 billion over the last decade and the equivalent cost of droughts is estimated at $3 billion (EM–DAT, 2012). Flooding in particular has been high on the global agenda with the number of reported flood events having risen over the last three decades and with trends in urbanisation and climate fluctuations contributing to increasing flood hazard, particularly in the developing nations where urban growth is highest. Flood losses have increased disproportionately, over and above the increase in flood hazard, as the value of assets at risk has grown through increased wealth and larger populations in areas at risk of flooding (Jha *et al.*, 2011).

However, these reported costs and disaster statistics are likely to be the tip of the iceberg in terms of the real economic, ecological and social impact of extreme events related to the water cycle. Many small-scale floods and water shortages are not reported as disasters and therefore not included in the disaster statistics (Satterthwaite *et al.*, 2007); rather they are accepted as part of the natural seasonal fluctuations and dealt with by local communities along with all the other challenges they face. Large-scale flooding has implications in terms of social and economic effects on the flooded areas and wider economy that are felt for many years after a large disaster and the effect of multiple smaller events can be just as corrosive to local development goals. Water shortages probably affect an even greater population indirectly and the implications on food supply chains of severe drought could be global (Stern, 2006).

While it is noted that the effect of a changing climate will exacerbate the extreme weather conditions affecting the prevalence of floods and water scarcity, many countries are already facing a cycle of flood and drought, driven by urbanisation and associated human demands

Water Resources in the Built Environment: Management Issues and Solutions, First Edition.
Edited by Colin A. Booth and Susanne M. Charlesworth.

on the ecosystem (Intergovernmental Panel on Climate Change (IPCC), 2007). In few places is this interdependence clearer than in the situation in the Asian deltas, such as the Mekong Delta, the Bay of Bengal and the Ciliwung (World Bank, 2010). This chapter will describe the complexities facing management of flooding in delta cities within the context of wider water management and then go on to illustrate the challenges within the specific case study of Jakarta in Indonesia.

28.2 Characteristics of Flooding in Asia

In terms of disaster statistics for flooding the largest losses are suffered by nations in the Asia-Pacific and the size of populations involved in flood events in Asia demonstrates the importance of this region in reducing global flood risk. During the last 30 years the total number of flood events that occurred in Asia represented about 40% of global events (EM–DAT, 2011). The worst affected country is China, followed by India, Indonesia, Bangladesh, Vietnam, Thailand and Pakistan. Figure 28.1 shows the spread of flood events worldwide, clearly demonstrating the prevalence of flood events across Asia.

Asia is subject to all kinds of flooding. The seismic vulnerability of the Asian region is associated with a high risk of tsunami events, for example the 2011 tsunami in Japan and the 2004 Indian Ocean tsunami. Monsoon and other intense rainfall events are a seasonal reality in cities such as Mumbai and can lead to events such as the 2005 Mumbai floods. Characteristics of Asian flooding are also determined by weather phenomena such as the El Niño South Oscillation (ENSO). Tropical cyclones bring heavy precipitation and storm surges in coastal areas (Jha and Brecht, 2011). Floods are often magnified by the massive urbanisation of the region due to the impervious surfaces, clogged drainage and poor land use controls.

Asian delta regions are among the most highly stressed areas in the world. They experience fierce competition for land and water use, and accumulation of impacts from events and activities across the whole river basin. Deltas are naturally vulnerable to complex problems such as flooding, subsidence and salt water intrusion. Delta cities, often important national or regional ports, are made more vulnerable by the human activities and associated demands on resources and land use. As a result many Asian delta cities and urbanised areas are at risk from flooding from the sea, from the confluence of multiple rivers, sea level rise, pluvial flash flooding, inadequate drainage and failure of man-made systems (IPCC, 2007).

Intense rainfall is a feature of tropical weather patterns that leads to large quantities of water flowing on to the surfaces of delta cities without the possibility of upstream interception or redirection. The detailed pattern of rainfall is often unpredictable, making accurate prediction difficult and reducing the possibility of deploying protection or implementing evacuation procedures. In a low-lying delta region it is often not possible to rely on gravity drainage to mitigate against intense rainfall and widespread surface water flooding is commonly experienced.

Coastal location at the output point of major river systems makes delta areas particularly vulnerable to fluvial and coastal flooding (Aerts *et al.*, 2009). The Mekong River serves as an illustration of the issue; rising in Tibet the Mekong runs through China, Myanmar, Laos, Thailand and Cambodia and is the cause of flooding along its length such that the nations have set up a commission, the Mekong River Commission, to track and control the flow (Mekong River Commission, 2009). In Vietnam the Mekong discharges to the sea via a delta and despite the mainly rural nature of the delta this is one of the most

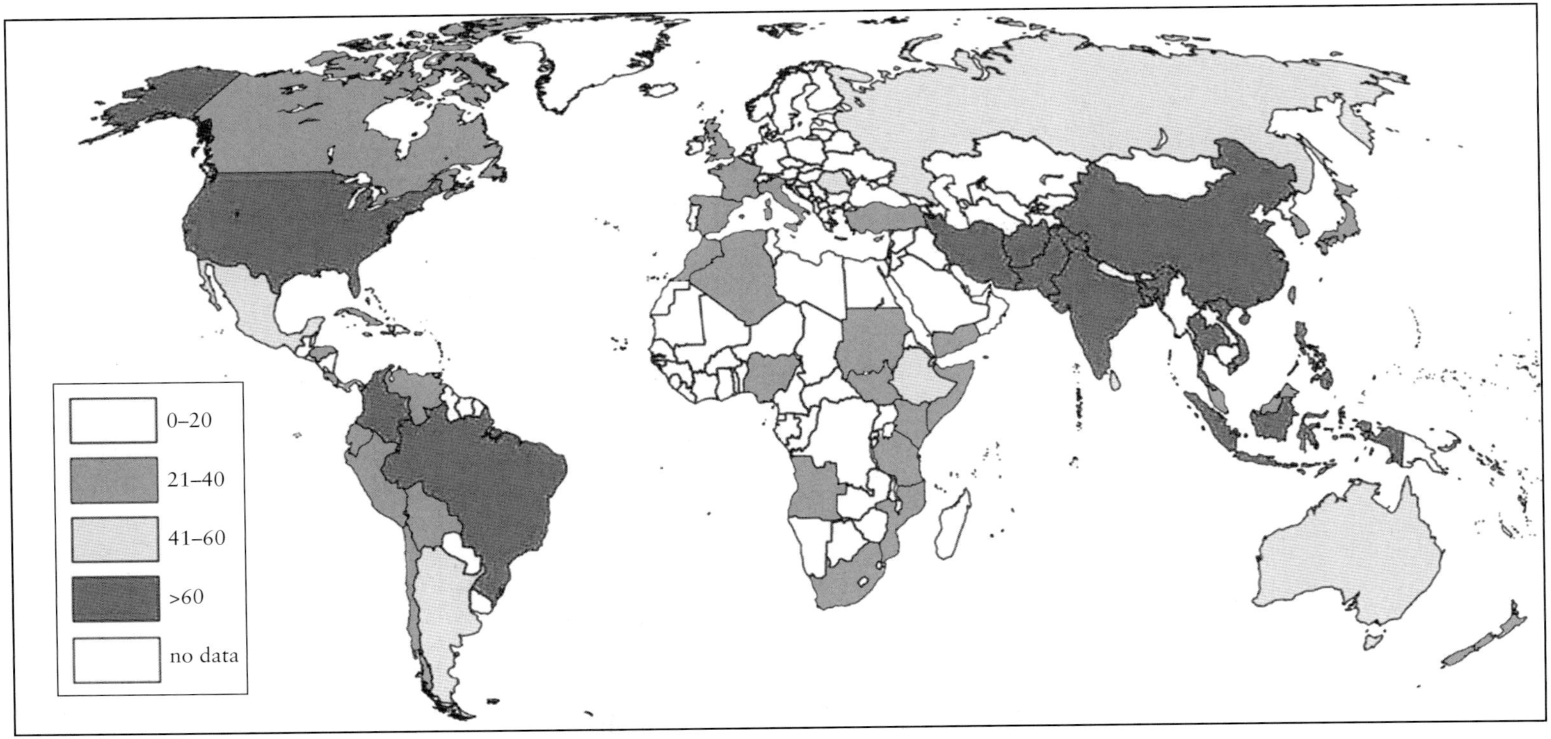

Figure 28.1 Map showing the spread of flood events worldwide, clearly demonstrating the prevalence of flood events across Asia.

densely populated parts of Vietnam, with Ho Chi Minh City sitting on the edge of the delta.

Coastal erosion processes associated with a rising sea level are already an issue in Asia and the projected sea level rise in Asia is predicted to result in severe shoreline retreat (IPCC, 2007; Yang *et al.*, 2006) (see Chapter 5). Declining sediment transport in monsoonal Asia is also implicated in coastal erosion and in some cases this is related to the construction of dams in upstream districts (Yang *et al.*, 2006). Millions of people in South East and East Asia are vulnerable to flooding caused by coastal erosion and sea-level rise (Stern, 2006). In Tokyo, Osaka and Nagoya vast tracts of land are already located below the high-water level. Coastal lowlands below the 1000 year event are widely distributed in Asia (Nicholls and Lowe, 2004) with Bangladesh, Japan, Vietnam and Thailand highly at risk (IPCC, 2007). In 2011 flooding in Ho Chi Minh City was caused by a combination of extreme tidal surge and release of water from hydropower reservoirs (Chinh, 2008).

A major finding from a recent World Bank report on flooding and other climate impacts on Asian Coastal Megacities was that the issue of land subsidence was sometimes more important than climate-related factors. For example, in Bangkok there has been a doubling of flood-related costs attributable to land subsidence (World Bank, 2010). Large deltas are seen to be sinking at rates up to 30 centimetres per year, with the Pearl River and Mekong Deltas particularly vulnerable (Fuchs, 2010). This is supported by more recent evidence from Jakarta that the projected impact of sea-level rise will be dwarfed by the impact of subsidence over the next years (Brinkmann, 2011; World Bank, 2012).

The exact causes of land subsidence are difficult to determine but it is thought that land compaction and extraction of groundwater are major contributors (Nicholls *et al.*, 2007). Excessive groundwater pumping leads to compaction of sediments (Taniguchi *et al.*, 2009a) and the scale of the impact is dependent on the physical characteristics of the aquifer and urbanisation of the surface (Yoshikoshi *et al.*, 2009; Taniguchi *et al.*, 2009b). Other geological processes may also be involved. Land compaction is a partly natural process but this is speeded up by urbanisation and the levels of groundwater extraction is driven by the needs of residential, agricultural and industrial users of water and their alternative supply sources. In deltas there is also a vicious circle effect, where subsidence leads to increased sea water incursion into fresh water supplies and causes increased need for groundwater extraction.

Water demand is influenced by the population and in major Asian delta cities and conurbations the sheer scale of the population puts demands on the water resources and the supply of piped water is often low. Therefore normal practice is to extract water from the ground via the sinking of boreholes or wells. Urbanisation of the delta areas also reduces the infiltration capabilities of the land, reducing the replenishment of the groundwater. Furthermore, measures to reduce flooding such as channellization of water courses may further contribute to low levels of groundwater as they minimise infiltration, moving water quickly through and out of the city and into the sea.

28.3 Approaches to Risk Reduction

Section 28.2 showed that the causes of flooding within the Asian deltas are complex; equally there are also many alternative and complementary measures that can be adopted to reduce flood risk. These measures are often categorised as structural (hard engineered) or nonstructural (soft) measures, or alternatively as those measures designed to control the water, keeping away from settlements or management measures that are designed to minimise the impact of water that reaches the urban environment. Traditionally, the natural desire to keep water away from settled areas has meant that structural measures, such as barriers and diversion channels, have been seen as the first recourse. However, structural measures are not always the most appropriate choice as they can be ineffective, expensive and move the flooding problem elsewhere (Jha *et al.*, 2012). It is vitally important to recognise that the risk from flooding can never be completely removed. Even with the most comprehensive structural protection systems in place there will remain a residual risk. Early warning systems and evacuation plans, resilient construction, emergency and recovery plans, and financial measures such as insurance are designed to manage the damage from flooding. Therefore, to tackle the challenges associated with multiple sources of hazard, an integrated approach to risk management is indicated, combining structural and nonstructural measures to progressively reduce the expected human and financial damage. Within the Asian deltas a core element of the integrated approach often relates to management of the wider water cycle to prevent land subsidence.

The management of water crosses the categories of structural and nonstructural measures, including structural measures such as storage and conveyance through environmental approaches such as wetlands to regulation of extraction. Optimising storage is likely to be one component in an overall strategy for flood risk reduction while also contributing to the supply of water and reducing the risk of drought.

There is storage in all parts of the natural water cycle, which can be enhanced by creating additional opportunities for storage within a catchment or water basin. Storage, preferably semi-natural storage, is also a key to modern urban drainage design. Storage occurs naturally in a catchment, for example within the floodplain or, more locally, in ponds. In heavily urbanised catchments the issue of construction within these natural storage areas necessitates careful planning to maintain the catchment capacity to absorb flows. Regulations can be used to minimise the loss of storage associated with construction, as has been implemented in India (Jha *et al.*, 2012) and Jakarta (Tanaka, 2011). Where such regulations are rigorously enforced they have the potential to limit significantly the impact of urbanisation on flood risk and could be particularly important in the Asian deltas where the expansion of urban development is rapid.

Storage can be enhanced by the creation of artificial facilities including flood storage reservoirs, retention ponds and detention ponds. Such storage facilities may often be part of the water supply network. Temporary flood storage capacity can also be utilised by planning the deliberate flooding of farmland or urban areas like playing fields or car parks. An example of the temporary use of playing fields and recreational areas for flood storage is seen in Japan in the Tsurumi Retarding Basin (Tanaka, 2011). Here the planned inundation of the area is supported by careful design, which allows controlled flooding and emptying of the area, with elevated access to the main stadia allowing safe egress during an event. The planning of sacrificial areas for temporary flood storage can be highly effective in protecting urban centres but is also controversial, as was seen in the Mississippi floods in 2011 (Lovett, 2011; Wynne, 2011). The role of negotiation, compensation and continuous communication in such circumstances is critical. Temporary facilities could provide an opportunity to harvest the stored water for greywater supply or treatment.

Structural measures to control the flow of water through the river system such as dams may be an integral part of the systems for water supply and may also contribute to energy generation as in Mozambique (van Ogtrop *et al.*, 2005), Latin America (Tucci, 2007) and China (Yang *et al.*, 2006). These structures also facilitate the management of fluvial flooding as peak flows can be attenuated to a certain extent through online and offline storage reservoirs (Tucci, 2007). A catchment wide perspective is important in these circumstances to ensure coordination of flows; if attenuated flows from separate tributaries are allowed to join the main river at the same time, the resulting flood downstream may be even worse. For flood control purposes the traditional approach to managing reservoir operations, based on purely hydraulic considerations, may need to be widened and placed within a context of environmentally sensitive integrated water resources management. Moreover, the ability to control river levels has often led to a false sense of security and the feeling that floods are preventable. This in turn leads to a feeling of being let down or neglected, and a culture of blame, when floods do occur (Rose *et al.*, 2010). Another important consideration concerns sediment flows, as there is a risk that reservoirs can lose capacity as a result of long-term deposition and for delta areas the lack of sediment transport can contribute to land loss.

Rain and stormwater harvesting have been proposed as a multipurpose approach to reduce the risk of urban flooding while providing a source of water suitable for reducing demand on the piped or extracted sources. As discussed above, water from communal temporary storage has the potential to be harvested and directed into the supply system. The 'linear' approach of managing large amounts of water in times of peak rainfall by diverting the water through a line of conveyance towards discharge out of the area is replaced by an 'areal' approach of managing rainwater at source by collecting and then storing it. The water thus stored can be used for nondrinking purposes or, properly treated, it can also be used for drinking purposes.

In heavily populated areas the use of individual rainwater collection systems could also be a green and sustainable option for increasing the supply of water in areas of water scarcity where the conventional water supply has failed to meet the demand of the community. Rainwater harvesting has been used informally since ancient times but in many cases has declined in importance as piped water supplies improved and because of the lack of information and technology within rapidly expanding urban environments. As a contribution to the management of urban flooding, a concerted approach would be needed that generated a large combined capacity in locally distributed facilities. Such a system has proved feasible in Seoul and in Thailand (United Nations Environment Programme (UNEP), 2008).

In the flood control context, the major design considerations must ensure sufficient storage capacity at times of peak flow in order to reduce runoff and prevent flooding. This capacity then needs to be quickly restored to accommodate the next peak event. These aims are complementary to, but could also conflict with, the aims of a harvesting system solely for water supply. In water supply systems overflowing may not cause too many problems but collection of heavily polluted runoff, such as is common in urban overland flood flows, would not be acceptable. Therefore, if a rainwater harvesting system includes flood mitigation as a primary part of the purpose these priorities have to be balanced and the system should accommodate or treat polluted runoff.

Two advantages of the use of rainwater harvesting for flood control lie in the flexibility and incremental nature of the installations and in the relatively low cost for construction, operation and maintenance. The dispersed nature of the water storage also implies that failure of part of the system may have lower consequences than failure of part of a hard engineered defence.

In many Asian deltas, where land subsidence has become problematic and is implicated in flood risk, control of groundwater extraction can be instrumental in slowing the growth in hazard and may, potentially, help to reduce it. Regulation of groundwater extraction has been introduced in order to slow subsidence in many places (Taniguchi *et al.*, 2009b) and there is some evidence that it can be effective. In Japan, for example, as far back as the 1950s groundwater regulation was instituted as a result of 17–18 cm annual land subsidence in Tokyo. The success of this has been seen in the greatly reduced subsidence of 2 cm observed today (Jago-on *et al.*, 2009; Tanaka, 2011). Extraction control must be seen as a long-term measure, not least because there is a time lag between reduction of pumping and slowing of subsidence: in Osaka the time lag was seen to be 8 years (Taniguchi *et al.*, 2009b). Further, the introduction of regulation can take time and the need for enforcement procedures to be put in place may delay implementation. It is often necessary to phase in restrictions in order to allow for alternative supply sources to be operationalised. Water pricing can be an alternative or complementary strategy in reducing the abstraction of groundwater. Charging for groundwater has been implemented, for example, in Jakarta, Bangkok, Bandung and Tianjin. In Bangkok the successful introduction of extraction controls and water pricing has been a long-term commitment, starting in 1969 as a response to significant land subsidence causing damage to infrastructure and increasing flood risk. The Groundwater Act of 1977 introduced licensing, critical zones were identified and drilling of wells banned in these areas. In 1984 a groundwater tariff was imposed, which was more than trebled in 1994 and then increased yet again to more than 8 times the original charge 10 years later. This pricing mechanism, introduced over a long period and coupled with expansion of a public piped supply, has resulted in groundwater being more expensive than piped water. A consequent reduction in demand and reduction in subsidence has followed (World Bank, 2010; Babel, 2008). Furthermore, funds generated from the groundwater charge have been directed into water supply and conservation projects.

The regulation of groundwater abstraction can be a thorny legislative issue, involving the consideration of who currently owns the rights to abstraction and how these rights may be restricted, as well as pricing issues (Institute for Global Environment Strategies (IGES), 2008). In arid areas, the demand for food production, for home consumption and export, has led to the exemption of agriculture from extraction controls and this limits their effectiveness. In other cases the effect of pricing of the urban and industrial water supply is undermined by low rates. Therefore provision of alternative water sources is an essential element of any programme to prevent groundwater extraction. An increase in the number of water treatment facilities and the encouragement of rainwater harvesting, or other surface water management techniques, can assist somewhat here. Addressing water demand is another alternative strategy.

A further concern of groundwater management is the stability of subsurface structures. As groundwater levels decline during pumping, rigid infrastructure is often threatened. Equally, as controls take effect, the infrastructure constructed during periods of high pumping may suffer from increased groundwater levels. In extreme cases restriction of pumping could lead to groundwater flooding (Foster and Garduño, 2002).

28.4 Flood and Water Management in the Delta City of Jakarta, Indonesia

Jakarta is the capital city and main economic centre of Indonesia with a population of about 9 million in central Jakarta and 27 million in the greater Jakarta metropolitan area (World Bank, 2012). The city is situated on the Ciliwung Delta with about 40% of the land area lying significantly below sea level (Handhayani, 2009). Localised flooding is a regular occurrence, especially during the rainy season, and larger events periodically disrupt the whole city, as, for example, in 1993, 1996, 2002, 2007 and 2008 (Abidin *et al.*, 2009). The 2007 flood event affected about 36% of the city, caused over 70 deaths and displaced 340 000 people and cost an estimated US$900 million. Flood risk in Jakarta is due to 13 rivers, monsoon rainfall, sea-level rise, sea storms, typhoons and tsunamis (Handhayani, 2009).

Flood control measures installed in the past to control river, coastal and pluvial flooding have mostly consisted of heavily engineered structures such as the Western and Eastern Floodway (Nasir, 2008; Handhayani, 2009). These channels were designed to intercept flows and divert them away from the city and into the sea. Sea walls protect Jakarta from coastal flooding and reservoirs and pumping stations or gravity drainage control the water levels in the drainage zones downstream of the floodways. As a result of these investments, dating back many decades and still ongoing, the incidence of major flood events in Jakarta was judged to have been reduced up until 2002 (Tucci, 2009). However, these measures, while extensive, have not reduced the flood risk to an acceptable level, as the 2002 and 2007 events showed, and their full implementation has been delayed and impeded by various administrative and logistical issues such as the need to acquire land and resettle displaced populations.

Factors depleting the effectiveness of Jakarta flood control systems include the fact that massive population growth has expanded the city limits and increased the density of development within the existing city. Since 1980 the population of Greater Jakarta has more than doubled and an estimated 250 000 people relocate to the city each year. Infilling with dense development has resulted in the destruction of retention lakes and building on green spaces (Abidin *et al.*, 2009). This new development has not been planned or regulated to conform to flood-sensitive design and therefore the density has increased peak flows within the existing city and new areas of the city have also added to runoff to levels that were not accounted for in the design of the channels.

The operations and maintenance of flood control systems has also been a contributing factor to their poor functioning. Primary watercourses within the city with a design capacity for a 1 in 25 year flood have been reduced to less than 1 in 5 years through sediment and waste build-up. Lack of dredging, coupled with poor waste management in the city, ensure that blocked drains not only cause flooding but also compromise quality of life through poor water quality and increased disease. Lack of resources for continued maintenance of the system partially explain the situation, but there is also a lack of recognition of the implications of the issue and lack of clear ownership among the multiple agencies engaged in managing the systems that may be more urgent to address (Tucci, 2009; Tanaka, 2011).

However, in common with many of the Asian delta areas, a major change that has affected the operation of Jakarta flood control systems is the issue of land subsidence. Recent evidence confirms that areas of Jakarta are subsiding rapidly at rates of up to 20 cm per year (Delinom *et al.*, 2009), with some areas worse affected than others and rates of subsidence variable across time and related to extraction and construction factors (Abidin *et al.*, 2009). There is strong evidence that extraction of groundwater is a major contributory factor both by the shallow wells of the general population and the deeper drilled wells used by industry (Abidin *et al.*, 2009). Future subsidence is predicted to be at least 5 to 10 cm per year; this projection is contingent on the successful introduction of measures to slow the extraction of groundwater.

The impact of subsidence on the city is seen in the cracking of constructed facilities, changes in rivers and canals, increased flood risk and malfunctioning drainage (Abidin *et al.*, 2009). The subsidence has also caused an increase in the coastal inundation risk and sea walls are continually raised and extended into new areas to combat this risk. Building higher walls offshore is made more problematic by the fact that the sea bed is also subsiding and the long-term feasibility of maintaining the sea defences in this way is questionable. Land subsidence also results in the reduction of clearance gaps for structures such as bridges above the canals. During a flood these low-lying structures form obstructions, constrict flows and can become blocked, thus increasing flooding. Subsidence also contributes to the impact from pluvial flooding as the speed of conveyance is affected by the topography of the land and the relative sea levels.

Plans for the future to tackle the particularly complex flood risk in Jakarta include both structural and nonstructural approaches and short- and long-term measures. Some immediate emergency actions are necessary; for example action was taken to elevate the road to the airport after it was cut off in the 2008 event – an initiative financed on the expectation that road tolls would cover the cost of construction over the long term – and strengthening of sea walls to prevent frequent coastal inundation is ongoing. Simultaneously Jakarta has commissioned detailed flood risk assessment under multiple climate and development scenarios and the short-term measures will be coupled with the design of a sustainable long-term plan for sea defences and other flood management initiatives (Brinkmann, 2011).

These initiatives include the Jakarta Urgent Flood Mitigation Project/Jakarta Emergency Dredging Initiative Project (JUFMP/JEDI Project) designed to improve the operation and maintenance of the existing flood management systems (Fook, 2011); the Jakarta Comprehensive Flood Management Plan (JCFMP), which is tackling risk upstream of the flood affected areas; and the Jakarta Coastal Defense Strategy (JCDS), which is tackling the risk at the city–coastal interface (Brinkmann, 2011). Other more local initiatives are also cutting across these strategic projects and tackle issues such as waste management and ecological issues (Handhayani, 2009). The challenges faced in implementation of these plans and in predicting their individual and joint contribution to flood risk reduction illustrates the importance of integrating flood risk management not only across sources and causes of flooding but also within urban management as a whole.

In this context the JUFMP is a useful illustration. The dredging programme within the JUFMP has been delayed for several years by the need to conduct safe resettlement programmes for those residents needing to be moved from the sides of the channels. Resettlement can be an opportunity to improve the well-being of inhabitants by moving them to an area with reduced flood risk and higher quality housing, but it is often resisted by residents who are keen to retain their local communities and employment opportunities. Once dredging is underway the plan is for disposal of waste material dredged from the watercourses to contribute to flood risk reduction by being used in land reclamation elsewhere (Fook, 2011). In the future the management and control of waste within and upstream of the city will reduce the need for resources to keep the watercourses clear. Local initiatives to retrieve and recycle waste have provided some income for city dwellers but the impact of this activity on total waste levels is small (Mercycorps, 2011). Drain clearance activity can also be undertaken by local communities, but the utility of clearing downstream areas is questionable in the knowledge that upstream areas, unaffected by flood issues, continue to dispose of their waste into the watercourses.

The JCDS Project demonstrates the need to integrate the plans for structural and nonstructural measures. The project aims to develop a strategy for Jakarta that can reduce the risk of flooding from the dual impact of sea-level rise and land subsidence in the long term (Brinkmann, 2011). The strategy recognises the need to continue with short-term protective activities around the existing defences and coastal pumping. However, in the longer term the projections of sea-level rise and subsidence suggest that other options will have to be considered. The options include: major offshore coastal defences, new and more stringent groundwater extraction controls and the building of new and larger retention areas in the city. These measures will all require substantial lead times and the success of abstraction control may affect the design criteria of other measures such as the coastal defence as the levels of future subsidence are dependent upon extraction rates.

Groundwater extraction controls have to be introduced slowly in Jakarta. Industry is dependent on extraction due to the lack of surface water sources. Data suggest that surface water supplies only 1% of the required volume (Delinom, 2007); therefore industrial extraction is a major contributor to depletion of resources and much of it may be unlicensed. Few incentives exist to drive more responsible water management among companies even though licensed water tariffs for hotels and businesses have been increased (Delinom, 2007). Forbidding extraction would leave large parts of the population without any supply due to the low levels of public infrastructure available to replace the extracted supply.

The design of the built environment is tackled by the JCFMP. This JICA-funded project is aimed at spatial planning, building regulations and construction of infiltration schemes upstream of the flood affected areas. It will be vital for the planned schemes to consider how retention areas can contribute to the water supply in order to maximise their impact on flood risk.

The case of Jakarta shows that a combination of measures is necessary to tackle flood risk in the city and metropolitan area and that these measures are linked closely to other urban issues such as water supply and waste management. For Jakarta the linkage between too much and too little water is particularly apparent because of the impact of land subsidence on flood risk. The challenge of integrating flood risk management into urban management is increased by the number and diversity of organisations and stakeholders involved in the processes. The sheer scale of the population in Jakarta means that scaling-up of solutions is also problematic. A comprehensive integrated solution combining downstream, midstream and upstream retentions could be effective in order to anticipate possible intense rainfall within this large catchment and to mitigate subsidence along the coast and sea-level rise.

28.5 Conclusions

Cities and urbanised areas in the Asian Deltas can be seen to encapsulate the dual problems of flooding and water shortages as their geographical location and rapidly growing populations place stress on the water cycle. The linkages between shortage of water supply and increasing flood risk are also demonstrated in delta cities as land subsidence and associated coastal flooding are driven by increased extraction of water when demand exceeds the local provision from other sources.

In these circumstances it is appropriate to consider innovative approaches to flood risk management that take into consideration the whole water cycle. Increase of semi-natural storage, rainwater harvesting and extraction control can slow the increase in flood risk while simultaneously improving water supply factors and enhancing environmental quality. Multiple benefits arising from these measures may add to the cost-effectiveness of schemes and increase the quality of life for inhabitants of delta areas. However, in many delta areas it will be necessary to implement these innovative schemes as part of an integrated plan that includes hard engineered defences and other structural and nonstructural measures.

The example of Jakarta in Indonesia is used to illustrate the real and present impacts of too much and too little water in the Asian deltas. Despite decades of investment and regular flooding over wide tracts of the city the flood risk in Jakarta continues to rise. Future plans for the defence of the city include increased water storage, extraction controls and measures to improve water supply to the region. However, the challenge of managing all these changes in a complex megacity is one that ensures that this is a long-term goal rather than a quick fix.

The conclusion highlighted by the Asian delta cities, but increasingly true everywhere, is that flood risk management must be viewed within the context of the wider water cycle and integrated within urban management as a whole. There are clear opportunities to benefit from the integration of water supply and flood management and clear risks in tackling one at the expense of the other.

References

Abidin, H.Z., Andreas, H., Gumilar, I., Gamal, M., Fukuda, Y. and Deguchi, T. (2009) Land subsidence and urban development in Jakarta (Indonesia). In: *7th FIG Regional Conference*, Hanoi, Vietnam.

Aerts, J.C.J.H., Major, D.C., Bowman, M.J., Dircke, P. and Marfai, M.A. (2009) *Connecting Delta Cities, Coastal Cities, Flood Risk Management and Adaptation to Climate Change*. VU University Press, Amsterdam.

Babel, D.M.S. (2008) *Groundwater Management in Asian Coastal Cities: A Proposal for Climate Change Impact and Adaptation Study*. Water Engineering and Management, AIT, Bangkok.

Brinkmann, J.J. (2011) Towards a public–private partnership for the coastal development and protection of Jakarta. *Urban Resilience Workshop*. World Bank, Jakarta.

Chinh, D.T. (2008) *Urban Flooding in Ho Chi Minh City: The Causes and Effects*. International GIS Summer School.

Delinom, R.M. (2007) Groundwater management issues in the Greater Jakarta area, Indonesia. In: *International Workshop on Integrated Watershed Management for Sustainable Water Use in a Humid Tropical Region*, JSPS–DGHE Joint Research Project. University of Tsukuba, Tsukuba.

Delinom, R.M., Assegaf, A., Abidin, H.Z., Taniguchi, M., Suherman, D., Lubis, R.F. and Yulianto, E. (2009) The contribution of human activities to subsurface environment degradation in Greater Jakarta Area, Indonesia. *Science of the Total Environment*, 407, 3129–3141.

EM–DAT May (2011) *The OFDA/CRED International Disaster Database*, May. Université Catholique de Louvain, Brussels, Belgium.

EM–DAT (2012) *The OFDA/CRED International Disaster Database*, July. Université Catholique de Louvain, Brussels, Belgium.

Fook, C.E. 2011. Overview of Jakarta Urgent Flood Mitigation Project (JUFMP). In: *Workshop – East Asia Urban Resilienc Study*, Jakarta, Indonesia.

Foster, S. and Garduño, H. (2002) *Argentina: Mitigation of Groundwater Drainage Problems in the Buenos Aires Conurbation Technical and Institutional Way Forward*. GW Mate/World Bank.

Fuchs, R.J. (2010) Cities at risk: Asia's coastal cities in an age of climate change. *Asia Pacific Issues, Analysis from the East West Center*. East West Center.

Handhayani, S. (2009) Ciliwung–Jakarta: delta diaogue. In: *Second Forum on Delta and Coastal Development*, Amsterdam.

IGES (2008) *Groundwater and Climate Change: No Longer the Hidden Resource*. IGES White Paper.

Intergovernmental Panel on Climate Change (IPCC) (2007) *Climate Change 2007: Impacts, Adaptation and Vulnerability. Contribution of Working Group II to the Fourth Assessment Report of the Intergovernmental Panel on Climate Change*. Cambridge University Press, Cambridge, UK.

Jago-on, K.A.B., Kanekob, S.,Fujikurac, R.,Fujiwarab, A.,Imaid, T.,Matsumotoe, T., Zhangb, J. Tanikawaf, H., Tanakab, K., Leeg, B. and Taniguchia, M. (2009) Urbanization and subsurface environmental issues: an attempt at DPSIR model application in Asian cities. *Science of the Total Environment*, 407, 3089–3104.

Jha, A.K. and Brecht, H. (2011) *Building Urban Resilience in East Asia. An Eye on East Asia and Pacific*. World Bank, Washington, DC.

Jha, A., Lamond, J., Bloch, R., Bhattacharya, N., Lopez, A., Papachristodoulou, N., Bird, A., Proverbs, D., Davies, J. and Barker, R. (2011) *Five Feet High and Rising – Cities and Flooding in the 21st Century*, Policy Research Working Paper 5648. World Bank, Washington, DC.

Jha, A., Lamond, J. and Bloch, R. (2012) *Cities and Flooding: A Guide to Integrated Urban Flood Risk Management for the 21st Century*. GFDRR/World Bank, Washington, DC.

Lovett, R.A. (2011) Why Mississippi floods were expected. *Nature News*, *13* May 2011.

Mekong River Commission (2009) *Best Practise Guidelines for Structural Measures and Flood Proofing. The Flood Management and Mitigation Programme, Component 2: Structural Measures and Flood Proofing in the Lower Mekong Basin.*

Mercycorps (2011) Community based flood risk management initiatives. *Workshop on Global Flood Risk Management*, Jakarta, Indonesia.World Bank.

Nasir, H. (2008) *Flood since Batavia until Jakarta (1614–2008) problems and solutions*. Available at: http://www.eu-changde.com/english/download.asp.

Nicholls, R.J. and Lowe, J.A. (2004) Benefits of mitigation of climate change for coastal areas. *Global Environmental Change Part A*, **14**, 229–244.

Nicholls, R.J., Wong, P.P., Burkett, V., Codignotto, J., Hay, J., McLean, R., Rogoonaden, S. and Woodroffe, C.D. (2007) Coastal systems and low-lying areas. In: Parry, M.L., Canziani, O.F., Palutikof, J.P. *et al*. (eds), *Climate Change 2007: Impacts, Adaptation and Vulnerability. Contribution of Working Group II to the Fourth Assessment Report of the Intergovernmental Panel on Climate Change*. Cambridge University Press, Cambridge.

Rose, C.B., Proverbs, D.G., Manktelow, K. and Booth, C.A. (2010) Psychological factors affecting flood coping strategies. In: de Wrachien, D., Proverbs, D.G., Brebbia, C.A. and Mambretti, S. (eds), *Flood Recovery Innovation and Response II*. WIT Press, Southampton, pp. 305–312.

Satterthwaite, D., Huq, S., Pelling, M., Reid, H. and Romero Lankao, P. (2007) *Adapting to Climate Change in Urban Areas. The Possibilities and Constraints in Low- and Middle-Income Nations*. Human Settlements Discussion Paper Series. IIED International Institute for Environment and Development.

Stern, N. (2006) *Stern Review: The Economics of Climate Change* London: H M Treasury.

Tanaka, T. (2011) The Project for Capacity Development of Jakarta Comprehensive Flood Management (JCFM). In: *Presentation by JICA at Workshop on Global Flood Risk Management*, Jakarta, Indonesia.

Taniguchi, M., Burnett, W.C. and Ness, G. (2009a) Human impacts on urban subsurface environments. *Science of the Total Environment*, **407**, 3073–3074.

Taniguchi, M., Shimada, J., Fukuda, Y., Yamano, M., Onodera, S.-i., Kaneko, S. and Yoshikoshi, A. (2009b) Anthropogenic effects on the subsurface thermal and groundwater environments in Osaka, Japan and Bangkok, Thailand. *Science of the Total Environment*, **407**, 3153–3164.

Tucci, C.E.M. (2007) *Urban Flood Management*. WMO/Capnet.

Tucci, C.E.M. (2009) LAC Water Bean Retreat Presentation. *LAC Water Bean Retreat*. World Bank, Washington, DC.

UNEP (2008) *Rainwater Harvesting and Disaster Management*. UNEP.

van Ogtrop, F.F., Hoekstra, A.Y. and van der Meulen, F. (2005) Flood management in the Lower Incomati River Basin, Mozambique: two alternatives. *Journal of the American Water Resources Association*, June.

World Bank (2010) *Climate Risks and Adaptation in Asian Coastal Megacities*. World Bank, Washington, DC.

World Bank (2012) *Tools for Building Urban Resilience: Integrating Risk Information into Investment Decisions. Pilot Cities Report – Jakarta and Can Tho*. World Bank/Australia Aid, Washington, DC.

Wynne, A. (2011) Engineers fight Mississippi floods. *New Civil Engineer* [Online]. Available: http://www.nature.com/news/2011/110513/full/news.2011.289.html [Accessed May 19, 2011].

Yang, S.L., Li, M., Dai, S.B., Liu, Z., Zhang, J. and Ding, P.X. (2006) Drastic decrease in sediment supply from the Yangtze River and its challenge to coastal wetland management. *Geophysics Research Letters*, **33**, L06408.

Yoshikoshi, A., Adachi, I., Taniguchi, T., Kagawa, Y., Kato, M., Yamashita, A., Todokoro, T. and Taniguchi, M. (2009) Hydro-environmental changes and their influence on the subsurface environment in the context of urban development. *Science of the Total Environment*, **407**, 3105–3111.

Section 8 Summary of the Book

29

Water Resources Challenges – Penury and Peace

Susanne M. Charlesworth and Colin A. Booth

29.1 Introduction

Water is a fundamental requirement for an array of environmental, economic and social needs, from ecosystems, habitats, biodiversity and agricultural necessities through to industrial, commercial and recreational activities. It also offers wealth and political stability, but none is more important than its requirement for human quality of life (e.g. drinking, cooking and hygiene). Unfortunately, water is not universally abundant and for many nations the uneven distribution of water and human settlement continues to create growing problems of fresh water availability and accessibility. When it is abundant, disaster can ensue, as evidenced by the number of global disasters attributable to floods, for example Pakistan 2010 and China 2010. With more than 80% of the global population living on land prone to flooding, it is a problem that will undoubtedly worsen with climate change. The built environment has become more prone to flooding because urbanisation has meant that landscapes, which were once porous and allowed surface water to infiltrate, have been stripped of vegetation and soil and have been covered with impermeable roads, pavements and buildings. As a consequence, precipitation, which would have once soaked away, now increasingly collects as surface runoff, enhancing the hydrograph storm peak, reducing the lag-time between peak rainfall and peak river discharge, and therefore promoting flooding of the immediate surroundings and/or those downstream.

Unfortunately, those circumstances portrayed above will further deteriorate when they are coupled with predicted climate change because it is anticipated that global rainfall regimes will dramatically alter. Some areas will experience wetter winters and drier summers but may also experience changes in the patterns of rainfall frequency and intensity (Intergovernmental Panel on Climate Change (IPCC), 2007). Therefore, more than ever before, water supply and urban water management have become a necessary requirement for the attention of governments, local authorities, relevant organisations and property owners.

Water Resources in the Built Environment: Management Issues and Solutions, First Edition.
Edited by Colin A. Booth and Susanne M. Charlesworth.

Various approaches have been outlined and solutions proposed in this book that address the main water resources challenges surrounding both water supply shortages and property flooding challenges. This chapter provides an overview of the information and insights gathered from the chapters collated. Based on these findings, the chapter finishes with a discussion of the complex political agenda surrounding international water security issues.

29.2 Insights Gathered from the Chapters of this Book

The authors of the chapters in this book have addressed between them the dual problems of water scarcity and flooding. The reasons underlying these interrelated issues are complex, but one thing is certain: without the proactive involvement of governments and individuals, the situation will get worse, exacerbation by global climate change is simply *in addition to* current trends of increasing abstraction, increasing urbanisation and increasing water demand.

This book therefore offers various means to engage with water, treating it as a valuable, invaluable resource. The reasons why water has become problematic is outlined in chapters dealing with the impacts of urbanisation and the propensity of urban areas to flood, the impacts of dams on society and the environment and issues around water quality. It offers approaches from policy-driven means to meet the demand for water through to costing the water supply. Big technological fixes such as desalination and dams are covered, but water conservation measures and harvesting of both greywater and rainwater are shown to have the potential to play an increasingly important role. As was clearly explained in the introductory chapter, scarcity is but one side of the water coin: too much water causes human misery, the destruction of infrastructure and substantial environmental impacts. Whether fluvial, pluvial or coastal, measures to provide resilience and resistance to flooding are given in these pages: from hard infrastructure along coasts and rivers to *back-to-nature* methods, such as SUDS, and from building-scale to whole communities. Society cannot fight water, but also societies cannot fight each other over water supplies. The following section charts some recent history around the world regarding the unequal distribution of water, which some seriously thought brought various countries to the brink of war.

29.3 Dialogue on International Water Resources Challenges

The quote 'whiskey is for drinking, but water is for fighting over' is often attributed to Mark Twain, and whatever the truth of the source, the latter half of the statement has certainly reflected fears of impending 'water wars' over inequity of distribution, the impacts on water supplies of large-scale technical fixes such as dams and the transboundary abstraction of water such as from the Saharan aquifers or the River Nile and its tributaries. Indeed, in 1979 Anwar Sadat, then President of Egypt, threatened Ethiopia by stating that 'The only matter that could take Egypt to war again is water' (Kameri-Mbote, 2007). By 1988, Boutros Boutros-Ghali, who was Egyptian Foreign Minister at the time, later to become United Nations Secretary-General, predicted that the next war in the Middle East would be fought over the River Nile as a source of water rather than a conflict brought on by politics. Even by 1999 the President of Libya, Muammar Al-Gaddafi, was asserting that 'the next Middle East war would be over dwindling water supplies'. Whilst the Pacific Institute (http://www.worldwater.org/conflict/map/) can plot 68 water–related conflicts that have occurred globally between 2000 and 2010 on its 'Water Conflict Chronology Map', illustrating that there is no doubt whatsoever that conflicts over water have occurred and continue to occur, nonetheless, the all-out war predicted by some has yet to come to pass. Some authors (e.g. Wolf *et al.*, 2003) argue that situations where the protagonists cooperate outnumber those where conflict ensues by more than 2 to 1 and that, in general, the conflicts that do occur tend not to be extreme. In fact, the only true 'war' found by Wolf (1998) was 4500 years ago between the city-states of Lagash and Umma on the Tigris–Euphrates. So, rather than 'water wars' there would appear to be 'water peace' (Trottier, no date). Authors such as Baskin (1994) proposed that the issue of water scarcity was approached in a rational way by states who found that cooperation was simply the most logical thing to do. Davis and Hirji (2005) go so far as to say that water wars are a myth, stating in fact that: 'Countries have attempted to use transboundary waters as vehicles for enhancing cooperation, regional integration, and development rather than as a source of conflict' (p. 120).

Nonetheless, they quote projections of 50 countries amounting to a total population of one billion people who will be classified as waterscarce and a further three billion people living with water stress by 2025. They also highlight global climate change as affecting flooding and droughts, water quality and water demand. So, the question posed by the UK Adaption Sub-Committee's 2012 Progress Report: *Climate Change – Is the UK Prepared for Flooding and Water Scarcity?* (Adaptation Sub-Committee, 2012) seems entirely appropriate to the water war/water peace dialogue as well as when applied to the discussions presented throughout this book.

To some extent the two halves of the question of water scarcity and flooding can be answered together: reduce demand, recycle and reuse. Thus, technologies such as rainwater harvesting, greywater recycling and water conservation can go some way to replacing some potable supplies and also provide a means of attenuating the storm peak and storing excess water. However, when the affected populations are in the billions and water is seen variously as a commodity to be bought and sold, a global common and a basic human right (Grover, 2006), Rosemann (in Grover, 2006) asks the question of whether water is becoming so commercialised that it is becoming a luxury affordable by the well-off rather than by all. If the answer to the question is in the affirmative, then the Millenium Development Goals, containing the human right to water through which the international community committed to halve the numbers of people who did not have access to safe drinking water by 2015, are not working.

In terms of the problem of flooding, with the projection that by 2050 over two-thirds of the world population will be urban (United Nations Children's Fund (UNICEF), 2012) and that most urban growth will occur in less developed countries (United Nations (UN), 2012), Wang *et al.* (in Grover, 2006) highlight reduced infiltration of water in general due to the increase in impermeable surfacing installed during the urbanisation process. Exacerbation of this effect by the potential for higher intensity, longer duration rainfall as the climate undergoes global change can only lead to the potential for more flooding. Jones and Macdonald (2007) call water a 'troublesome substance' whose 'unruly' behaviour humans need to discipline in some way.

King Cnut was a Dane who ruled England between 1016 and 1035, and is often credited with the arrogance of thinking he could hold back the tide by ordering it not to flow over his land. In fact, he was deliberately illustrating the futility of his human power against that of nature. Nearly 980 years after Cnut's death, human beings are still attempting to fight water, forcing it to conform to society's needs rather than changing their behaviour for the sake of water.

29.4 Conclusions

Without doubt, the next generations of water resources managers have real and complex engineering, environmental, social, economic and political issues to consider in their decision making, which requires the forward-thinking integration of governments and industry with an informed and steadfast society. Ultimately, their roles will influence the sustainability of the built environment, lifestyle quality and, moreover, may also contribute to global security.

Rather than remaining reliant on the government and relevant industries to address water resources challenges, it is apparent that society must also take responsibility to change its frivolous behaviour and attitudes to water and, where appropriate, embrace alternative water supply and/or property flooding technologies. For instance, as a nation where there are water shortages, and in a world where energy use and carbon footprints are measurable, there is no justifiable reason why society should use potable water for flushing toilets – people should embrace water saving and harvesting approaches. Similarly, in the current economic climate, there is no sizeable reason why the insurance industry should be subsidised by the government (or by the premiums of other policyholders) to pay for the repairs and replacements of properties that have been flooded, where the owner has chosen not to install flood resistance and/or resilience measures. Therefore, although no mandatory policy exists for builders or homeowners (yet), future flooding policy and legislation should have minimal freedom of choice and be intrinsically linked to re-educating people to change and promoting awareness campaigns (e.g. Environment Agency flood pamphlets).

References

Adaptation Sub-Committee (2012) *Climate Change – Is the UK Prepared for Flooding and Water Scarcity?* 102 pp. Available at: http://www.preventionweb.net/files/27607_cccascreport2012bookmarked2.pdf.

Baskin, G. (1994) The clash over water: an attempt at demystification. *Palestine–Israel Journal*, **3**, 27–35.

Davis, J.R. and Hirji, R. (2005) The myth of water wars. *Georgetown Journal of International Affairs, Science and Technology*, **6** (1), 115–124.

Grover, V.I. (ed.) (2006) *Water: Global Common and Global Problems*. United Nations Environment Programme. Science Publishers, 548 pp.

Intergovernmental Panel on Climate Change (IPCC) (2007) In: Solomon, S., Qin, D., Manning, M. et al. (eds), *Contribution of Working Group I to the Fourth Assessment Report of the Intergovernmental Panel on Climate Change*. Cambridge University Press, Cambridge, United Kingdom and New York, USA.

Jones, P. and Macdonald, N. (2007) Making space for unruly water: sustainable drainage systems and the disciplining of surface runoff. *Geoforum*, 38(3), 534–544.

Kameri-Mbote, P. (2007) *Water, Conflict, and Cooperation: Lessons from the Nile River Basin. Navigating Peace*, Volume 4. Environmental Change and Security Program, Woodrow Wilson International Center for Scholars, 6 pp. Available at: http://www.wilsoncenter.org/sites/default/files/NavigatingPeaceIssuePKM.pdf.

Trottier, J. (no date) *Water Wars: The Rise of a Hegemonic Concept. Exploring the Making of the Water War and Water Peace Belief within the Israeli–Palestinian Conflict*. UNESCO, 20 pp.

United Nations (UN) (2012) *World Urbanization Prospects: The 2011 Revision*. United Nations Department of Economic and Social Affairs/Population Division iii, 50 pp.

United Nations Children's Fund (UNICEF) (2012) *The State of the World's Children 2012: Children in an Urban World*. UNICEF, 156 pp.

Wolf, A. (1998) Conflict and cooperation along international waterways. *Water Policy*, **1** (2), 251–265.

Wolf, A.T., Yoffe, S.B. and Giordano, M. (2003) International waters: identifying basins at risk. *Water Policy*, 5, 29–60.

Index

Figures are indicated by *italic* page numbers; Tables, Examples and Case Studies by **bold** page numbers.

Water Resources in the Built Environment: Management Issues and Solutions, First Edition.
Edited by Colin A. Booth and Susanne M. Charlesworth.

WILEY Blackwell

Books also available from Wiley-Blackwell

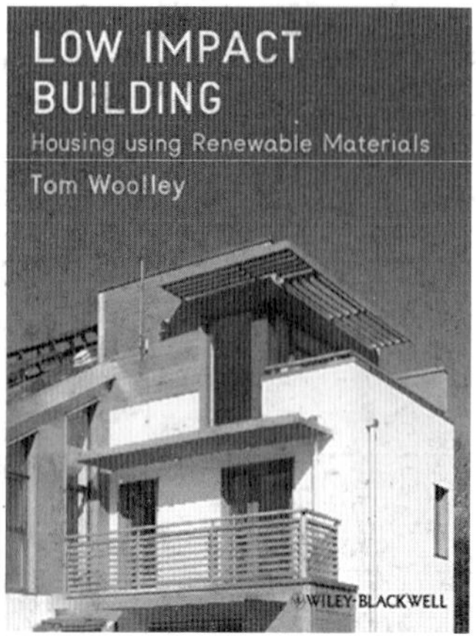

Low Impact Building
Woolley
9781444336603

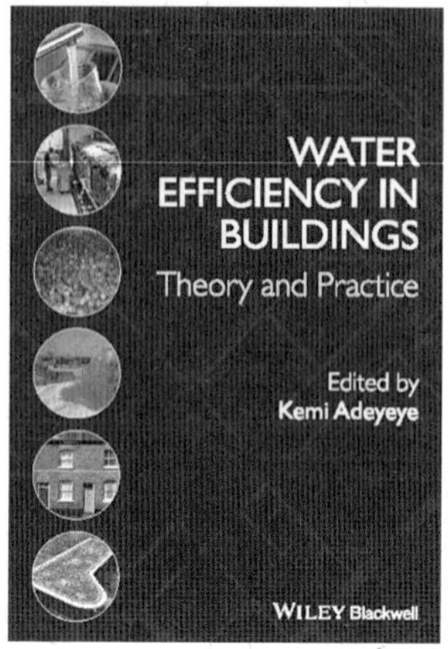

Water Efficiency in Buildings: theory and practice
Adeyeye
9781118456576

Ecosystem Services Come To Town
Grant
978-1-4051-9506-5

Solutions to Climate Change: Challenges in the Built Environment
Booth et al
978-1-4051-9507-2

Residential Landscape Sustainability
Smith et al
978-1-4051-5873-2

Flood Damaged Property
Proverbs
978-1-4051-1616-9

other/forthcoming Books of Interest

Sustainable Building Adaptation
Wilkinson
9781118477106

Sustainable Refurbishment
Shah
9781405195089

www.wiley.com/go/construction